Operator Methods for Optimal Control Problems

LECTURE NOTES

IN PURE AND APPLIED MATHEMATICS

1. *N. Jacobson,* Exceptional Lie Algebras
2. *L. -Å. Lindahl and F. Poulsen,* Thin Sets in Harmonic Analysis
3. *I. Satake,* Classification Theory of Semi-Simple Algebraic Groups
4. *F. Hirzebruch, W. D. Newmann, and S. S. Koh,* Differentiable Manifolds and Quadratic Forms (out of print)
5. *I. Chavel,* Riemannian Symmetric Spaces of Rank One (out of print)
6. *R. B. Burckel,* Characterization of C(X) Among Its Subalgebras
7. *B. R. McDonald, A. R. Magid, and K. C. Smith,* Ring Theory: Proceedings of the Oklahoma Conference
8. *Y.-T. Siu,* Techniques of Extension on Analytic Objects
9. *S. R. Caradus, W. E. Pfaffenberger, and B. Yood,* Calkin Algebras and Algebras of Operators on Banach Spaces
10. *E. O. Roxin, P.-T. Liu, and R. L. Sternberg,* Differential Games and Control Theory
11. *M. Orzech and C. Small,* The Brauer Group of Commutative Rings
12. *S. Thomeier,* Topology and Its Applications
13. *J. M. Lopez and K. A. Ross,* Sidon Sets
14. *W. W. Comfort and S. Negrepontis,* Continuous Pseudometrics
15. *K. McKennon and J. M. Robertson,* Locally Convex Spaces
16. *M. Carmeli and S. Malin,* Representations of the Rotation and Lorentz Groups: An Introduction
17. *G. B. Seligman,* Rational Methods in Lie Algebras
18. *D. G. de Figueiredo,* Functional Analysis: Proceedings of the Brazilian Mathematical Society Symposium
19. *L. Cesari, R. Kannan, and J. D. Schuur,* Nonlinear Functional Analysis and Differential Equations: Proceedings of the Michigan State University Conference
20. *J. J. Schäffer,* Geometry of Spheres in Normed Spaces
21. *K. Yano and M. Kon,* Anti-Invariant Submanifolds
22. *W. V. Vasconcelos,* The Rings of Dimension Two
23. *R. E. Chandler,* Hausdorff Compactifications
24. *S. P. Franklin and B. V. S. Thomas,* Topology: Proceedings of the Memphis State University Conference
25. *S. K. Jain,* Ring Theory: Proceedings of the Ohio University Conference
26. *B. R. McDonald and R. A. Morris,* Ring Theory II: Proceedings of the Second Oklahoma Conference
27. *R. B. Mura and A. Rhemtulla,* Orderable Groups
28. *J. R. Graef,* Stability of Dynamical Systems: Theory and Applications
29. *H.-C. Wang,* Homogeneous Branch Algebras
30. *E. O. Roxin, P.-T. Liu, and R. L. Sternberg,* Differential Games and Control Theory II
31. *R. D. Porter,* Introduction to Fibre Bundles
32. *M. Altman,* Contractors and Contractor Directions Theory and Applications
33. *J. S. Golan,* Decomposition and Dimension in Module Categories
34. *G. Fairweather,* Finite Element Galerkin Methods for Differential Equations
35. *J. D. Sally,* Numbers of Generators of Ideals in Local Rings
36. *S. S. Miller,* Complex Analysis: Proceedings of the S.U.N.Y. Brockport Conference
37. *R. Gordon,* Representation Theory of Algebras: Proceedings of the Philadelphia Conference
38. *M. Goto and F. D. Grosshans,* Semisimple Lie Algebras
39. *A. I. Arruda, N. C. A. da Costa, and R. Chuaqui,* Mathematical Logic: Proceedings of the First Brazilian Conference

40. *F. Van Oystaeyen,* Ring Theory: Proceedings of the 1977 Antwerp Conference
41. *F. Van Oystaeyen and A. Verschoren,* Reflectors and Localization: Application to Sheaf Theory
42. *M. Satyanarayana,* Positively Ordered Semigroups
43. *D. L. Russell,* Mathematics of Finite-Dimensional Control Systems
44. *P.-T. Liu and E. Roxin,* Differential Games and Control Theory III: Proceedings of the Third Kingston Conference, Part A
45. *A. Geramita and J. Seberry,* Orthogonal Designs: Quadratic Forms and Hadamard Matrices
46. *J. Cigler, V. Losert, and P. Michor,* Banach Modules and Functors on Categories of Banach Spaces
47. *P. T. Liu and J. G. Sutinen,* Control Theory in Mathematical Economics: Proceedings of the Third Kingston Conference, Part B
48. *C. Byrnes,* Partial Differential Equations and Geometry
49. *G. Klambauer,* Problems and Propositions in Analysis
50. *J. Knopfmacher,* Analytic Arithmetic of Algebraic Function Fields
51. *F. Van Oystaeyen,* Ring Theory: Proceedings of the 1978 Antwerp Conference
52. *B. Kedem,* Binary Time Series
53. *J. Barros-Neto and R. A. Artino,* Hypoelliptic Boundary-Value Problems
54. *R. L. Sternberg, A. J. Kalinowski, and J. S. Papadakis,* Nonlinear Partial Differential Equations in Engineering and Applied Science
55. *B. R. McDonald,* Ring Theory and Algebra III: Proceedings of the Third Oklahoma Conference
56. *J. S. Golan,* Structure Sheaves over a Noncommutative Ring
57. *T. V. Narayana, J. G. Williams, and R. M. Mathsen,* Combinatorics, Representation Theory and Statistical Methods in Groups: YOUNG DAY Proceedings
58. *T. A. Burton,* Modeling and Differential Equations in Biology
59. *K. H. Kim and F. W. Roush,* Introduction to Mathematical Consensus Theory
60. *J. Banas and K. Goebel,* Measures of Noncompactness in Banach Spaces
61. *O. A. Nielson,* Direct Integral Theory
62. *J. E. Smith, G. O. Kenny, and R. N. Ball,* Ordered Groups: Proceedings of the Boise State Conference
63. *J. Cronin,* Mathematics of Cell Electrophysiology
64. *J. W. Brewer,* Power Series Over Commutative Rings
65. *P. K. Kamthan and M. Gupta,* Sequence Spaces and Series
66. *T. G. McLaughlin,* Regressive Sets and the Theory of Isols
67. *T. L. Herdman, S. M. Rankin, III, and H. W. Stech,* Integral and Functional Differential Equations
68. *R. Draper,* Commutative Algebra: Analytic Methods
69. *W. G. McKay and J. Patera,* Tables of Dimensions, Indices, and Branching Rules for Representations of Simple Lie Algebras
70. *R. L. Devaney and Z. H. Nitecki,* Classical Mechanics and Dynamical Systems
71. *J. Van Geel,* Places and Valuations in Noncommutative Ring Theory
72. *C. Faith,* Injective Modules and Injective Quotient Rings
73. *A. Fiacco,* Mathematical Programming with Data Perturbations I
74. *P. Schultz, C. Praeger, and R. Sullivan,* Algebraic Structures and Applications Proceedings of the First Western Australian Conference on Algebra
75. *L. Bican, T. Kepka, and P. Nemec,* Rings, Modules, and Preradicals
76. *D. C. Kay and M. Breen,* Convexity and Related Combinatorial Geometry: Proceedings of the Second University of Oklahoma Conference
77. *P. Fletcher and W. F. Lindgren,* Quasi-Uniform Spaces
78. *C.-C. Yang,* Factorization Theory of Meromorphic Functions
79. *O. Taussky,* Ternary Quadratic Forms and Norms
80. *S. P. Singh and J. H. Burry,* Nonlinear Analysis and Applications
81. *K. B. Hannsgen, T. L. Herdman, H. W. Stech, and R. L. Wheeler,* Volterra and Functional Differential Equations

82. *N. L. Johnson, M. J. Kallaher, and C. T. Long,* Finite Geometries: Proceedings of a Conference in Honor of T. G. Ostrom
83. *G. I. Zapata,* Functional Analysis, Holomorphy, and Approximation Theory
84. *S. Greco and G. Valla,* Commutative Algebra: Proceedings of the Trento Conference
85. *A. V. Fiacco,* Mathematical Programming with Data Perturbations II
86. *J.-B. Hiriart-Urruty, W. Oettli, and J. Stoer,* Optimization: Theory and Algorithms
87. *A. Figa Talamanca and M. A. Picardello,* Harmonic Analysis on Free Groups
88. *M. Harada,* Factor Categories with Applications to Direct Decomposition of Modules
89. *V. I. Istrăţescu,* Strict Convexity and Complex Strict Convexity: Theory and Applications
90. *V. Lakshmikantham,* Trends in Theory and Practice of Nonlinear Differential Equations
91. *H. L. Manocha and J. B. Srivastava,* Algebra and Its Applications
92. *D. V. Chudnovsky and G. V. Chudnovsky,* Classical and Quantum Models and Arithmetic Problems
93. *J. W. Longley,* Least Squares Computations Using Orthogonalization Methods
94. *L. P. de Alcantara,* Mathematical Logic and Formal Systems
95. *C. E. Aull,* Rings of Continuous Functions
96. *R. Chuaqui,* Analysis, Geometry, and Probability
97. *L. Fuchs and L. Salce,* Modules Over Valuation Domains
98. *P. Fischer and W. R. Smith,* Chaos, Fractals, and Dynamics
99. *W. B. Powell and C. Tsinakis,* Ordered Algebraic Structures
100. *G. M. Rassias and T. M. Rassias,* Differential Geometry, Calculus of Variations, and Their Applications
101. *R.-E. Hoffmann and K. H. Hofmann,* Continuous Lattices and Their Applications
102. *J. H. Lightbourne, III, and S. M. Rankin, III,* Physical Mathematics and Nonlinear Partial Differential Equations
103. *C. A. Baker and L. M. Batten,* Finite Geometries
104. *J. W. Brewer, J. W. Bunce, and F. S. Van Vleck,* Linear Systems Over Commutative Rings
105. *C. McCrory and T. Shifrin,* Geometry and Topology: Manifolds, Varieties, and Knots
106. *D. W. Kueker, E. G. K. Lopez-Escobar, and C. H. Smith,* Mathematical Logic and Theoretical Computer Science
107. *B.-L. Lin and S. Simons,* Nonlinear and Convex Analysis: Proceedings in Honor of Ky Fan
108. *S. J. Lee,* Operator Methods for Optimal Control Problems
109. *V. Lakshmikantham,* Nonlinear Analysis and Applications

Other Volumes in Preparation

Operator Methods for Optimal Control Problems

Edited by
Sung J. Lee
University of South Florida
Tampa, Florida

MARCEL DEKKER, INC. New York and Basel

Library of Congress Cataloging-in-Publication Data

Operator methods for optimal control problems.

(Monographs and textbooks in pure and applied mathematics ; v. 108)
Papers from the proceedings of a special session held at the Annual Meeting of the American Mathematical Society, New Orleans, 1986.
Includes index.
1. Control theory--Congresses. 2. Mathematical optimization--Congresses. 3. Operator theory.
I. Lee, Sung J. II. American Mathematical Society. Meeting (92nd ; 1986 : New Orleans, La.) III. Series
QA402.3.064 1987 629.8'312 87-13451
ISBN 0-8247-7811-1

MARCEL DEKKER, INC.
270 Madison Avenue, New York, New York 10016

Current printing (last digit):
10 9 8 7 6 5 4 3 2 1

PRINTED IN THE UNITED STATES OF AMERICA

Preface

The present volume is an outgrowth of the proceedings of a special session, Operator Methods of Optimal Control Problems, held at the Annual Meeting of the American Mathematical Society in New Orleans in 1986. It includes full length original research papers and state-of-the-art survey articles as well as research announcements by invited speakers to the session.

The purpose of the conference was to feature recent advances in the operator-theoretical approaches to optimal control problems arising from deterministic or stochastic differential equations. The following topics were discussed at the conference: necessary conditions for nonlinear problems, identification problems, wave equations, computational alogorithms, geometry of Riccati equations, operator extremal problems, stabilization problems, stochastic problems, delay equations, beam equations, differential games and singular differential equations.

I wish to thank the Program Committee of the American Mathematical Society for making this session possible. Special thanks are due to Ms. Vickie Kearn, Executive Editor, for inviting me to publish this volume in the Lecture Notes Series, and the staff of Marcel Dekker, Inc., for their cooperation.

Sung J. Lee

Contents

Contributors

N. U. AHMED, University of Ottawa, Ottawa, Ontario, Canada

S. BALAKUMAR, Texas Technical University, Lubbock, Texas

STEPHEN L. CAMPBELL, North Carolina State University, Raleigh, North Carolina

G. CHEN, The Pennsylvania State Unviersity, University Park, Pennsylvania

ETHELBERT N. CHUKWU, University of Tennessee, Knoxville, Tennessee

RICHARD DATKO, Mathematical Reviews, Ann Arbor, Michigan

H. O. FATTORINI, University of California, Los Angeles, California

S. G. KRANTZ*, The Pennsylvania State University, University Park, Pennsylvania

I. LASIECKA+, University of Florida, Gainesville, Florida

SUNG J. LEE, University of South Florida, Tampa, Florida

D. W. MA*, The Pennsylvania State University, University Park, Pennsylvania

JACK W. MACKI, University of Alberta, Edmonton, Alberta, Canada

C. R. MARTIN, Texas Technical University, Lubbock, Texas

M. Z. NASHED, University of Delaware, Newark, Delaware

A. N. V. RAO, University of South Florida, Tampa, Florida

R. TRIGGIANI+, University of Florida, Gainesville, Florida

C. E. WAYNE, The Pennsylvania State University, University Park, Pennsylvania

H. H. WEST, The Pennsylvania State University, University Park, Pennsylvania

PIETRO ZECCA, University of Florence, Florence, Italy

Current affiliations

*Washington Universtiy, St. Louis, Missouri

+University of Virginia, Charlottesville, Virginia

Operator Methods for Optimal Control Problems

Identification of Linear Operators in Differential Equations on Banach Space

N. U. Ahmed
Department of Electrical Engineering and Mathematics
University of Ottawa
Ottawa, Ontario, Canada

INTRODUCTION

We consider the problem of approximate identification of parameters, and, in general, operators in systems governed by parabolic, hyperbolic, and structurally damped hyperbolic evolution equations. The questions of existence of approximating elements in admissible parameter and operator spaces are discussed for the above systems. The method used in the proof of existence is based on arguments leading to sequential compactness of the set of admissible operators (or parameters), and lower semicontinuity of cost functionals with respect to strong operator topologies. Also necessary conditions for optimal identification, using variational methods, are presented. This requires Gateaux differentiability of solutions with respect to the admissible operators, which is proved using a technique similar, in content, to the Lax-Milgram theorem.

Scientists or engineers working in practical fields often encounter the problem of modelling physical systems. On the basis of fundamental physical laws and some idealizing assumptions, the analyst conjectures a certain type of model equation in which the structure of the system operator is partially known or the parameters in the model operator are approximately known. The analyst must determine the unknown elements on the basis of available field data.

The problem of identification of unknown parameters in distributed systems has received considerable attention in the past as documented in the survey paper [1]. The question of exact identifiability for parabolic systems have been considered in [3-5]. Using the uniqueness of Dirichlet series, several results on identifiability and non-identifiability of constant and spatially varying parameters are established in these papers. Exact identifiability conditions for all the eigenvalues of an operator, and for parameters in linear first and second order evolution equations in Hilbert spaces have been discussed by Nakagiri in [6]. Trotter-Kato theorem has been used in [7] to prove convergence of an approximating sequence of parameters which is obtained on the basis of finite dimensional approximation of the distributed system. Techniques of optimal control thoery have also been utilized [8-12] in parameter identification problems. Necessary conditions of optimality for both controls and parameters combined together in a class of distributed systems arising from stochastic differential equations are presented in [8]. Optimality conditions for controls in the coefficients of a class of hyperbolic systems are developed in [9]. Identification of parameters for certain first and second order partial differential equations have been discussed by Lions [10]. A method, based on semigroup theory, for identification of system parameters has been presented by Balakrishnan [2]. Numerical techniques based on optimal control theory, have been suggested by Chavent [11,12].

In this paper, we consider some general problems of identification for systems governed by first and second order linear evolution equations of the form

$$\begin{aligned} &\dot{x} + A(q,t)x + Bx = f, \\ &\ddot{x} + 2\gamma A(q,t)\dot{x} + A(q,t)x + Bx = f, \\ &\ddot{x} + A(q,t)x + Bx = f. \end{aligned} \tag{1}$$

In section 3, we consider the questions of identification of the operator B from an admissible class with the identification errors given by functionals of the forms

$$J(B) = \int_I g(t, x(t))dt \qquad \text{(for parabolic systems)}, \tag{2}$$

and

$$J(B) = \int_I g(t, x(t), \dot{x}(t))dt \qquad \text{(for hyperbolic systems)}. \tag{3}$$

In this paper, we allow the operator B (to be identified) to be a differential operator or an itegro-differential operator in contrast to [2] where B is assumed to be a bounded operator in Hilbert space. Under this set up, the problems of identification of constant as well as spatially varying parameters could be treated as special cases.

We present existence results in Theorems 3.1, 3.3 and 3.4. Note that the question of existence here is important as it is in optimal control theory. In section 4, we consider the extension of these results towards identification of the pair $\{q, B\}$ with the error functionals $J(q, B)$ of the same basic form as given by (2) and (3) and conclude the section with a final result that includes identification of the initial states, q, B and the free term.These results are presented in Theorems 4.1 and 4.2.

In sections 5 and 6, we present several necessary conditions of optimality for identification of the operator B with the function g chosen as quadratic. These results are given in Theorem 5.1 for the parabolic system, Theorem 5.2 for the structurally damped hyperbolic system and Theorem 5.3 for the undamped hyperbolic system. In Theorem 6.1, we present necessary conditions of optimality for the pair $\{q, B\}$ for all the three classes of systems.

2. NOTATIONS

Let H be a real separable Hilbert space and V a linear subspace of H carrying the structure of a reflexive Banach space with V dense in H and the injection $V \hookrightarrow H$ continuous. Then identifying H with its dual, we have $V \hookrightarrow H \hookrightarrow V'$, where V' is the topological dual of V. Let $|\cdot|_H, \|\cdot\|_V$ and $\|\cdot\|_{V'}$ denote the norms in H, V and V', respectively, with scalar products in H denoted by $(\cdot,\cdot)_H$ or $(\cdot,\cdot)$, and the duality pairings for the pair $\{V, V'\}$ denoted by $< \cdot,\cdot >_{V,V'}$ or $< \cdot,\cdot >_{V',V}$. For any Banach space E and a bounded interval $I \subset R$, let $L_2(I, E)$ denote the equivalence classes of strongly measurable functions f on I with values in E such that $\int_I |f(t)|_E^2 dt < \infty$. The space $L_2(I, E)$ furnished with the norm topology $\|f\| = (\int_I |f(t)|_E^2 dt)^{1/2}$ is a Banach space. In case E is Hilbert, $L_2(I, E)$ is also a Hilbert space. We use $\mathcal{L}(E, F)$ to denote the space of bounded linear operators from a Banach space E to a Banach space F.

Let $\tau_{so}(\tau_{wo})$ denote the strong (weak) operator topology on $\mathcal{L}(E, F)$. Then $\mathcal{L}_s(E, F) \equiv (\mathcal{L}(E, F), \tau_{so})$ and $\mathcal{L}_w(E, F) \equiv (\mathcal{L}(E, F), \tau_{wo})$ are locally convex linear topological vector spaces. In this paper, we shall often make use of the space $\mathcal{L}(V, V')$ furnished with the strong or weak operator topologies. For each $b > 0$ and $a \in R$, define the set

$$\begin{aligned} \mathcal{P}_{a,b} \equiv \{B \in \mathcal{L}(V, V') : \|B\|_{\mathcal{L}(V,V')} \leq b, \\ < B\xi, \xi >_{V',V} + a|\xi|_H^2 \geq 0 \quad \text{for all} \quad \xi \in V\}. \end{aligned}$$

Let Q_m denote any metrizable compact Hausdorff space, and let $I \equiv (0, T), T < \infty$, and $\{A(t), t \in \bar{I}\}$ or $\{A(q, t), q \in Q_m, t \in \bar{I}\}$ denote a family of linear operators with values $A(t)$ or $A(q, t) \in \mathcal{L}(V, V')$. Further notations will be introduced in the sequel as required.

3. EXISTENCE THEORY IN IDENTIFICATION I:

In this section we present existence results in identification for an operator B in systems governed by the first and second order evolution equations of the form (1).

We assume that the operator A is known and satisfies the general formalism of Lions :

A1) $t \to < A(t)\phi, \psi >$ is measurable for each $\phi, \psi \in V$ and that there exists a constant c such that

$$|< A(t)\phi, \psi >| \leq c\|\phi\|_V \|\psi\|_V \qquad \text{for all} \quad t \in [0,T] \equiv \bar{I}.$$

A2) A satisfies the Garding's inequality, i.e. there exists an $\alpha > 0$, and $\beta \in R$ such that

$$< A(t)\phi, \phi > +\beta|\phi|_H^2 \geq \alpha\|\phi\|_V^2 \qquad \text{for all} \quad t \in \bar{I}.$$

3.1 Parabolic Systems

Problem P1 : Consider the first order evolution equation given by

$$\begin{aligned} &\dot{x} + A(t)x + Bx = f, \\ &x(0) = x_0, \quad B \in P_{a,b}, \end{aligned} \tag{4}$$

and let $g : I \times H \to \bar{R} \equiv [-\infty, \infty]$. The identification problem is to find a $B^0 \in P_{a,b}$ so that $J(B^0) \leq J(B)$ for all $B \in P_{a,b}$ where

$$J(B) = \int_I g(t, x(t))dt.$$

Our main concern in this section is to show that the identification problem P1 has a solution (not necessarily unique). For this purpose we will need certain preparatory results.

Lemma 3.1

Consider the system (4) and suppose that the operator A satisfies the assumptions A1) and A2) and $B \in P_{a,b}$. Then for every $x_0 \in H$ and $f \in L_2(I, V')$ the system has a unique (weak) solution $x \in L_\infty(I, H) \cap L_2(I, V)$ and further, neglecting a set of measure zero on I, $x \in C(\bar{I}, H)$.

Proof: See Ahmed and Teo [13], Lions and Megenes [14]. ■

Consider the space of operators $\mathcal{L}(V, V')$ and suppose that it is given the strong operator topology which we denote by τ_{so}. Given this topology, $\mathcal{L}_s(V, V') \equiv (\mathcal{L}(V, V'), \tau_{so})$ is locally convex linear topological vector space which is sequentially complete. Similarly, $\mathcal{L}_w(V, V') \equiv (\mathcal{L}(V, V'), \tau_{wo})$ with the weak operator topology τ_{wo} is also a sequentially complete locally convex topological space.

With this background we can prove the following result.

Lemma 3.2

Consider the identification problem P1 and suppose the assumptions of Lemma 3.1 hold and the function $t \to g(t, x)$ is measurable for each $x \in H$; and $x \to g(t, x)$ is lower semicontinuous on H for almost all t. Then the functional $B \to J(B)$ is also lower semicontinuous on $\mathcal{L}_s(V, V')$.

Proof:

Let $\{B^n\} \in \mathcal{L}_s(V, V')$ and suppose $B^n \xrightarrow{\tau_{so}} B^0$. Clearly, $B^0 \in \mathcal{L}_s(V, V')$ and hence it follows from Lemma 3.1 that the system

$$\dot{x} + A(t)x + B^0 x = f,$$

$$x(0) = x_0,$$

has a unique solution $x^0 \in L_\infty(I, H) \cap L_2(I, V) \cap C(\bar{I}, H)$. Similarly, corresponding to each B^n, the system

$$\dot{x} + A(t)x + B^n x = f,$$

$$x(0) = x_0,$$

has a unique solution $x^n \in L_\infty(I, H) \cap L_2(I, V) \cap C(\bar{I}, H)$.

Defining $y^n \equiv (x^n - x^0)$, one observes that y^n is the solution of the problem

$$\dot{y}^n + A(t)y^n + B^n y^n = (B^0 - B^n)x^0, \qquad t \in I,$$

$$y^n = 0.$$

Scalar multiplying the first equation on either side by y^n and using the elementary inequality $ab \leq \frac{1}{2\varepsilon}a^2 + \frac{\varepsilon}{2}b^2$ for all $a, b \in R$ and $\varepsilon > 0$, one can easily verify (using $\varepsilon = \alpha$) that

$$|y^n(t)|_H^2 + \alpha \int_0^t \|y^n(\theta)\|_V^2 d\theta \leq \frac{1}{\alpha} \int_0^t \|(B^0 - B^n)x^0\|_{V'}^2 d\theta + 2(a + \beta) \int_0^t |y^n(\theta)|_H^2 d\theta. \tag{5}$$

Defining

$$\phi^n(t) \equiv |y^n(t)|_H^2 + \alpha \int_0^t \|y^n(\theta)\|_V^2 d\theta, \tag{6}$$

it follows from the above inequality that

$$\phi^n(t) \leq \frac{1}{\alpha} \int_0^T \|(B^0 - B^n)x^0\|_{V'}^2 d\theta + 2|(a+\beta)| \int_0^t \phi^n(\theta) d\theta. \tag{7}$$

Using Gronwall's lemma, one concludes that

$$\phi^n(t) \leq (\exp 2|(a+\beta)|T)(\frac{1}{\alpha}) \int_0^T \|(B^0 - B^n)x^0\|_{V'}^2 d\theta, \tag{8}$$

for all $t \in I$.

Since $B^n \to B^0$ in the strong operator topology in $\mathcal{L}_s(V, V')$ and $x^0 \in L_2(I, V)$ it is clear that $\|(B^0 - B^n)x^0(t)\|_{V'} \to 0$ almost everywhere on I and also there exists a finite number γ such that $\|(B^0 - B^n)x^0\|_{V'} \leq \gamma \|x^0(t)\|_V$. Hence by Lebesgue's dominated convergence theorem it follows that $\phi^n(t) \to 0$ as $n \to \infty$ uniformly on I. Hence one may conclude from (6) and (8) that $x^n \to x^0$ in $C(\bar{I}, H)$ as well as in $L_2(I, V)$ and in particular $x^n(t) \to x^0(t)$ in H for all $t \in \bar{I}$.

Define

$$J(B^n) \equiv \int_I g(t, x^n(t)) dt,$$

and

$$J(B^0) \equiv \int_I g(t, x^0(t)) dt,$$

where x^n and x^0 are the solutions of the system (4) corresponding to B^n and B^0, respectively. Since, by assumption, for almost all $t \in I, x \to g(t, x)$ is lower semicontinuous on H, we have

$$g(t, x^0(t)) \leq \underline{\lim}_n g(t, x^n(t)) \qquad \text{a.e. on } I,$$

and consequently, by Fatou's lemma

$$\int_I g(t, x^0(t)) dt \leq \underline{\lim}_n \int_I g(t, x^n(t)) dt. \tag{9}$$

Clearly, this is equivalent to

$$J(B^0) \leq \underline{\lim}_n J(B^n).$$

This completes the proof of the lemma . ■

Ideally one would like to solve the problem P1 with respect to the class $\mathcal{P}_{a,b}$. However, for technical reasons which will be clear from the following arguments, the problem P1 may not have a solution. So we restrict ourselves to a smaller class $\mathcal{P}^0_{a,b} \subset \mathcal{P}_{a,b}$. Let W be another reflexive Banach space with $V \hookrightarrow W \hookrightarrow H \hookrightarrow W' \hookrightarrow V'$ and suppose that the injection $V \hookrightarrow W$ is compact. Define for any $b > 0$ and $a \in R$,

$$\mathcal{P}^0_{a,b} \equiv \{B \in \mathcal{L}(V, W') : \|B\|_{\mathcal{L}(V,W')} \leq b,$$
$$< B\xi, \xi > +a|\xi|^2_H \geq 0 \quad \text{for all} \quad \xi \in V\}.$$

Lemma 3.3

The set $\mathcal{P}^0_{a,b}$ considered as a subset of $\mathcal{L}(V, V')$ is sequentially compact in the strong operator topology τ_{so}.

Proof:

Since W is reflexive Banach space, so also is W'. Hence $\mathcal{P}^0_{a,b}$ is compact in the weak operator topology of $\mathcal{L}(V, W')$. Therefore, corresponding to every sequence $\{B^k\} \in \mathcal{P}^0_{a,b}$, there exists a subsequence relabeled as $\{B^k\}$ and a $B^0 \in \mathcal{P}^0_{a,b}$ such that $B^k\xi \xrightarrow{w} B^0\xi$ in W' for every $\xi \in V$ with $\|\xi\|_V \leq 1$. But the injection $W' \subset V'$ is compact and hence, for a subsequence, $B^k\xi \xrightarrow{s} B^0\xi$ in V'. Since ξ is arbitrary this implies that $B^k \to B^0$ in the strong operator topology of $\mathcal{L}(V, V')$, i.e. in $\mathcal{L}_s(V, V')$. This proves the assertion. ■

With this background, we can now prove our main result of this subsection which claims that the identification problem P1 has a solution.

THEOREM 3.1

Suppose $g(t, x) > -\infty$ for $(t, x) \in I \times H$. Then under the assumptions of Lemmas 3.1, 3.2 and 3.3, there exists a $B^0 \in \mathcal{P}^0_{a,b}$ such that $J(B^0) \leq J(B)$ for all $B \in \mathcal{P}^0_{a,b}$.

Proof:

Define $\nu \equiv \inf\{J(B) \ , \ B \in \mathcal{P}^0_{a,b}\}$. Since $g(t, x) > -\infty$ for $(t, x) \in I \times H$, the infimum is well defined and $\nu > -\infty$. Let $\{B^k\}$ be a minimizing sequence from $\mathcal{P}^0_{a,b}$, i.e. $\lim_k J(B^k) = \nu$. Then by Lemma 3.3, there exists $\{B^{k_\iota}\} \in \{B^k\}$ relabeled as $\{B^k\}$ and a $B^0 \in \mathcal{P}^0_{a,b}$ such that $B^k \xrightarrow{\tau_{so}} B^0$. Since $B \to J(B)$ is lower semicontinuous with respect to the topology τ_{so} (see Lemma 3.2) and $B^k \xrightarrow{\tau_{so}} B^0$, we have

$$\nu \leq J(B^0) \leq \underline{\lim}_k J(B^k) \leq \lim_k J(B^k) = \nu.$$

Hence $J(B^0) = \nu$ implying that $J(\cdot)$ attains its infimum on $\mathcal{P}^0_{a,b}$. This completes the proof. ■

3.2 Hyperbolic Systems

In this section, we present similar results for structurally damped second order evolution equations. The proofs are essentially similar.

Problem P2 : Consider the second order evolution equation given by

$$\begin{aligned} &\ddot{x} + 2\gamma A(t)\dot{x} + A(t)x + Bx = f, \quad t \in (0,T) \equiv I, \\ &x(0) = x_0, \quad \dot{x}(0) = x_1, \quad B \in \mathcal{P}^0_{a,b}, \end{aligned} \tag{10}$$

and let $g : I \times V \times H \to \bar{R}$. The identification problem is to find a $B^0 \in \mathcal{P}^0_{a,b}$ so that $J(B^0) \leq J(B)$ for all $B \in \mathcal{P}^0_{a,b}$, where

$$J(B) = \int_I g(t, x(t), \dot{x}(t))dt. \tag{11}$$

For the solution of this problem, we need the following results.

THEOREM 3.2

Suppose the operator A is self adjoint and satisfies the basic assumptions A1), A2) and $A \in C^1(\bar{I}, \mathcal{L}(V, V')), \gamma > 0$ and B any element of $\mathcal{P}_{a,b} \subset \mathcal{L}(V, V')$. Then for every $f \in L_2(I, V'), x_0 \in V$ and $x_1 \in H$, the system (10) has a unique solution $x \equiv x(B)$ satisfying the following properties :

(I) $x \in L_\infty(I, V) \subset L_2(I, V)$,

(II) $\dot{x} \in L_\infty(I, H) \cap L_2(I, V)$,

(III) $\ddot{x} \in L_2(I, V')$,

(IV) $x \in C(\bar{I}, V)$, and $\dot{x} \in C(\bar{I}, H)$.

Lemma 3.4

Suppose the hypotheses of Theorem 3.2 hold and consider the identification problem P2. Suppose $t \to g(t, \xi, \eta)$ is measurable for each $\{\xi, \eta\} \in V \times H$, and for almost all $t \in I$, $\{\xi, \eta\} \to g(t, \xi, \eta)$ is lower semicontinuous on $V \times H$ and that $g(t, \xi, \eta) > -\infty$ for all $(t, \xi, \eta) \in I \times V \times H$. Then the functional $B \to J(B)$, (equation (11)), is lower semicontinuous in the strong operator topology of $\mathcal{L}(V, V')$.

Proof:

The proof is essentially similar to that of Lemma 3.2 and follows from the fact that whenever $B^n \xrightarrow{\tau_{so}} B^0, x^n \to x^0$ in $C(\bar{I}, V)$ and $\dot{x}^n \to \dot{x}^0$ in $C(\bar{I}, H)$. The later follows from similar arguments as given in Lemma 3.2 with the only exception that for a priori estimates the equation (10) is now scalar multiplied by $\dot{x}$ instead of x. ■

THEOREM 3.3

Under the assumptions of Theorem 3.2, Lemma 3.3 and Lemma 3.4, the problem P2 has a solution.

Proof: Similar to that of Theorem 3.1. ■

3.3 Undamped Hyperbolic System

We consider now the undamped system

$$\begin{aligned} &\ddot{x} + A(t)x + Bx = f, \quad t \in (0, T) \equiv I, \\ &x(0) = x_0, \quad \dot{x}(0) = x_1, \quad B \in \mathcal{P}_{a,b}. \end{aligned} \tag{12}$$

It is known (see Lions and Magenes [14], vol. 1) that for $f \in L_2(I, H), x_0 \in V$ and $x_1 \in H$, this system has a unique solution; and that $x \in L_\infty(I, V) \subset L_2(I, V)$ and $\dot{x} \in L_\infty(I, H) \subset L_2(I, H)$. Hence $\ddot{x} \in L_2(I, V')$ and consequently it follows from intermediate dervative theorem that $\dot{x} \in C(\bar{I}, H)$ and $x \in C(\bar{I}, [V, H]_{1/2})$. However, by parabolic regularization (which is equivalent to adding a small damping term to the evolution equation and after analysis letting the parameter go to zero) one can verify that $x \in C(\bar{I}, V)$. Thus the results corresponding to the structurally damped system (10) can be applied with slight modification. In fact we shall prove the following result.

Lemma 3.5

Consider the undamped system (12) with the operators A, B as in the prceeding section. Let $f \in L_2(I, H), x_0 \in V$ and $x_1 \in H$, and suppose g satisfies the assumptions of Lemma 3.4. Then $B \to J(B)$ is lower semicontinuous in the strong operator topology of $\mathcal{L}(V, V')$.

Proof:

We give an outline of the proof. We introduce a damping factor $\gamma > 0$ to the system (12) so as to convert it into the system (10). Let $B^n \xrightarrow{T_{so}} B^0$ in $\mathcal{L}(V, V')$ and suppose that x_γ^n and x_γ^0 are the solutions of the regularized problem (see (10)) corresponding to $B = B^n$ and $B = B^0$, respectively. Then it follows from Lemma 3.4 that as $n \to \infty$,

$$\begin{aligned} x_\gamma^n &\to x_\gamma^0 \quad \text{in} \quad C(\bar{I}, V), \\ \dot{x}_\gamma^n &\to \dot{x}_\gamma^0 \quad \text{in} \quad C(\bar{I}, H). \end{aligned} \tag{13}$$

At this point we use the arguments of regularization (see [14], Vol. 1, Theorem 8.3, P. 281) and prove the existence of an x^0 such that as $\gamma \downarrow 0$,

$$\begin{aligned} x_\gamma^0 &\to x^0 \quad \text{in} \quad C(\bar{I}, V), \\ \dot{x}_\gamma^0 &\to \dot{x}^0 \quad \text{in} \quad C(\bar{I}, H), \end{aligned} \tag{14}$$

where x^0 is the solution of the udamped system (12) corresponding to $B = B^0$.

Letting x^n denote the solution of the undamped system with $B = B^n$, one can again show that

$$\begin{aligned} x_\gamma^n &\to x^n \quad \text{in} \quad C(\bar{I}, V), \\ \dot{x}_\gamma^n &\to \dot{x}^n \quad \text{in} \quad C(\bar{I}, H), \end{aligned} \tag{15}$$

as $\gamma \downarrow 0$. Using (13), (14) and (15), we can then verify that as $n \to \infty$,

$$\begin{aligned} x^n &\to x^0 \quad \text{in} \quad C(\bar{I}, V), \\ \dot{x}^n &\to \dot{x}^0 \quad \text{in} \quad C(\bar{I}, H). \end{aligned} \tag{16}$$

The rest of the proof follows immediately from (16) and our assumption on g. ■

Problem P3:Consider the undamped hyperbolic system (12). Our problem P3 is : find a $B \in P_{a,b}^0 \subset \mathcal{L}(V, W') \subset \mathcal{L}(V, V')$ that imparts a minimum to

$$J(B) = \int_I g(t, x(t), \dot{x}(t))dt,$$

subject to the dynamic constraint (12).

THEOREM 3.4

Under the assumption of Lemma 3.5 the identification problem P3 has a solution.

Proof:

The proof is similar to that of Theorem 3.1 and makes use of Lemma 3.5 in place of Lemma 3.3. ■

Remark 3.1

We can also allow the operator B to be function of time by taking for the admissible class, the set

$$\mathcal{P}^0_{a,b} = \{B \in L_\infty(I, \mathcal{L}(V, W')) : ess\ \sup\{\|B(t)\|_{\mathcal{L}(V,W')}, t \in I\} \leq b,$$
$$\text{and} \quad < B(t)\xi, \xi >_{V',V} + a|\xi|^2_H \geq 0 \text{ a.e. on } I\},$$

where $b > 0$ and $a \in R$. The space $L_\infty(I, \mathcal{L}(V, W'))$ may be given the $w^\star-$strong operator $(w^\star - \tau_{so})$ toplogy in the sense that $B^n \overset{w^\star - \tau_{so}}{\longrightarrow} B^0$, whenever

$$\int_I < B^n(t)v, f(t) >_{W',W} dt \to \int_I < B^0(t)v, f(t) >_{W',W} dt,$$

for every $v \in V$ and $f \in L_1(I, W)$. With respect to this topology $\mathcal{P}^0_{a,b}$ is compact.

4. EXISTENCE THEORY IN IDENTIFICATION II:

In section 3 we considered identification problems assuming that the operator A is known. However, in certain situations this operator may also depend on the unknown parameters which is to be identified along with the operator B. In other words, we wish to consider similar identification problems for the following class of systems :

$$\dot{x} + A(q,t)x + Bx = f, \tag{17}$$
$$\ddot{x} + 2\gamma A(q,t)\dot{x} + A(q,t)x + Bx = f, \tag{18}$$
$$\ddot{x} + A(q,t)x + Bx = f, \tag{19}$$

where both q and B are unknown and must be identified on the basis of certain related observations and measurements.

Let Q be a compact metric space and τ_m denote the metric topology on Q and denote the corresponding topological space by Q_m. As before, we denote by $\mathcal{L}_s(V,V')$ the locally convex sequentially complete linear topological vector space $(\mathcal{L}(V,V'),\tau_{so})$. Then we consider the product space $Q_m \times \mathcal{L}_s(V,V')$ with the product topology and our general problem, as discussed, is to identify the elements (q,B) from $Q_m \times \mathcal{P}^0_{a,b} \subset Q_m \times \mathcal{L}_s(V,V')$, where $\mathcal{P}^0_{a,b}$ is a subset of $\mathcal{L}_s(V,V')$ as defined in section 3 (see Lemma 3.3).

The basic assumptions we use in this section are as follows :

A3) The family of operators $\{A(q,\cdot), q \in Q_m\}$ satisfy the assumptions A1) and A2) with certain constants c, α, β independent of $q \in Q_m$.

A4) The mapping $q \to A(q,t)$ from Q_m to $\mathcal{L}_s(V,V')$ is continuous (in the sense that whenever $q^n \xrightarrow{\tau_m} q^0$ in Q_m, $A(q^n,t) \xrightarrow{\tau_{so}} A(q^0,t)$ for almost all $t \in \bar{I} \equiv [0,T]$). For the hyperbolic systems, A is assumed to be selfadjoint and C^1 in t on I, uniformly in $q \in Q_m$.

Let $x \equiv x(q,B)$ denote the solutions of any of the systems (17), (18) or (19), as described above corresponding to $(q,B) \in Q_m \times \mathcal{L}_s(V,V')$ and consider the functionals

$$J(q,B) = \int_I g(t,x(t))dt, \tag{20}$$

for the parabolic case (17) and

$$J(q,B) = \int_I g(t,x(t),\dot{x}(t))dt, \tag{21}$$

for the hyperbolic cases (18) and (19).

We can prove the following results using similar procedure as in section 3.

Lemma 4.1

Consider any of the systems (17), (18) and (19) and suppose that the operator A satisfies the assumptions A3) and A4) with $x_0 \in H$ and $f \in L_2(I,V')$ for the parabolic system (17); and $x_0 \in V, x_1 \in H$ and $f \in L_2(I,V')$ for the hyperbolic system (18); and $x_0 \in V, x_1 \in H$ and $f \in L_2(I,H)$ for the undamped hyperbolic system (19). Let g satisfy the assumptions of Lemma 3.2 for the parabolic system (17) and Lemma 3.4 for the hyperbolic systems (18) and (19). Then the mapping $(q,B) \to J(q,B)$ is lower semicontinuous on $Q_m \times \mathcal{L}_s(V,V')$.

Proof:

We shall give an outline of the proof for the parabolic case. The rest is similar. Suppose $q^n \xrightarrow{\tau_m} q^0$ in Q_m and $B^n \xrightarrow{\tau_{so}} B^0$ in $\mathcal{L}_s(V,V')$. Let $\{x^n\}$ and x^0 denote the solutions for the system (17) corresponding to $\{(q^n, B^n)\}$ and (q^0, B^0), respectively with $x^n(0) = x^0(0) = x_0$. Define $y^n \equiv (x^0 - x^n)$ and ϕ^n with values

$$\phi^n(t) \equiv |y^n(t)|_H^2 + \alpha \int_0^t \|y^n(\theta)\|_V^2 d\theta, \quad t \in I. \tag{22}$$

Then under the assumption A3) one shows, as in section 3, that

$$\begin{aligned} \phi^n(t) \leq & \frac{2}{\alpha}\{\int_0^t \|(A(q^n,\theta) - A(q^0,\theta))x^0(\theta)\|_{V'}^2 d\theta \\ & + \int_0^t \|(B^n - B^0)x^0(\theta)\|_{V'}^2 d\theta\} + 2|(a+\beta)| \int_0^t \phi^n(\theta)d\theta, \quad t \in I. \end{aligned} \tag{23}$$

Hence by Gronwall's lemma

$$\phi^n(t) \leq \eta^n \exp 2|(a+\beta)|t \qquad \text{for all} \quad t \in I,$$

where

$$\begin{aligned} \eta^n = \frac{1}{\alpha}\{ & \int_0^T \|(A(q^n,t) - A(q^0,t))x^0(t)\|_{V'}^2 dt \\ & + \int_0^T \|(B^n - B^0)x^0(t)\|_{V'}^2 dt\}. \end{aligned}$$

Since $B^n \xrightarrow{\tau_{so}} B^0$ in $\mathcal{L}_s(V, V')$ and, by assumption A4), $A(q^n, t) \xrightarrow{\tau_{so}} A(q^0, t)$ in $\mathcal{L}_s(V, V')$ whenever $q^n \xrightarrow{\tau_m} q^0$ in Q_m, it follows from dominated convergence theorem that $\eta^n \to 0$ as $n \to \infty$. Hence $\phi^n(t) \to 0$ as $n \to \infty$ uniformly on I. The rest of the proof is concluded as in Lemma 3.2. ■

THEOREM 4.1

Suppose the hypotheses of Lemma 4.1 hold. Then the functional $J(q, B)$ attains its minimum on $Q_m \times \mathcal{P}_{a,b}^0$, ($\mathcal{P}_{a,b}^0$ as in Lemma 3.3), and hence the systems (17), (18) and (19) are identifiable.

Proof:

The proof follows from lower semicontinuity of J on $Q_m \times \mathcal{L}_s(V, V')$ and compactness of the set $Q_m \times \mathcal{P}_{a,b}^0$ in the product topology $\tau_m \times \tau_{so}$. ■

The preceeding results can be further generalized. In general one may consider that all the variables $\{x_0, x_1, f, B, A(q, \cdot)\}$ appearing in the system equations (17)-(19) are to be identified.

In the following, we present a result for the parabolic system. For hyperbolic systems similar results can be deduced following the same procedure.

For each $(q, B) \in Q_m \times \mathcal{L}_s(V, V')$, define

$$L_{q,B}\psi \equiv \dot{\psi} + A(q,t)\psi + B\psi,$$

and

$$X_{q,B} \equiv \{\phi \in L_2(I,V) : \phi(T) = 0, L^*_{q,B}\phi = \xi, \xi \in L_2(I,V')\}.$$

The set $X_{q,B}$ is the space described by the solutions of

$$L^*_{q,B}\phi \equiv -\dot{\phi} + A^*(q,t)\phi + B^*\phi = \xi, \qquad \phi(T) = 0,$$

as ξ describes the space $L_2(I,V')$. We assume that $X_{q,B}$ is independent of $(q,B) \in Q_m \times \mathcal{P}^0_{a,b}$ and write $X_{q,B} \equiv X$ and furnish X with the norm topology

$$\|\phi\|_X = \|L^*_{q,B}\phi\|_{L_2(I,V')} = \|\xi\|_{L_2(I,V')},$$

and note that the norm, however, may depend on (q,B). Clearly, X furnished with the above graph norm is a Banach space and for each $(q,B) \in Q_m \times \mathcal{P}^0_{a,b}$, $L^*_{q,B}$ is an isomorphism of X onto $L_2(I,V')$.

Let C_0 be a weakly compact subset of H, $\mathcal{F}$ a weakly compact subset of $L_2(I,V')$ and Q_m and $\mathcal{P}^0_{a,b}$ are as given before. For the identification problem we consider $\mathcal{A} \equiv C_0 \times \mathcal{F} \times Q_m \times \mathcal{P}^0_{a,b}$ to be the admissible set. Let $x \equiv x(x_0, f, q, B)$ denote the solution of the problem (17) corresponding to $x_0 \in C_0, f \in \mathcal{F}, q \in Q_m$ and $B \in \mathcal{P}^0_{a,b}$. For a given $g : I \times H \to \bar{R}$, the problem is to find $(x^0_0, f^0, q^0, B^0) \in \mathcal{A}$ so that

$$J(x^0_0, f^0, q^0, B^0) \leq J(x_0, f, q, B), \quad \text{for all} \quad (x_0, f, q, B) \in \mathcal{A},$$

where

$$J(x_0, f, q, B) \equiv \int_0^T g(t, x(t))dt \qquad \text{with} \quad x = x(x_0, f, q, B).$$

We introduce the following assumption:

A5) :

$$\left.\begin{matrix} A(q^n,t) \\ A^*(q^n,t) \end{matrix}\right\} \xrightarrow{\tau_{so}} \left\{\begin{matrix} A(q^0,t) \\ A^*(q^0,t) \end{matrix}\right. \qquad \text{in } \mathcal{L}_s(V,V'),$$

whenever $q^n \xrightarrow{\tau_m} q^0$ in Q_m, and

$$\left.\begin{matrix} B^n \\ (B^n)^* \end{matrix}\right\} \xrightarrow{\tau_{so}} \left\{\begin{matrix} B^0 \\ (B^0)^* \end{matrix}\right. \qquad \text{in } \mathcal{L}_s(V,V'),$$

simultaneously.

Lemma 4.2

Suppose the hypotheses A3), A4) and A5) hold. Let $t \to g(t,\xi)$ be a measurable function on $I \equiv (0,T)$ for each $\xi \in H$; and $\xi \to g(t,\xi)$ be weakly lower semicontinuous on H for almost all $t \in I$ and that $g(t,\xi) > -\infty$ for all $(t,\xi) \in I \times H$.

Then the functional $(x_0, f, q, B) \to J(x_0, f, q, B)$ is lower semicontinuous on $\mathcal{A} \equiv \mathcal{C}_0 \times \mathcal{F} \times \mathcal{Q}_m \times \mathcal{P}^0_{a,b}$ with respect to the product topology.

Proof:

Let

$$
\begin{aligned}
x_0^n &\xrightarrow{w} x_0^0 && \text{in } \ \mathcal{C}_0 \subset H, \\
f^n &\xrightarrow{w} f^0 && \text{in } \ L_2(I, V'), \\
q^n &\xrightarrow{\tau_m} q^0 && \text{in } \ \mathcal{Q}_m, \\
B^n &\xrightarrow{\tau_{so}} B^0 && \text{in } \ \mathcal{P}^0_{a,b} \subset \mathcal{L}_s(V, V').
\end{aligned}
\tag{24}
$$

Let $x^n \equiv x(x_0^n, f^n, q^n, B^n)$ denote the solution of the problem (17) with $x_0 = x_0^n, f = f^n, q = q^n$ and $B = B^n$ and similarly $x^0 = x(x_0^0, f^0, q^0, B^0)$. We will show that $x^n(t) \xrightarrow{w} x^0(t)$ for each $t \in I$. First we show that $x^n \xrightarrow{w} x^0$ in $L_2(I, V)$. Defining $y^n \equiv (x^n - x^0)$ we have, in $L_2(I, V)$,

$$
\begin{aligned}
\dot{y}^n + A(q^n, t)y^n + B^n y^n &= f^n - f^0 + (B^n - B^0)x^0 \\
&\quad + (A(q^n, t) - A(q^0, t))x^0, \qquad t \in I, \\
y^n(0) &= (x_0^n - x_0^0),
\end{aligned}
$$

or equivalently

$$
\begin{aligned}
\dot{y}^n + A(q^0, t)y^n + B^0 y^n &= (f^n - f^0) + (B^n - B^0)x^0 + (A(q^n, t) - A(q^0, t))x^0 \\
&\quad + (B^0 - B^n)y^n + (A(q^0, t) - A(q^n, t))y^n, \\
y^n(0) &= (x_0^n - x_0^0).
\end{aligned}
\tag{25}
$$

Defining the operator $L_0 \equiv L_{\{q^0, B^0\}}$ and scalar multiplying (25) by $\phi \in X$ and integrating over I, we obtain

$$
\begin{aligned}
\int_I < y^n, L_0^* \phi > dt &= \ell^n(\phi) \\
&\equiv \ell_1^n(\phi) + \ell_2^n(\phi) + \ell_3^n(\phi) + \ell_4^n(\phi) + \ell_5^n(\phi) + \ell_6^n(\phi),
\end{aligned}
\tag{26}
$$

where

$$
\begin{aligned}
\ell_1^n(\phi) &\equiv (y^n(0), \phi(0))_H, \\
\ell_2^n(\phi) &\equiv \int_I < f^n - f^0, \phi >_{V',V} \ dt, \\
\ell_3^n(\phi) &\equiv \int_I < (B^n - B^0)x^0, \phi >_{V',V} \ dt, \\
\ell_4^n(\phi) &\equiv \int_I < (A(q^n, t) - A(q^0, t))x^0, \phi >_{V',V} \ dt,
\end{aligned}
\tag{27}
$$

$$\ell_5^n(\phi) \equiv \int_I < y^n, (B^0 - B^n)^\star \phi >_{V,V'} \ dt,$$

$$\ell_6^n(\phi) \equiv \int_I < y^n, (A(q^0,t) - A(q^n,t))^\star \phi >_{V,V'} \ dt.$$

Note that by virtue of Lemma 3.1, $X \subset L_2(I,V) \cap C(\bar{I},H)$ and hence $\phi(0)$ is well defined as an element of H. Consequently, it follows from (24) and (27) that for each $\phi \in X, \ell_i^n(\phi) \to 0$, as $n \to \infty$ for $i = 1,2,3,4$. For the terms $\ell_5^n(\phi)$ and $\ell_6^n(\phi)$, note that

$$\begin{aligned} \ell_5^n(\phi) &\le (\int_I \|y^n(t)\|_V^2 dt)^{1/2} (\int_I \|(B^0 - B^n)^\star \phi\|_{V'}^2 dt)^{1/2}, \\ \ell_6^n(\phi) &\le (\int_I \|y^n(t)\|_V^2 dt)^{1/2} (\int_I \|(A(q^0,t) - A(q^n,t))^\star \phi\|_{V'}^2 dt)^{1/2}. \end{aligned} \tag{28}$$

Hence by virtue of assumption A5), dominated convergence theorem and the fact that $\{y^n\}$ lies in a bounded subset of $L_2(I,V)$ (due to assumption A3), and boundedness of the sets $\mathcal{P}_{a,b}, \mathcal{C}_0$ and $\mathcal{F}$) it follows from the above inequalities that $\ell_5^n(\phi) \to 0$ and $\ell_6^n(\phi) \to 0$ for each $\phi \in X$. Therefore, we conclude that

$$\lim_n \int_I < y^n, L_0^\star \phi > dt = 0, \qquad \text{for each} \quad \phi \in X.$$

Since $L_0^\star$ is an isomorphism of X onto $L_2(I,V')$, this means that $y^n \xrightarrow{w} 0$ in $L_2(I,V)$ or equivalently $x^n \xrightarrow{w} x^0$ in $L_2(I,V)$ and certainly in $L_2(I,H)$. Recalling that $\{x^n, x^0\} \in C(\bar{I},H)$ and that

$$\sup_n \{\sup_{t \in \bar{I}} |x^n(t)|_H^2\} < \infty,$$

we may conclude that, for each $t \in \bar{I} = [0,T]$,

$$x^n(t) \xrightarrow{w} x^0(t) \qquad \text{in} \quad H.$$

Thus we have proved that as $x_0^n \xrightarrow{w} x_0^0$ in $H, f^n \xrightarrow{w} f^0$ in $L_2(I,V'), B^n \xrightarrow{\tau_{so}} B^0$ and $A(q^n,t) \xrightarrow{\tau_{so}} A(q^0,t), t \in I$, in $\mathcal{L}_s(V,V')$,

$$x^n(t) = x(x_0^n, f^n, q^n, B^n)(t) \xrightarrow{w} x(x_0^0, f^0, q^0, B^0)(t) = x^0(t),$$

in H for all $t \in \bar{I}$. Since g is assumed to be weakly lower semicontinuous on H and bounded away from $-\infty$, we obtain

$$\int_I g(t, x^0(t)) dt \le \int_I \underline{\lim}_n\, g(t, x^n(t)) dt \le \underline{\lim}_n \int_I g(t, x^n(t)) dt,$$

which is equivalent to

$$J(x_0^0, f^0, q^0, B^0) \leq \underline{\lim}_n J(x_0^n, f^n, q^n, B^n).$$

This completes the proof of the lemma . ■

THEOREM 4.2

Let $\mathcal{A} \equiv C_0 \times \mathcal{F} \times Q_m \times P_{a,b}^0$ with C_0 and $\mathcal{F}$ being weakly compact subset of H and $L_2(I, V')$, respectively, and Q_m and $P_{a,b}^0$ as defined before. Let g satisfy the assumptions as given in Lemma 4.2. Then the functional $(x_0, f, q, B) \rightarrow J(x_0, f, q, B)$ attains its infimum on $\mathcal{A}$.

Proof: The proof follows immediately from Lemma 4.2 using similar arguments as in Theorem 3.1. ■

Remark 4.1

Following similar procedure, one can prove existence results for the hyperbolic systems with all the parameters $\{x_0, x_1, f, q, B\}$ considered unkown.

Remark 4.2

In all the Lemmas 3.2, 3.4, 3.5, 4.1 and 4.2, the Lebesgue measure used for the cost integrals can be replaced by any bounded positive Radon measure ν on I, provided the integrand g is continuous in t on I. That is, the error functionals

$$J = \int_I g(t, x(t))dt,$$

and

$$J = \int_I g(t, x(t), \dot{x}(t))dt,$$

can be replaced by

$$J = \int_I g(t, x(t))\nu(dt),$$

and

$$J = \int_I g(t, x(t), \dot{x}(t))\nu(dt),$$

respectively. This allows for pointwise observations with weights attached to each observation as dseired; for example, in case of pure pointwise observation, we have

$$J = \sum_i \beta_i g(t_i, x(t_i)) \quad \text{or} \quad J = \sum_i \beta_i g(t_i, x(t_i), \dot{x}(t_i)),$$

where $\nu(dt) = \sum_i \beta_i \delta_{t_i}(dt)$ with δ_{t_i} denoting the Dirac measure with support $\{t_i\}$ and $\beta_i \geq 0$, such that $\sum \beta_i < \infty$.

In the preceeding results we have used $\mathcal{P}^0_{a,b}$ as the admissible class for identification of the operator B. In that we have used the conditions :

(i) for all $\xi \in V$,

$$< B\xi, \xi >_{V',V} + a|\xi|^2_H \geq 0 \ , \quad a \in R,$$

and

(ii) $\|B\|_{\mathcal{L}(V,W')} \leq b < \infty$.

Condition (i) is essentially used for apriori estimates and can be eleminated if the constant b in (ii) is taken sufficiently small. This is formally presented in the following result.

Proposition 4.1

All the preceeding results remain valid if the set $\mathcal{P}^0_{a,b}$ is replaced by a ball $\mathcal{B}_{b_0} \subset \mathcal{L}(V, W')$ of sufficiently small radius b_0.

Proof:

Let η_0 be the embedding constant for the (compact) injection $W' \subset V'$. That is

$$\eta_0 = \inf\{\eta \geq 0 : \|\xi\|_{V'} \leq \eta \|\xi\|_{W'} \quad \text{for all} \quad \xi \in W'\}.$$

Then η_0 is also the embedding constant for the injection $\mathcal{L}(V, W') \subset \mathcal{L}(V, V')$. Choose $b_0 > 0$ such that $b_1 \equiv \eta_0 b_0 < \alpha$, where α is the (coercivity) constant appearing in the assumption A1) (section 3) and A3) (section 4). Under this condition, apriori estimates are again available since $\|B\|_{\mathcal{L}(V,V')} \leq \eta_0 \|B\|_{\mathcal{L}(V,W')} \leq \eta_0 b_0 = b_1 < \alpha$, for all $B \in \mathcal{B}_{b_0}$. Also our requirment for compactness is satisfied since the set $\mathcal{B}_{b_0}$ is compact in the weak operator topology of $\mathcal{L}(V, W')$ and hence, by assumption of compactness of the injection $W' \subset V'$, $\mathcal{B}_{b_0}$ considered as a subset of $\mathcal{L}(V, V')$, is compact in its strong operator topology. ■

In case $\{A(q), q \in Q\}$ are merely generators of C_0−semigroups in a Hilbert space H, we can prove similar existence results for the following problem.

$$\dot{x} = A(q)x,$$
$$J(q) = \int_I g(t, x(t))dt = \text{min.}, \tag{29}$$

with g satisfying the hypotheses of Lemma 3.2.

We use the standard symbol $R(\lambda, A) \equiv (\lambda I - A)^{-1}$ to denote the resolvent of the operator A corresponding to $\lambda \in \rho(A) \equiv$ resolvent set.

THEOREM 4.3

Let Q be a compact metric space and suppose for each $q \in Q, A(q)$ is the generator of a strongly continuous contraction semigroup in H and suppose

$$R(\lambda, A(q^n)) \to R(\lambda, A(q^0)),$$

in the strong operator topology for each $\lambda > 0$, whenever $q^n \to q^0$ in Q. Then the problem (29) has a solution.

Proof:

The proof follows from Trotter-Kato theorem and the lower semicontinuity of g. ■

5. NECESSARY CONDITIONS FOR OPTIMALITY IN IDENTIFICATION

In this section we present several necessary conditions of optimality for the identification problems as stated in the previous sections. First we note that usually the mapping $B \to x(B)$ from $\mathcal{L}(V, V')$ to $L_2(I, V)$ is unique, however, for $\xi \in Z \equiv \{x(B) : B \in \mathcal{P}_{a,b}\}$ the set $x^{-1}(\xi) \equiv \{B \in \mathcal{P}_{a,b} : x(B) = \xi\}$ is a closed subset of $\mathcal{P}_{a,b}$ and may contain infinitely many points. Hence a unique solution to the identification problem is not expected and in many instances is not essential.

Let $\mathcal{H}$ denote the Hilbert space of observations and $\tilde{y} \in C(\bar{I}, \mathcal{H})$ the observed data or the response of the natural system. We assume that the observation equation for the model system is given by

$$y = Cx, \tag{30}$$

where $C \in \mathcal{L}(H, \mathcal{H})$ and x is the response of any of the model systems (4) (parabolic), (10) (hyperbolic), and (12) (undamped hyperbolic) corresponding to $B \in \mathcal{P}_{a,b}$. We seek a $B^0 \in \mathcal{P}_{a,b}$ that minimizes the mean square error between the model output y and the observed data $\tilde{y}$. That is

$$\begin{aligned} J(B^0) &\equiv \frac{1}{2}\int_I \|Cx(B^0)(t) - \tilde{y}(t)\|^2_{\mathcal{H}} dt \\ &\le \frac{1}{2}\int_I \|Cx(B)(t) - \tilde{y}(t)\|^2_{\mathcal{H}} dt = J(B), \end{aligned} \tag{31}$$

for all $B \in \mathcal{P}_{a,b}$. Defining

$$g(t,\xi) \equiv \frac{1}{2}\|C\xi - \tilde{y}(t)\|_{\mathcal{Y}}^2, \quad \xi \in H, \quad t \in \bar{I},$$

we note that this is a special case of the general identification problems, P1, P2 and P3, considered in the preceeding sections and hence the corresponding existence theorems, Theorem 3.1, Theorem 3.3 and Theorem 3.4 hold with $\mathcal{P}_{a,b} = \mathcal{P}^0_{a,b}$.

For the proof of necessary conditions of optimality, we shall make use of the Gateaux differential of $x(B)$ with respect to the operator B. Indeed we show that the Gateaux differential of x at B^0 in the direction B,

$$\hat{x}(B^0, B) = \text{w-}\lim_{\varepsilon \to 0}((x(B^0 + \varepsilon B) - x(B^0))/\varepsilon), \tag{32}$$

exists and that it is the solution of a related differential equation.

Throughout the rest of the paper, we assume without further notice that the operators A and B and their adjoints satisfy the basic assumptions of the previous sections.

5.1 Parabolic System

We consider the parabolic system (4) and present in the following lemma the Gateaux differentiability of $B \to x(B)$ in the weak sense.

Lemma 5.1

Let $x(B)$ denote the (weak) solution of the Cauchy problem (4) corresponding to $B \in \mathcal{P}_{a,b}$. Then at each point $B^0 \in \mathcal{P}_{a,b}$, the function $B \to x(B)$ has a weak Gateaux differential in the direction $B - B^0$, denoted $\hat{x}(B^0, B - B^0)$, and it is the solution of the Cauchy problem

$$\begin{aligned} &\dot{e} + A(t)e + B^0 e = (B^0 - B)x(B^0), \\ &e(0) = 0\ . \end{aligned} \tag{33}$$

Proof:

Let $B^0, B \in \mathcal{P}_{a,b}$. Since $\mathcal{P}_{a,b}$ is a closed convex subset of $\mathcal{L}_w(V, V')$,

$$B^\varepsilon \equiv B^0 + \varepsilon(B - B^0) \in \mathcal{P}_{a,b}\ , \quad \text{for} \quad 0 \le \varepsilon \le 1.$$

Define

$$\phi^\varepsilon \equiv (x(B^\varepsilon) - x(B^0))/\varepsilon.$$

Then using the differential equation (4), one obtains

$$\dot{\phi}^\varepsilon + A(t)\phi^\varepsilon + B^\varepsilon\phi^\varepsilon = (B^0 - B)x(B^0) \ ,$$
$$\phi^\varepsilon(0) = 0 \ . \tag{34}$$

We show that the Gateaux differential given by the weak limit (in $L_2(I,V)$) of ϕ^ε or a subsequence thereof exists and it is the (weak) solution of (33). Since $(B^0 - B)x(B^0) \in L_2(I,V')$, it follows from (34), as in Lemma 3.2, that the set $\{\phi^\varepsilon \ , \ \varepsilon \in [0,1]\}$ is contained in a bounded subset of $L_2(I,V) \cap L_\infty(I,H)$. Hence from every sequence $\phi^n \equiv \phi^{\varepsilon_n}$, with $\varepsilon_n \in [0,1]$ and $\varepsilon_n \to 0$, one can extract a subsequence relabeled as $\{\phi^n\}$ and a $\phi^0 \in L_2(I,V) \cap L_\infty(I,H)$ such that $\phi^n \xrightarrow{w} \phi^0$ in $L_2(I,V)$. Hence the Gateaux diifferential of x exists and it is given by $\hat{x}(B^0, B - B^0) = \phi^0$. It remains to show that ϕ^0 is the solution of (33).

Indeed, since $A\phi^n \xrightarrow{w} A\phi^0$ in $L_2(I,V')$, (weak and weak* convergence being equivalent in reflexive Banach spaces) and

$$B^n\phi^n = B^0\phi^n + \varepsilon_n(B - B^0)\phi^n \xrightarrow{w} B^0\phi^0, \quad \text{in} \quad L_2(I,V') \ ,$$

it follows from (34) that $\dot{\phi}^n \in L_2(I,V')$ and $\dot{\phi}^n \xrightarrow{w} \eta$ in $L_2(I,V')$ for a suitable $\eta \in L_2(I,V')$ and that η is the distributional derivative of ϕ^0. Hence ϕ^0 satisfies the equality

$$\dot{\phi}^0 + A(t)\phi^0 + B^0\phi^0 = (B^0 - B)x(B^0) \ ,$$

in the sense of vector-valued distributions in V'.

Since $\phi^0 \in L_2(I,V)$ and $\dot{\phi}^0 \in L_2(I,V')$, it is clear that $\phi^0 \in C(\bar{I},H)$ and $\phi^0(0)$ is well defined and equals $\phi^n(0) = 0$, for all n. Hence ϕ^0 satisfies the differential equation (33) and one may identify ϕ^0 as e. This completes the proof. ■

With the help of the above lemma, we prove the following necessary conditions for optimality.

THEOREM 5.1

Consider the parabolic system (4) and the identification problem P1 with $g(t,\xi) \equiv \frac{1}{2}\|C\xi - \tilde{y}(t)\|^2_{\mathcal{H}}$ and $J(B) = \int_I g(t,x(t))dt$. Then the best approximation B^0 for the unknown operator is determined by the simultaneous solution of the system equation

$$\dot{x} + A(t)x + B^0x = f \ ,$$
$$x(0) = x_0 \ , \tag{35}$$

the adjoint equation

$$-\dot{z} + A^*(t)z + (B^0)^* z = C^* \Lambda_{\mathcal{H}}(Cx(B^0) - \tilde{y}) , \quad z(T) = 0 , \tag{36}$$

and the inequality

$$\int_I < B^0 x(B^0), z >_{V',V} dt \geq \int_I < Bx(B^0), z >_{V',V} dt , \tag{37}$$

for all $B \in \mathcal{P}^0_{a,b}$.

Proof:

Since $B \to J(B)$ has (weak) Gateaux differential on $\mathcal{P}_{a,b}$, it follows that J, as defined above, also has a Gateaux differential. Then, in order that J attains its minimum at $B^0 \in \mathcal{P}^0_{a,b}$ (whose existence is assured by Theorem 3.1), it is necessary that

$$J'_{B^0}(B - B^0) \equiv \lim_{\varepsilon \to 0}\{(J(B^0 + \varepsilon(B - B^0)) - J(B^0))/\varepsilon\} \geq 0 , \tag{38}$$

for all $B \in \mathcal{P}^0_{a,b}$. Using the result of Lemma 5.1, it follows from the above that

$$J'_{B^0}(B - B^0) = \int_I (C\hat{x}(B^0, B - B^0), Cx(B^0) - \tilde{y})_{\mathcal{H}} \, dt \geq 0 , \tag{39}$$

for all $B \in \mathcal{P}^0_{a,b}$, where $\hat{x}$ is the Gateaux differential as given by Lemma 5.1. Using the canonical isomorphism $\Lambda_{\mathcal{H}}(\ (\Lambda_{\mathcal{H}}(x), x) \ = \|x\|^2_{\mathcal{H}}, \|\Lambda_{\mathcal{H}}(x)\|_{\mathcal{H}'} \ = \|x\|_{\mathcal{H}}$ for all $x \in \mathcal{H})$, we can rewrite (39) as

$$J'_{B^0}(B - B^0) = \int_I (\hat{x}(B^0, B - B^0), C^* \Lambda_{\mathcal{H}}(Cx(B^0) - \tilde{y}))_{\mathcal{H}} \, dt \geq 0, \tag{40}$$

for all $B \in \mathcal{P}^0_{a,b}$, where C^* is the adjoint of C and belongs to $\mathcal{L}(\mathcal{H}', H)$. The inequality (40) can be further simplified by introducing the so called adjoint variable z, which is the solution of the following equation :

$$-\dot{z} + A^*(t)z + (B^0)^* z = C^* \Lambda_{\mathcal{H}}(Cx(B^0) - \tilde{y}) , \quad z(T) = 0 . \tag{41}$$

Then, since $C^* \Lambda_{\mathcal{H}}(Cx(B^0) - \tilde{y}) \in L_2(I, H) \subset L_2(I, V')$, reversing the flow of time $t \to T - t$, it follows from Lemma 3.2 that the system (41) also has a unique weak solution $z \in L_2(I, V) \cap L_\infty(I, H) \cap C(\bar{I}, H)$.

Utilizing (41) into the inequality (40) and integrating by parts, one obtains

$$\int_I < \dot{\hat{x}} + A(t)\hat{x} + B^0 \hat{x}, z >_{V',V} \quad dt \ \geq 0 \ . \tag{42}$$

The necessary inequality (37) now follows from (33) and (42). This completes the proof. ■

5.2 Damped Hyperbolic System

In this section, we consider the structurally damped second order evolution equation (10) as in section 3.2. The proofs of the necessary conditions of optimality are essentially similar and hence omitted.

Lemma 5.2

Suppose the assumptions of Theorem 3.2 hold, and let $x(B)$ denote the solution of the evolution equation (10) corresponding to $B \in \mathcal{L}(V,V')$. Then at each point $B^0 \in \mathcal{P}_{a,b}$, the function $B \to x(B)$ has a (weak) Gateaux differential in the direction $B - B^0 \in \mathcal{L}(V,V')$, denoted $\hat{x}(B^0, B - B^0)$, and it is the solution of the Cauchy problem :

$$\begin{aligned} &\ddot{e} + 2\gamma A(t)\dot{e} + A(t)e + B^0 e = (B^0 - B)x(B^0) , \\ &e(0) = 0 , \\ &\dot{e} = 0 . \end{aligned} \tag{43}$$

THEOREM 5.2

Consider the system (10) and suppose the assumptions of Theorem 3.2 hold and let the functional $J(B)$ (equation (11)) be given by

$$J(B) = \int_I g(t, x(B)(t))dt , \qquad B \in \mathcal{P}^0_{a,b} ,$$

with $g(t,\xi)$ as defined in Theorem 5.1. Then, in order that $B^0 \in \mathcal{P}^0_{a,b}$ be the best approximation to the unknown operator $B \in \mathcal{L}(V,V')$, it is necessary that the pair $\{x, z\}$ satisfy

$$\begin{aligned} &\ddot{x} + 2\gamma A(t)\dot{x} + A(t)x + B^0 x = f , \\ &x(0) = x_0 , \\ &\dot{x}(0) = x_1 , \end{aligned} \tag{44}$$

the adjoint system

$$\begin{aligned} &\ddot{z} - 2\gamma A^*(t)\dot{z} + A^*(t)z + (B^0)^* z = C^* \Lambda_{\mathcal{Y}}(Cx(B^0) - \tilde{y}) , \\ &z(T) = 0 , \\ &\dot{z}(T) = 0 , \end{aligned} \tag{45}$$

and the inequality

$$\int_I < B^0 x(B^0), z >_{V',V} \quad dt \geq \int_I < Bx(B^0), z >_{V',V} \quad dt , \tag{46}$$

for all $B \in \mathcal{P}^0_{a,b}$.

Proof: Similar to that given for Theorem 5.1. ■

5.3 Undamped Hyperbolic System

In this section, we present the necessary conditions of optimality for the undamped system (12) of section 3.3.

Lemma 5.3

Suppose the operator A satisfy the assumption of Theorem 3.2 with $\beta = 0$ (see A2)). Then, the solution x of the system (12) has a weak Gateaux differential at each point $B^0 \in \mathcal{P}_{a,b}$ in the direction $(B - B^0)$, denoted $\hat{x}(B^0, B - B^0)$, and is given by the weak solution of

$$\begin{aligned} &\ddot{e} + Ae + B^0 e = (B^0 - B)x(B^0), \\ &e(0) = 0 , \\ &\dot{e}(0) = 0 , \end{aligned} \tag{47}$$

where $x(B^0)$ is the response of the system (12) (see section 3.3) corresponding to $B = B^0 \in \mathcal{P}_{a,b}$. Further, $\hat{x} \in L_2(I, H) \cap C(\bar{I}, H)$, and $\dot{\hat{x}} \in L_2(I, V') \cap C(\bar{I}, V')$.

Proof:

Let $x(B^\varepsilon)$ and $x(B^0)$ denote the solutions of the problem (12) corresponding to $B^\varepsilon = B^0 + \varepsilon(B - B^0)$ and B^0, respectively with $0 \leq \varepsilon \leq 1$ and $B, B^0 \in \mathcal{P}_{a,b} \subset \mathcal{L}(V, V')$. Defining

$$\phi^\varepsilon \equiv (x(B^\varepsilon) - x(B^0)/\varepsilon,$$

one obtains

$$\begin{aligned} &\ddot{\phi}^\varepsilon + A(t)\phi^\varepsilon + B^\varepsilon \phi^\varepsilon = (B^0 - B)x(B^0), \\ &\phi^\varepsilon(0) = 0 , \qquad \dot{\phi}^\varepsilon(0) = 0 . \end{aligned} \tag{48}$$

Since $x(B^0) \in L_2(I, V), (B^0 - B)x(B^0) \in L_2(I, V')$, hence the standard techniques based on energy estimates can not be directly applied. However, by the method of transposition (see [14], Vol. 1, Theorem 9.3, P.288), which is essentially equivalent to the well known Lax-Milgram Theorem, one can show that both (47) and (48) have solutions satisfying

$$e\ ,\ \phi^{\varepsilon} \in L_2(I,H)\cap C(\bar{I},H)\ ,$$
$$\dot{e}\ ,\ \dot{\phi}^{\varepsilon} \in L_2(I,V')\cap C(\bar{I},V') \quad \text{for all} \quad \varepsilon\in[0,1]\ . \tag{49}$$

Further, e solves problem (47) in the sense that

$$\int_I (e,\ddot{\psi}+A^*(t)\psi+(B^0)^*\psi)_H\ dt=\int_I <(B^0-B)x(B^0),\psi>_{V',V}\ \ dt, \tag{50}$$

for all $\psi\in X\equiv\{\psi\in L_2(I,V): L_0^*\psi\in L_2(I,H),\psi(T)=0,\dot{\psi}(T)=0\}$, where

$$L_0\equiv(\frac{d^2}{dt^2}+A(t)+B)$$

with L_0^* being the adjoint. Similarly, for each $\varepsilon\in[0,1]$, ϕ^{ε} solves the problem (48) in the sense that

$$\int_I (\phi^{\varepsilon},\ddot{\psi}+A^*(t)\psi+(B^{\varepsilon})^*\psi)_H\ dt=\int_I <(B^0-B)x(B^0),\psi>_{V',V}\ \ dt\ , \tag{51}$$

for all $\psi\in X$ such that $L_{\varepsilon}^*\psi\equiv\ddot{\psi}+A^*(t)\psi+(B^{\varepsilon})^*\psi\in L_2(I,H)$.

Note that X furnished with the norm topology given by $\|\psi\|_X=\|L_0^*\psi\|_{L_2(I,H)}$, is a Banach space and L_0^* (as well as L_{ε}^*) is an isomorphism of X onto $L_2(I,H)$. Since both $L_0^*\psi$ and $L_{\varepsilon}^*\psi\in L_2(I,H)$, it is clear that $\varepsilon(B-B^0)\psi\in L_2(I,H)$ and converges strongly to zero as $\varepsilon\to 0$. Letting $\varepsilon\to 0$ in (51) and denoting the weak limit of ϕ^{ε} by ϕ^0, one obtains

$$\int_I (\phi^0,\ddot{\psi}+A^*(t)\psi+(B^0)^*\psi)_H\ dt=\int_I <(B^0-B)x(B^0),\psi>_{V',V}\ \ dt\ , \tag{52}$$

for all $\psi\in X$. Comparing (50) with (52), we obtain

$$\int_I (e-\phi^0\ ,\ L_0^*\psi)\ dt\ =0\ , \tag{53}$$

for all $\psi\in X$. Since L_0^* is an isomorphism of X onto $L_2(I,H)$, it follows from (53) that $\phi^0=e$. Hence $\hat{x}(B^0,B-B^0)=\phi^0=e$. This completes the proof. ■

Using Lemma 5.3, we can prove the following result.

<u>THEOREM 5.3</u>

Consider the undamped system (12) and suppose the assumptions of Lemma 5.3 hold and $x_0\in V, x_1\in H, f\in L_2(I,H)$ and

$$J(B)=\int_I g(t,x(B)(t))dt\ ,$$

with $g(t,\xi)=\frac{1}{2}\|C\xi-\tilde{y}(t)\|_{\mathcal{H}}^2, \tilde{y}\in L_2(I,\mathcal{H})$ and $C\in\mathcal{L}(H,\mathcal{H})$. Then in order that $B^0\in P_{a,b}^0$ be the best approximation to the unknown operator B, it is necessary that the pair $\{x,z\}$ satisfy

$$\ddot{x}+A(t)x+B^0x=f\ ,$$
$$x(0)=x_0\ , \tag{54}$$
$$\dot{x}(0)=x_1\ ,$$

and

$$\ddot{z} + A^*(t)z + (B^0)^*z = C^*\Lambda_{\mathcal{H}}(Cx(B^0) - \tilde{y}) \; ,$$
$$z(T) = 0 \; , \tag{55}$$
$$\dot{z}(T) = 0 \; ,$$

and the inequality

$$\int_I < B^0 x(B^0), z >_{V',V} \;\; dt \geq \int_I < Bx(B^0), z >_{V',V} \;\; dt \; , \tag{56}$$

for all $B \in \mathcal{P}^0_{a,b}$.

6. NECESSARY CONDITIONS OF OPTIMALITY FOR $\{q, B\}$

In this section, we present the necessary conditions for optimum identification of both $\{q, B\} \in Q_m \times \mathcal{P}^0_{a,b}$ for systems (17), (18) and (19) with the same error functional

$$J(q, B) \equiv \frac{1}{2} \int_I \|Cx(q, B) - \tilde{y}\|^2_{\mathcal{H}} \; dt \; . \tag{57}$$

Define $f_0 \equiv C^*\Lambda_{\mathcal{H}}(Cx(q^0, B^0) - \tilde{y}), C \in \mathcal{L}(H, \mathcal{H}), \tilde{y} \in C(\bar{I}, \mathcal{H})$, and the adjoint systems corresponding to (17), (18) and (19) by

$$-\dot{z} + A^*(q^0, t)z + (B^0)^*z = f_0 \; , \tag{17)*$$
$$z(T) = 0 \; ,$$
$$\ddot{z} - 2\gamma A^*(q^0, t)\dot{z} + A^*(q^0, t)z + (B^0)^*z = f_0 \; , \tag{18)*$$
$$z(T) = \dot{z}(T) = 0 \; ,$$
$$\ddot{z} + A^*(q^0, t)z + (B^0)^*z = f_0 \; , \tag{19)*$$
$$z(T) = \dot{z}(T) = 0 \; ,$$

respectively. Following similar procedure as in the preceeding sections, we have the following necessary conditions for optimality.

THEOREM 6.1

Suppose Q is algebraically contained in a linear topological vector space and that Q is convex and $Q_m = (Q, \tau_m)$ is compact. Let A satisfy the assumptions A3) and A4) and further $q \to A(q, t)$ is once Gateaux differentiable in the weak operator topology of $\mathcal{L}(V,V')$ and that the Gateaux differential is strongly (strong operator topology) measurable on I. Then, in order that $\{q^0, B^0\} \in Q_m \times \mathcal{P}^0_{a,b}$ be optimal, it is necessary that there exists a pair

$$\{x^0, z^0\} \equiv \{x(q^0, B^0), z(q^0, B^0)\} \in C(\bar{I}, H) \times C(\bar{I}, H),$$

such that x^0 satisfies (17), (respectively (18), (19)) and z^0 satisfies (17)* (respectively (18)*, (19)*) and that

$$\int_I < (B^0 - B)x^0, z^0 > dt + \int_I < A'(q^0, t; q^0 - q)x^0, z^0 > dt \geq 0, \qquad \text{(for (17) and (19))}$$

$$\int_I < (B^0 - B)x^0, z^0 > dt + \int_I < A'(q^0, t; q^0 - q)x^0, z^0 > dt \qquad (58)$$

$$+ 2\gamma \int_I < A'(q^0, t; q^0 - q)\dot{x}^0, \dot{z}^0 > dt \geq 0, \quad \text{(for (18))},$$

for all $B \in \mathcal{P}^0_{a,b}$ and $q \in Q_m$, where A' denotes the weak Gateaux differential of A at q^0 in the direction $q^0 - q$.

7. COMPUTATIONAL PROCEDURE

Based on the necessary conditions presented in the preceeding sections, an iterative procedure was developed for determining the optimal operators or parameters to approximate the unknown. The algorithm has been used for both hyperbolic and parabolic problems with constant as well as spatially varying parameters, and the results are reported elsewhere. For simplicity of presentation, we discuss the contents of the algorithm with reference to the parabolic system; but the same algorithm holds for hyperbolic problems if references are made to the appropriate system equations.

Rewriting the inequality (37) as

$$J'_{B^0}(B - B^0) = -\int_I << B - B^0, x^0(B^0) \otimes z(B^0) >> dt \geq 0\ , \qquad (59)$$

we can identify the product $x(B^0) \otimes z(B^0)$ as an element of the tensor product space $V \otimes V$ and consider duality pairing $<< \cdot, \cdot >>$ as given by

$$<< B, x \otimes y >> \equiv < Bx, y >_{V',V}\ ,$$

for each $B \in \mathcal{L}(V,V')$. Clearly, this defines a continuous linear functional on $\mathcal{L}(V,V')$. Since the operator B is time invariant, (59) is equivalent to

$$J'_{B^0}(B - B^0) = << B - B^0, -\int_I x(B^0) \otimes z(B^0) dt >> \ \geq 0\ .$$

Thus for each $B \in \mathcal{L}(V, V')$, the gradient of the cost function, denoted J'_B, given by

$$J'_B = -\int_I x(B) \otimes z(B)\, dt \ , \tag{60}$$

is an operator with values in $\mathcal{L}(V', V)$, where x and z are solutions of the system equation (35) and the adjoint equation (36), respectively, with B replacing B^0. Using this gradient, we can use the following algorithm to identify the unknown operator.

Algorithm

1. Guess $B^1 \in \mathcal{P}_{a,b}$ for B^0 and set $i = 1$.
2. Solve the system equation (35) with $B^0 = B^i$.
3. Solve the adjoint equation (36) with $B^0 = B^i$.
4. Compute the gradient J'_{B^i} as given in (60).
5. Compute search direction
   a) Gradient Method : $S^i = -J'_{B^i}$, or,
   b) Conjugate gradient method : $S^i = -J'_{B^i} + \beta^i S^{i-1}$, where
   $$\beta^i = \frac{\|J'_{B^i}\|^2_{\mathcal{L}(V',V)}}{\|J'_{B^{i-1}}\|^2_{\mathcal{L}(V',V)}}.$$
6. Update the operator by setting
   $$B^{i+1} = B^i + \varepsilon S^i \ ,$$
   where $\varepsilon(> 0)$ is chosen sufficiently small so that
   $$J(B^{i+1}) \leq J(B^i) \qquad \text{and} \quad B^{i+1} \in \mathcal{P}_{a,b}.$$
7. Set $i = i + 1$ and repeat from step 2 until a convergence criterion is satisfied, for example, $|J(B^{i+1}) - J(B^i)| \leq \rho$, where $\rho > 0$, is chosen suitably small.

Remark 7.1

In case the structure of the operator is given with the parameters unknown, the above algorithm becomes relatively simple. In particular, the gradient of the cost function can be computed easily from the solution of the system equation and the corresponding adjoint equation.

CONCLUSION

In this paper, we have presented a general abstract theory for identification of linear systems in infinite dimensional spaces. Currently, similar results are being developed for nonlinear systems involving monotone operators, accretive operators and nonlinear semigroup, and stochastic systems [15].

ACKNOWLEDGMENTS

(i) This work was supported by the Natural Sciences and Engineering Research Council of Canada under grant no. A7109.

(ii) The author would like to thank the organizer Prof. S.J. Lee for the invitation to present this paper at the annual meeting of the A.M.S. held in New Orleans, Jan. 1986.

REFERENCES

[1] C.S. KUBRUSLY, "Distributed Parameter Sytem Identification : A Survey", Int. Journal of Control, Vol. 26, 1977, P.509-535.

[2] A.V. BALAKRISHNAN, "Identification-Inverse Problem for Partial Differential Equations : A Stochastic Formulation", 6th IFIP Conference on Optimization, Novosibirsk, 1974, Proceedings in Lecture Notes in Computer Science, Springer, 1975.

[3] S. KITAMURA and S.NAKAGIRI, "Identification of Spatially - varying and Constant Parameters in Distributed Systems of Parabolic Type", SIAM Journal on Control and Optimization, Vol. 15, 1977, P. 785-802.

[4] A. PIERCE, "Unique Identification of Eigenvalues and Coefficients in a Parabolic Problem", SIAM Journal on Control and Optimization, Vol. 17, 1979, P. 494-499.

[5] T. SUZUKI, "Uniqueness and Nonuniqueness in an Inverse Problem for the Parabolic Equation", Journal of Differential Equations, Vol. 47, 1983, P. 296-316.

[6] S. NAKAGIRI, "Identifiability of Linear Systems in Hilbert Spaces", SIAM Journal on Control and Optimization, Vol. 21, 1983, P. 501-530.

[7] H.T. BANKS and K. KUNISCH, "An Approximation Theory for Nonlinear Partial Differential Equations with Applications to Identification and Control", SIAM Journal on Control and Optimization, Vol. 20, 1982, P. 815-848.

[8] N.U. AHMED and K.L. TEO, "Optimal Control of Stochastic Differential Systems with Fixed Terminal Time", Advances in Applied Probability, Vol. 7, 1975, P. 154-178.

[9] N.U. AHMED, "Necessary Conditions of Optimality for a Class of Second order Hyperbolic Systems with Spatially Dependent Controls in the Coefficients", Journal of Optimization Theory and Applications, Vol. 38, 1982, P. 423-446.

[10] J.L. LIONS, "Some Aspects of Modelling Problems in Distributed Parameter Systems : Modelling and Identification", Ed. A. Ruberti, Springer - Verlag, 1978.

[11] G. CHAVENT, "Identification of Functional Parameters in Partial Differential Equations", in Identification of Parameters in Distributed Systems, Eds. R.E. Goodson and M. Polis, American Socity of Mech. Engrs., 1974.

[12] G. CHAVENT, "On the Identification of Distributed Parameter Systems", Proc. 5th IFAC Symposium on Identification and System Parameter Estimation, Darmstadt, Federal Republic of Germany, 1979.

[13] N.U. AHMED and K.L. TEO, "Optimal Control of Distributed Parameter Systems", North Holland, New York, Oxford, 1981.

[14] J.L. LIONS and E. MAGENES, "Nonhomogeneous Boundary Value Problems and Applications," Vol. 1,2, Springer-Verlag, New York, 1972.

[15] N.U. AHMED and T.E. DABBOUS, "Parameter Identification of Diffusion Processes Using Nonlinear Filtering", Submitted.

Two-Point Boundary Value Problems and the Matrix Riccati Equations

S. Balakumar C. Martin

Department of Mathematics
Texas Tech University
Lubbock, Texas

Abstract: The geometry of two-point boundary value problems is discussed and the associated Riccati equations are derived. Necessary and sufficient conditions are given for the existence and uniqueness of solutions to a two-point boundary value problem in terms of the finite escape time for the associated Riccati equation.

1 INTRODUCTION

In this paper we study the geometry of two-point boundary value problems and the associated matrix Riccati equations. We begin with a review of a series of techniques for solving two-point boundary value problems and compare the associated Riccati equations. It is well known in the modern literature that any solution technique ultimately is equivalent to solving some Riccati equation or in some cases to solving a sequence of Riccati equations. In this paper we ultimately show that the existence and uniqueness of the solution to a two-point boundary value problem is equivalent to the problem of determining the existence of finite escape time for a certain associated Riccati equation. The major technique which is used in this paper is the geometry of the Grassmann manifolds. This is of course an established technique in the study of the Riccati equation.

There are many computational schemes available, such as the method of invariant imbedding, the box scheme, the multiple shooting techniques, the method of factorization and so forth, for the computation of solutions of two-point boundary value problems. In methods such as the invariant imbedding and the box scheme the boundary value problem is first transformed into an initial value problem via a suitable decoupling transformation. The standard decoupling transformation is the so called *Riccati transformation.* The transformed system is then resolved by forward and backward integration, with the initial values and terminal values. However, there is a certain amount of uncertainty in these procedures for, it may not be possible to solve the resulting Riccati equation within the given interval. This is discussed in detail in section II.

The factorization method of J. Taufer, [9], is a very useful and powerful method. The close relationship between this method and the method of invariant imbedding is discussed in some detail in section III.

In this paper, the existence and uniqueness of the solution of a two–point boundary alue problem is discussed using the Grassmannian manifold as the main tool. The two–point boundary value problem is described as a flow in the Grassmannian manifold and the study is carried out in section IV. For this purpose, it is shown that it suffices to consider a linear homogeneous two–point boundary value problem.

In section V, the Riccati differential equation is identified with a flow generated by a one–parameter group acting on the Grassmann manifold. The main result of this paper is to show that the existence and uniqueness of solutions of two–point boundary value problems is equivalent to the existence of the solution to a certain Riccati equation. This is proved by using the geometry of the Grassmann manifolds.

2 THE RICCATI TRANSFORMATION

In this section, the problem of decoupling in solving linear boundary value problems is discussed. In the process of the discussion, the Riccati transformation is derived. This section is primarily based on the exposition of Mattheij, [7]. At first, we shall discuss the basic idea behind the decoupling, with the help of a geometric method of Mattheij, [7].

Let X and Y be two independent vectors such that $\| X \|_2 \gg \| Y \|_2$, and $a = \alpha_1 X + \alpha_2 Y$, and $b = \beta_1 X + \beta_2 Y$, where α_1, α_2, β_1, and β_2 are constants. If the angle defined by a and b is small they do not form a useful basis for the purposes of numerical calculations. We can find a more useful basis by defining $t_1 = a/\gamma_1$ where γ_1 is of the same order of magnitude as $\| a \|_2$, and by letting $t_2 \in$ span (a,b) with $\| t_2 \|_2 \approx \| t_1 \|_2$ and such that t_1 and t_2 do not define a small angle. Then for some γ_2 and γ_3 we have $b = \gamma_2 t_1 + \gamma_3 t_2$. This brings us to the following matrix equation

$$(X,Y)\begin{pmatrix} \alpha_1 & \beta_1 \\ \alpha_2 & \beta_2 \end{pmatrix} = (a,b) = (t_1,t_2)\begin{pmatrix} \gamma_1 & \gamma_2 \\ 0 & \gamma_3 \end{pmatrix}$$

This geometric description indicates a way to transform the system matrix of a two-point boundary value problem into a block upper triangular matrix. This is the main idea of Riccati transformation. A similar transformation can be used to obtain a block lower triangular matrix.

Now, we shall consider a linear homogeneous two–point boundary value problem as follows:

$$X'(t) = A(t)X(t), for\ a \le t \le b \tag{1}$$

with boundary conditions

$$X_1(a) = X_a \tag{2}$$

and

$$X_2(b) = X_b. \tag{3}$$

Where A(t) is a continuous $N \times N$ matrix valued function and X(t) is an N–vector. More general boundary conditions are of course possible but for the purposes of this exposition these suffice.

To solve the above problem, we transform the system matrix A(t) by a time–dependent matrix T(t). We let

$$X(t) = T(t)Y(t). \tag{4}$$

So, we have, $Y(t) = T^{-1}(t)X(t)$ and we calculate

$$\begin{aligned} Y'(t) &= T^{-1}(t)X'(t) - T^{-1}(t)T'(t)T^{-1}(t)X(t) \\ &= T^{-1}(t)A(t)T(t)Y(t) - T^{-1}(t)T'(t)Y(t). \end{aligned} \tag{5}$$

Thus,

$$Y'(t) = W(t)Y(t), \tag{6}$$

where we wish $W(t) = T^{-1}(t)A(t)T(t) - T^{-1}(t)T'(t)$ to be upper or lower triangular.

We choose the transformation $T(t)$ in order to make $W(t)$ upper or lower triangular by solving the Lyapunov's equation $T'(t) = A(t)T(t) - T(t)W(t)$. The matrix form of equation (6) is

$$\begin{pmatrix} Y_1' \\ Y_2' \end{pmatrix} = \begin{pmatrix} W_{11} & W_{12} \\ 0 & W_{22} \end{pmatrix} \begin{pmatrix} Y_1 \\ Y_2 \end{pmatrix} \tag{7}$$

and thus

$$Y' = W_{11}Y_1 + W_{12}Y_2 \tag{8}$$

and

$$Y_2' = W_{22}Y_2. \tag{9}$$

Hence, the boundary value problem is transformed into the initial value problems given by the equations (8) and (9). Now, to get the solution we have to integrate the ordinary differential equation (9) for increasing t and then integrate the ordinary differential equation (8) for decreasing t.

The decoupling mentioned above is utilized in methods such as invariant imbedding, and is described by Mattheij, [7]. There the decoupling is carried out by a lower triangular matrix, T(t), containing identity matrices in the diagonal blocks and the remaining block of T(t) satisfies a matrix Riccati equation. That is,

$$T(t) = \begin{pmatrix} I_n & 0 \\ P(t) & I_{N-n} \end{pmatrix}_{N\times N}.$$

Where, P(t) is an $(N-n)\times n$ matrix satisfying a matrix Riccati equation. The inverse of T(t) can be obtained easily and is

$$T^{-1}(t) = \begin{pmatrix} I_n & 0 \\ -P(t) & I_{N-n} \end{pmatrix}_{N\times N}.$$

By matrix differentiation, we get,

$$T'(t) = \begin{pmatrix} 0 & 0 \\ P'(t) & 0 \end{pmatrix}_{N\times N}.$$

Now, if we apply the transformation $X(t) = T(t)Y(t)$, with the above T(t) to the same problem (1), we get $Y'(t) = [T^{-1}(t)A(t)T(t) - T^{-1}(t)T'(t)]Y(t)$. In matrix form we can write this equation as,

$$\begin{pmatrix} Y_1'(t) \\ Y_2'(t) \end{pmatrix} =$$

$$\begin{pmatrix} A_{11}+A_{12} & A_{12} \\ P'-A_{21}+A_{22}P-PA_{11}-PA_{12}P & -PA_{12}+A_{22} \end{pmatrix} \begin{pmatrix} Y_1(t) \\ Y_2(t) \end{pmatrix}. \tag{10}$$

Where

$$A(t) = \begin{pmatrix} A_{11} & A_{12} \\ A_{21} & A_{22} \end{pmatrix} \begin{matrix} \mathrm{n} \\ \mathrm{N\text{-}n} \end{matrix},$$

$$X(t) = \begin{pmatrix} X_1(t) \\ X_2(t) \end{pmatrix},$$

and

$$Y(t) = \begin{pmatrix} Y_1(t) \\ Y_2(t) \end{pmatrix}.$$

In order to force equation (10) to be of the form $Y'(t) = W(t)Y(t)$, with

$$W(t) = \begin{pmatrix} A_{11}+A_{12} & A_{12} \\ P'-A_{21}+A_{22}P-PA_{11}-PA_{12}P & -PA_{12}+A_{22} \end{pmatrix}$$

and upper triangular we set

$$P'-A_{21}+A_{22}P-PA_{11}-PA_{12}P = 0.$$

This is the classical *matrix Riccati equation.* We first integrate this Riccati equation with the terminal condition P(b) to obtain the function $P(t)$. Then we compute the solution to the boundary value problem with the given boundary conditions by forward and backward integration. That is, from equation (10), with $W(t)$ upper triangular, we get,

$$Y_1'(t) = [A_{11}+A_{12}P]Y_1(t) + A_{12}Y_2(t) \tag{11}$$

and

$$Y_2'(t) = [-PA_{12} + A_{22}]Y_2(t). \tag{12}$$

These ordinary differential equations can be integrated with the boundary conditions $Y_1(a)$ and $Y_2(b)$ respectively. The values of $Y_1(a)$ and$Y_2(b)$ can be computed in the following way. Using $X(t) = T(t)Y(t)$ in the matrix form we get

$$\begin{pmatrix} X_1(t) \\ X_2(t) \end{pmatrix} = \begin{pmatrix} I & 0 \\ P(t) & I \end{pmatrix} \begin{pmatrix} Y_1(t) \\ Y_2(t) \end{pmatrix}.$$

So, $Y_1(a) = X_1(a) = X(a)$, and $Y_2(b) = -P(b)Y_1(b) + X_2(b)$. If we choose $P(b) = 0$, we get $Y_2(b) = X_2(b) = X_b$. Finally, we can recover X(t) as T(t)Y(t), since the inverse transformation is needed only for $X_2(t)$ and $X_2(t) = P(t)Y_1(t) + Y_2(t)$, where, $X_1(t) = Y_1(t)$.

Based on the above procedure we can formulate the following algorithm to obtain the solution to a two–point boundary value problem of the form (1).

Theorem 2.1 (Basic algorithm) *The algorithm for solving the two point boundary value problem can be defined as follows:*

Step I: Use the transformation $X(t) = T(t)Y(t)$ where

$$T(t) = \begin{pmatrix} I & 0 \\ P(t) & I \end{pmatrix},$$

with known terminal condition T(b), to transform the given system into

$$Y'(t) = W(t)Y(t).$$

Where W(t) is a block upper triangular matrix.

Step II: Integrate the resulting Riccati equation with the terminal condition P(b).

Step III: For the initial conditions $Y_1(a)$ and terminal conditions $Y_2(b)$, integrate the ordinary differential equations (11) and (12) to obtain Y(t).

Step IV: Compute the solution X(t) as T(t)Y(t).

We illustrate the theorem with the following example.

Example 2.1 Let $X'(t) = AX(t)$, $1/2 \le t \le 1$, subject to the boundary conditions $X_1(1/2) = 1$ and $X_2(1) = 1$. Where

$$A = \begin{pmatrix} 1 & 1 \\ -1 & -1 \end{pmatrix}.$$

When step I is applied to the problem we get

$$\begin{pmatrix} Y_1'(t) \\ Y_2'(t) \end{pmatrix} = \begin{pmatrix} 1+P(t) & 1 \\ 0 & -1+P(t) \end{pmatrix} \begin{pmatrix} Y_1(t) \\ Y_2(t) \end{pmatrix} \tag{13}$$

provided $P'(t) = -1 + P(t)^2$. This is the Riccati equation. When we integrate this differential equation with the terminal condition $P(1) = 0$ we get, $P(t) = 1/t - 1$. This is step II. Now, to complete step III, from equation (13) we have

$$Y_1'(t) = 1 + P(t)Y_1(t) + Y_2(t) \tag{14}$$

and

$$Y_2'(t) = -1 + P(t)Y_2(t). \tag{15}$$

The equation (14) when integrated with the terminal condition $Y_2(1) = 1$ gives $Y_2(t) = 1/t$ and to integrate the equation (15) we use the initial condition $Y_1(1/2) = 1$. When the value of P(t) is substituted in the equation (14), it becomes

$$Y_1'(t) = (1/t)Y_1(t) + 1/t \tag{16}$$

This is a non–homogeneous problem and a solution to the corresponding homogeneous problem, satisfying the initial condition $Y_1(1/2) = 1$, is 2t. Hence, a particular solution to the non–homogeneous problem (16), is

$$2t \int_{1/2}^{t} \frac{1}{2s^2} ds = 2t - 1 \tag{17}$$

Thus, the complete solution to problem (16), is $Y_1(t) = 2tc + 2t - 1$. Where c is a constant. By using the initial condition $Y_1(1/2) = 1$, we get, $c = 1$. So, $Y_1(t) = 4t - 1$. Now, by step IV, we get $X_1(t) = 4t - 1$ and $X_2(t) = -4t + 5$. That is, the solution to the given problem is

$$X(t) = \begin{pmatrix} 4t - 1 \\ -4t + 5 \end{pmatrix}.$$

This method is very powerful but it cannot be applied successfully to all problems. The successful application of the method depends on the behavior of the Riccati solution in the given interval. The following example illustrates this fact.

Example 2.2 Let $X'(t) = AX(t)$, $0 \le t \le \pi$, subject to the boundary conditions $X_1(0) = 0$ and $X_2(\pi) = 1$. Where

$$A = \begin{pmatrix} 0 & 1 \\ -1 & 0 \end{pmatrix}.$$

The Riccati transformation, $X(t) = T(t)Y(t)$, results in the Riccati equation $P'(t) + P^2(t) + 1 = 0$. This equation can be integrated with the terminal condition $P(\pi) = 0$, to get the solution $P(t) = -\tan t$, for $0 \le t \le \pi$. The Riccati solution does not behave well over the entire interval $[0, \pi]$ i.e., at $t = \pi/2$, P(t) does not exist. So, the method breaks down at this point. But, the actual solution to the problem is $X(t) = -\sin t$, which behaves well in the given interval.

In general terms this transformation is a method of determining a geometric basis for the directions of the solutions. Mattheij, in his paper, [7], introduced a consistency concept to evaluate the effectiveness of such transformations.

3 THE METHOD OF FACTORIZATION FOR THE SOLUTION OF BOUNDARY VALUE PROBLEMS

The factorization method is another effective method to solve two–point boundary value problems. Here again, the problem is transformed into an initial value problem and thereby, the solution is obtained. An example of the factorization method is as follows.

Consider the boundary value problem,

$$y''(t) + a(t)y'(t) + b(t)y(t) = 0, a < t < b. \tag{18}$$

$$y(a) = y_a, \tag{19}$$

and

$$y'(b) = \alpha_b y(b) + \beta_b. \tag{20}$$

Where, α_b and β_b are constants. The second order differential equation (18) can be written as

$$[D^2 + a(t)D + b(t)]y(t) = 0 \tag{21}$$

Where, $D = d/dt$. The method of factorization consists of writing the above equation, in the factored form as,

$$\{D + q(t)\}\{D + p(t)\}y(t) = 0. \tag{22}$$

This equation when simplified, gives the equation

$$[D^2 + p(t) + q(t)D + p'(t) + p(t) + q(t)]y(t) = 0.$$

Then, by equating coefficients with equation (21), we have, $p(t) + q(t) = a(t)$ and $p'(t)+p(t)q(t) = b(t)$. Now, these two equations combine to form the differential equation

$$p'(t) - p^2(t) + a(t)p(t) - b(t) = 0. \tag{23}$$

This is the classical *scalar Riccati equation.* In equation (22), by setting $D + p(t) = z(t)$, we can form the following system of ordinary differential equations

$$z'(t) + q(t)z(t) = 0 \tag{24}$$

or

$$z'(t) + a(t) - p(t)z(t) = 0 \tag{25}$$

and

$$y'(t) + p(t)y(t) = z(t). \tag{26}$$

Now, to get a solution to the boundary value problem, we only need to solve the equations (23), (25) and (26) by integration. The boundary conditions are obtained by comparing (20) and (26). Thus, $p(b) = -\alpha_b$ and $z(b) = \beta_b$. After getting $p(t)$ from equation (23) and $z(t)$ from equation (24), we can integrate equation (26) with the initial condition $y(a) = y_a$, to obtain the solution of the given boundary value problem.

It can be shown easily that the method of factorization and the method of order reduction are closely related for the scalar case. To illustrate this fact, consider the same boundary value problem. That is,

$$y''(t) + a(t)y'(t) + b(t)y(t) = 0.$$

$$y(a) = y_a$$

and

$$y'(b) = \alpha_b y_b + \beta_b.$$

The method of order reduction is applied in the following manner. Let $y'(t) = x(t)$, then the given second order differential equation becomes,

$$x'(t) + a(t)x(t) + b(t)y(t) = 0.$$

These two equations can be written in the matrix form as

$$\begin{pmatrix} y'(t) \\ x'(t) \end{pmatrix} = \begin{pmatrix} 0 & 1 \\ -b(t) & -a(t) \end{pmatrix} \begin{pmatrix} y(t) \\ x(t) \end{pmatrix}.$$

Now, by the Riccati transformation

$$\begin{pmatrix} y(t) \\ x(t) \end{pmatrix} = \begin{pmatrix} 1 & 0 \\ -p(t) & 1 \end{pmatrix} \begin{pmatrix} v(t) \\ u(t) \end{pmatrix}$$

we get,

$$\begin{pmatrix} v'(t) \\ u'(t) \end{pmatrix} = \begin{pmatrix} -p(t) & 1 \\ 0 & p(t) - a(t) \end{pmatrix} \begin{pmatrix} v(t) \\ u(t) \end{pmatrix}$$

provided,

$$p'(t) - p^2(t) + a(t)p(t) - b(t) = 0.$$

This is the scalar Riccati equation (23), we obtained by the factorization method. This differential equation can be solved for p(t) with the terminal condition $p(b) = -\alpha_b$. Then by the algorithm discussed in section II, we can find the complete solution of the problem. Hence, the method of factorization and the method of order reduction are essentially the same.

The main purpose of this section is to study the method of factorization described by J. Taufer, [9], and find a relationship between this method and the method of invariant imbedding, considered in section II.

Taufer in his paper, [9], described the factorization method for a boundary value problem of the form $X'(t) + A(t)X(t) = f(t)$. Where, A(t) is an $N \times N$ matrix and X and f are N–vectors. The method of direct factorization is to be considered in this section and accordingly the necessary notations, definitions, lemmas and theorem are taken from Taufer, [9], and modified where necessary to suit our discussion.

Notations 3.1 *Let there be given*

(a) *a sequence of numbers* $M = \{\gamma_i\}_i^{r+1} = 0$ *be such that* $a = \gamma_0 < \gamma_1 < \ldots < \gamma_r < \gamma_{r+1} = b$;

(b) *a constant* $n \times N$ *matrix* U_1 *of rank n (where* $1 \leq n \leq N$*)*;

(c) *a constant* $(N - n) \times N$ *matrix* U_2 *of rank (N–n)*;

(d) *a constant n–vector* u_1;

(e) *a constant (N–n)–vector* u_2;

(f) *constant non-singular* $N \times N$ *matrices* W_i, *for* $i = 1, \ldots, r$;

(g) *constant N–vectors* w_i, *for* $i = 1, \ldots, r$;

(h) *an* $N \times N$ *matrix* $A(t) = A_{ij}(t)$ *such that for* $i = 1, \ldots, N$ *and* $j = 1, \ldots, N, A_{ij}(t)$ *are integrable in Lebesgue sense in the interval (a,b)*;

(i) *an N–vector* $f(t) = (f_1(t), \ldots, f_N(t))^T$ *such that for* $i = 1, \ldots, N$, *each* $f_i(t)$ *is Lebesgue integrable.*

Using these notations Taufer formulated the boundary value problem in the following manner.

Definition 3.1 : *The following problem is called 'problem Ψ'. An X(t) is to be found for which it holds that:*

(a) $X'(t) + A(t)X(t) = f(t)$ *almost everywhere on (a,b)*;

(b) $U_1X^+(a) = u_1$, $U_2X^-(b) = u_2$;

(c) *X(t) has discontinuities at most of the first type;*

(d) $X^-(\gamma_i) = W_iX^+(\gamma_i) + w_i$ *for* $i = 1, \ldots, r.$;

(e)

$$X_i(t) = \left\{ \begin{array}{ll} X(t) & \text{for } t \in (\gamma_i, \gamma_{i+1}) \\ X^-(t) & \text{for } t = \gamma_i \\ X^+(t) & \text{for } t = \gamma_{i+1} \end{array} \right\}$$

are absolutely continuous in the interval (γ_i, γ_{i+1}), *for* $i = 1, \ldots, r$.

Definition 3.2 : *Let* $D_i(t)_{i=0}$ *be a sequence of* $N \times N$ *matrices and* $d_i(t)_{i=0}^r$ *be a sequence of n–vectors. In the interval* (γ_i, γ_{i+1}), *let the matrices* $D_i(t)$ *and the vectors* $d_i(t)$ *be absolutely continuous solutions of the equations* $D_i'(t) = D_i(t)A(t)$ *and* $d_i'(t) = D_i(t)f(t)$ *with the initial conditions*

$$D_0(a) = U_1$$

$$d_0(a) = u_1$$

$D_i(\gamma_i) = D_{i-1}(\gamma_i)W_i$, *for* $i = 1, \ldots, r$ *and* $d_i(\gamma_i) = d_{i-1}(\gamma_i)w_i$ *for* $i = 1, \ldots, r$.

Definition 3.3 : *Let* $\sigma = (\sigma_1, \ldots, \sigma_n)$ *be an order of the numbers* $1, \ldots, N$. *Let* $P^\sigma = p_{ij}^\sigma$ *be a square permutation matrix of order N defined as follows:* $p_{\sigma j}^\sigma = \delta_i^j$ *for* $i, j = 1, \ldots, N$. *where*

$$\delta_i^j = \left\{ \begin{array}{ll} 0 & \text{for } i \neq j \\ 1 & \text{for } i = j \end{array} \right\}$$

is the Kronecker delta.

Definition 3.4 : *Let* $L_n = l_{ij}^n$ *be an* $N \times n$ *rectangular matrix defined as* $l_{ij}^n = \delta_i^j$ *for* $i = 1, \ldots, N$ *and* $j = 1, \ldots, n$ $(n < N)$

Definition 3.5 : *Let* $R_n = r_{ij}^n$ *be an* $N \times (N - n)$ *rectangular matrix defined as*

$$r_{ij} = \delta_{i-n}^j,$$

for $i = 1, \ldots, N$ *and* $j = 1, \ldots, N - n$.

Theorem 3.1 : *There exists a partition* $a = \theta_0 < \theta_1 < \ldots < \theta_{v+1} = b$, *of the interval such that,*

(1) for each interval (θ_i, θ_{i+1}), *there exists* $j(i)$ *so that*

$$(\theta_i, \theta_{i+1}) \in (\gamma_{j(i)}, \gamma_{j(i)+1})$$

for $i = 0, 1, \ldots, v$.

(2) for each interval (θ_i, θ_{i+1}), *there exists an ordering* σ_i *such that the matrix*

$$D_{j(i)} P^{\sigma i} L_n$$

is non-singular in the interval (θ_i, θ_{i+1}), *for* $i = 0, 1, \ldots, v$.

We omit the proof and refer the reader to Taufer, [9]. A detailed discussion about the ordering σ_i and the selection of j(i) is given by Keller and Lentini, [3]. It should also be noted that there is a close relationship to the so called *Property W* of Polya, [8]. In the paper by Martin and Smith, [5], there is some discussion of related material.

Definition 3.6 *Let D be an $n \times N$ matrix and d be an n–vector. Let σ be a fixed ordering of numbers $1, 2, \ldots, N$ and let the matrix $DP^\sigma L_n$ be non–singular. Then define the operators:*

(a) $G(D, \sigma, n) = (DP^\sigma L_n)^{-1} DP^\sigma R_n$

(b) $G^*(D, \sigma, n) = (DP^\sigma L_n)^{-1} D$

(c) $g(D, d, \sigma, n) = (DP^\sigma L_n)^{-1} d$

Notice that G is an $n \times (N - n)$ matrix, G^* is an $n \times N$ matrix and g is an n–vector. The matrix $DP^\sigma L_n$ is a selected non–singular sub matrix of the matrix D and the matrix $DP^\sigma R_n$ consists of those columns of the matrix D which do not occur in the matrix $DP^\sigma L_n$.

Definition 3.7 *Let $a = \theta_0 < \theta_1 < \ldots < \theta_{v+1} = b$, be a partition of the interval (a,b) and let σ_i be an ordering such that the assertions of the Theorem 3.1 are true. For each interval $(\theta_i, \theta_{i+1}), i = 0, 1, \ldots, v$, define the following:*

(a) Matrices $G_i(t) = G(D_{j(i)}, \sigma_i, n)$ *and* $G_i^*(t) = G^*(D_{j(i)}, \sigma_i, n)$ *for* $t \in (\theta_i, \theta_{i+1})$;

(b) Vectors $g_i(t) = g(D_{j(i)}, d_{j(i)}, \sigma_i, n)$ *for* $t \in (\theta_i, \theta_{i+1})$

(c) Sub matrices of the matrix A, for $t \in (\theta_i, \theta_{i+1})$

$$A_1^i(t) = L_n^T (P^{\sigma i})^T A P^{\sigma i} L_n,$$

which is $n \times n$

$$A_2^i(t) = L_n^T (P^{\sigma i})^T A P^{\sigma i} R_n,$$

which is $n \times (N - n)$,

$$A_3^i(t) = R_n^T (P^{\sigma i})^T A P^{\sigma i} L_n,$$

which is $(N - n) \times n$, and

$$A_4^i(t) = R_n^T (P^{\sigma i})^T A P^{\sigma i} R_n,$$

which is $(N - n) \times (N - n)$

(d) Vectors, for $t \in (\theta_i, \theta_{i+1})$

$$F_1^i(t) = L_n^T (P^{\sigma i})^T f,$$

the first n rows of f, and

$$F_2^i(t) = R_n^T (P^{\sigma i})^T f,$$

the last N–n rows of f. And,

(e) Vectors, for $t \in (\theta_i, \theta_{i+1})$

$$Y_i(t) = L_n^T (P^{\sigma i})^T X^{j(i)},$$

the first n rows of $X^{j(i)}$ *and*

$$Z^i(t) = R_n^T (P^{\sigma i})^T X^{j(i)},$$

the last N–n rows of $X^{j(i)}$*, where j(i) is an index such that* $(\theta_i, \theta_{i+1}) \in (\gamma_{j(i)}, \gamma_{j(i)+1})$

With all these basic definitions and notations, Taufer, in [9], derived the following factorization equations which are used to obtain the solution of the boundary value problem. The derivation of these equations are given in Taufer, [9].

The factorization equations are: For $i = 0, 1, \ldots, v$,

$$Y^i(t) + G_i(t)Z^i(t) = g_i(t), for\ t \in (\theta_i, \theta_{i+1}), \tag{27}$$

$$G_i'(t) = G_i(t)A_4^i(t) - A_1^i(t)G_i(t) - G_i(t)A_3^i(t)G_i(t) + A_2^i(t) \tag{28}$$

almost everywhere on (θ_i, θ_{i+1}),

$$g_i'(t) = -A_1^i(t) + G_i(t)A_3^i(t)g_i(t) + F_1^i(t) + G_i(t)F_2^i(t), \tag{29}$$

almost everywhere on (θ_i, θ_{i+1}).

$$Z_i'(t) = -A_4^i(t) - A_3^i(t)G_i(t)Z^i(t) + F_2^i(t) - A_3^i(t)g_i(t), \tag{30}$$

almost everywhere on (θ_i, θ_{i+1}).

The differential equation (28) is a Riccati equation and it should be solved as the first step in the method of direct factorization. The method of direct factorization is carried out in the following manner. First, we integrate the equation (28), for increasing t, with the initial conditions: for $i = 0$, $G_0(a) = G(U_1, \sigma_0, n)$; for $i = 0, 1, \ldots, v$,

$$G_i(\theta_i) = \left\{ \begin{array}{ll} G\{G_{i-1}(\theta_i), \sigma_i, n\}, & if\ \theta_i \notin M \\ G\{G_{i-1}^*(\theta_i)W_i, \sigma_i, n\}, & if\ \theta_i = \gamma_i \notin M \end{array} \right\}.$$

Similarly, we integrate equation (29) for increasing t, with the initial conditions: for $i = 0$, $g_0(a) = g(U_1, u_1, \sigma_0, n)$; for $i = 0, 1, \ldots, v$,

$$g_i(\theta_i) = \left\{ \begin{array}{ll} g\{G_{i-1}^*(\theta_i), g_{i-1}(\theta_i), \sigma_i, n\} & if\ \theta_i \notin M \\ g\{G_{i-1}^*(\theta_i)W_i, g_{i-1}(\theta_i) - G_{i-1}^*(\sigma_i)w_i, \sigma_i, n\} & if\ \theta_i = \gamma_i \notin M \end{array} \right\}.$$

Thus, we obtain the vector Z^i and the remaining components Y^i of X^i are calculated form the equation (27). Hence, the complete solution X^i of the boundary value problem is obtained.

We must take special care in integrating the Riccati equation (28), not to pass the point where the solution of the equation (28) has a singularity. If the solution to equation

(28) has a pole at a point $t = t_0$ then, $| G_i(t) | \to \infty$ when $t \to t_0$. This implies that $det[D_{j(i)}(t)PL_n] \to 0$ as $t \to t_0$ and thus the matrix $D_{j(i)}(t)P^{\sigma i}L_n$ is singular. Hence, the solution of equation (28), for fixed i, has a pole at the point $t = t_0$ if and only if the corresponding matrix $D_{j(i)}(t)P^{\sigma i}L_n$ is singular. The following lemmas give conditions for the matrix $D_{j(i)}(t)P^{\sigma i}L_n$ to be non–singular. These lemmas have been proved in Taufer, [9], and we only state the lemmas here.

Lemma 3.1 *Let D be an $n \times N$ matrix of rank n. Then there exists an ordering $\sigma = (\sigma_1, \sigma_2, \ldots, \sigma_N)$ of the numbers $1, 2, \ldots, N$ such that the matrix $DP^{\sigma}L_n$ is non–singular and the absolute values of the elements of the matrix $(DP^{\sigma}L_n)^{-1}D$ are not greater than one.*

Lemma 3.2 *Let $A_{ij}(t) \leq K$ for all the i and j. Let h and μ satisfy the inequality $h < 4/NK(N\mu + 2)$ and let $| G^i_{lk}(t) | \leq \mu$ be valid for all l and k, then the matrix $D_{j(i)}(\tau) :^{\sigma i} L_n$ is non–singular for all $\tau \in [t - h, t + h]$.*

The method of direct factorization of Taufer, mentioned in this section, can also be considered in terms of the method of invariant imbedding. The same set of factorization equations can be derived by making a suitable Riccati transformation of the given boundary value problem. Now, we shall show this.

Assuming the notations of Taufer, [9], the problem $X'(t) + A(t)X(t) = f(t)$, can be written in the matrix form as;

$$\begin{pmatrix} (Y^i)' \\ (Z^i)' \end{pmatrix} = \begin{pmatrix} A^i_1 & A^i_2 \\ A^i_3 & A^i_4 \end{pmatrix} \begin{pmatrix} Y^i \\ Z^i \end{pmatrix} = \begin{pmatrix} F^i_1 \\ F^i_2 \end{pmatrix}.$$

The appropriate transformation, for the problem is,

$$\begin{pmatrix} Y^i \\ Z^i \end{pmatrix} = \begin{pmatrix} I & -G_i \\ 0 & I \end{pmatrix} \begin{pmatrix} g_i \\ h_i \end{pmatrix}.$$

Notice that the $T(t)$ matrix we use here, is a block upper triangular matrix and the value of P(t), which is placed in the off diagonal block is $-G_i(t)$. In this case, the resulting W matrix will be block lower triangular. When we substitute the transformation we get,

$$\begin{pmatrix} g'_i \\ h'_i \end{pmatrix} = \begin{pmatrix} -A^i_1 - G_i A^i_3 & 0 \\ -A^i_3 & A^i_3 G_i - A^i_4 \end{pmatrix} \begin{pmatrix} g_i \\ h_i \end{pmatrix} \tag{31}$$

provided, $G'_i = G_i A^i_4 - A^i_1 G_i - G_i A^i_3 G_i + A^i_2$. This is the Riccati equation (28), obtained by Taufer, in [9], by the method of factorization. We can also obtain the other factorization equations (27), (29) and (30) from the matrix equation (31).i.e., From equation (31) we have

$$g'_i = -A^i_1 + G_i A^i_3 g_i$$

and

$$h_i' = -A_3^i Y^i + A_3^i G_i - A_4^i h_i.$$

Now, from the transformation we have,

$$Y^i = g_i - G_i h_i \tag{32}$$

and

$$Z^i = h_i. \tag{33}$$

By differentiating the above equations we get,

$$(Y^i)' = g_i' - G_i h_i' - G_i' h_i$$

and

$$(Z^i)' = h_i'$$

and from the matrix form of the given problem we have,

$$(Y^i)' = -A_1^i Y^i - A_2^i Z^i + F_1^i$$

and

$$(Z^i)' = -A_3^i Y^i - A_4^i Z^i + F_2^i.$$

Using these equations and by substitution we obtain the equation

$$g_I' = -A_1^i + G_i A_3^i g_i + F_1^i = G_i F_2^i,$$

which is equation (29), and

$$(Z^i)' = -A_4^i - A_3^i G_i Z^i + F_2^i - A_3 g_i,$$

which is equation (30). Finally, from equation (32) we get equation (27). Which is

$$Y^i + G_i Z^i = g_i.$$

We can also derive the same set of initial conditions as considered by Taufer in [9].

Thus we conclude that the method of factorization introduced by Taufer in [9], is basically the same as the method of invariant imbedding.

4 NECESSARY AND SUFFICIENT CONDITION FOR THE EXISTENCE OF A UNIQUE SOLUTION OF A BOUNDARY VALUE PROBLEM

In this section we derive a necessary and sufficient condition for the existence of a solution of a two–point boundary value problem. This is in principle well known but the technique we use is of some interest. The geometry of the Grassmannian manifold is the main tool used in this section. First, we describe a boundary value problem as a flow in the Grassmannian manifold and then derive the condition.

We begin by recalling some basic facts about the Grassmannian manifold and certain group actions. We refer the reader to the exposition given by Doolin and Martin, [2] and Brickell and Clark, [1].

We use the following notations and basic definitions in this section. Let M be a manifold and T(M) be its tangent space. Then, the differentiable mapping Y: M $\to$ T(M) is called a vector field. A curve, C(t), in a manifold, M, is defined to be a differentiable function, C: R $\to$ M, whose domain, (a,b), is an open interval of R. Then, locally $C'(t) = Y(C(t))$, is a differential equation of first order on M. Now, a curve, C(t), in M satisfying the equation $C'(t) = Y(C(t))$ and such that $C(a) = C_a$. Such a curve is also called an integral curve starting from the point C_a.

In order to understand the flow of a vector field, the action of a group on a manifold should be considered. For this purpose, the Lie Group–G will be considered a closed sub–group of the group of $N \times N$ invertible matrices, Gl(N).

Now, we shall state the following definition from Doolin and Martin, [2].

Definition 4.1 *The group G is said to act on the manifold, M, if there exists an infinitely differentiable function* $\Omega : G \times M \to M$ *with the following properties.*

(1) for all $m \in M$, $\Omega(e, m) = m$, *where e is the identity element.*

(2) for all g, $h \in G$ *and* $m \in M$, $\Omega g, \Omega(h, m) = \Omega(gh, m)$.

Let β be a one–parameter sub–group of G with the following properties: $\beta(t)\beta(h) = \beta(t+h)$ and $\beta(0) = e$. Using these properties we can show that $\beta^{-1}(t) = \beta(-t)$ and $\beta'(t) = A\beta(t)$. Where,

$$A = \lim_{h \to 0} (\beta(h) - \beta(0))/h.$$

The matrix A is called the infinitesimal generator of the sub–group β. Clearly, the solution of the differential equation $\beta'(t) = A\beta(t)$, is $\beta(t) = \exp At$.

At this point, consider a curve X(t), in M, formed by the action of a one–parameter sub–group $\beta(t)$ of G. That is, $X(t) = \beta(t)X(a)$. By differentiating this equation and (working locally) by using the equation $\beta(t) = A\beta(t)$, we get the equation, $X'(t) = AX(t)$. This is an ordinary differential equation. Hence, this construction produces the initial value problem, $X'(t) = AX(t)$, with the initial value X(a). The curve, X(t) is an integral curve originating from X(a). The vector field generated by the matrix A is called the Riccati vector field and the flow, $\beta(t) = \exp At$, on M, is called the Riccati flow, when M is one of the Grassmannian manifolds.

Now, consider a problem of the form, $X'(t) = AX(t)$, subject to the boundary conditions $X(a) = X_a$ and $X(b) = X_b$. The solution X(t) for $a \le t \le b$, to this problem, is a curve in M that originates at X_a and terminates at X_b. The major question here is the existence of such a curve. To answer this question, let us clarify some properties of manifolds and integral curves. For this purpose, we shall state the following Definitions and lemmas from Brickell and Clark [1].

Definition 4.2 *Let Y be a vector field on a hausdorff manifold, M. Let m be a given point of M and let J(m), be the union of the domains of all the integral curves of the vector field Y, which start from m. The integral curve, C_m, which is defined on the domain, J(m), and starting form the point m is called the maximal integral curve.*

Definition 4.3 *A vector field on a hausdorff manifold, M, is said to be complete if* $J(m) = R$, *for each* $m \in M$.

The following lemma gives a useful criterion for a vector field to be complete.

Lemma 4.1 *A vector field on a hausdorff manifold, M, is complete if and only if there exists a neighborhood of 0, in R, such that each maximal integral curve is defined on the neighborhood.*

The following lemma gives a condition for the manifold to have a complete vector field.

Lemma 4.2 *Every vector field on a compact hausdorff manifold is complete.*

Based on these lemmas and definitions we come to the conclusion that if C: [a,b] → M is an integral curve of a vector field, Y, on a compact manifold then the domain of the curve, C, can be extended to R. In other words the solutions to ordinary differential equations on compact manifolds, are well–behaved over every finite interval.

Now, we will show that the Riccati differential equation is naturally associated with a compact manifold. By our construction, the Riccati equation will be seen to be a complete vector field associated with a one–parameter sub–group of Gl(N) acting on the Grassmannian manifold of n–planes in N–dimensional Euclidean space. Here, we will recall the construction but refer the reader to Doolin and Martin, [2], for the details.

Let V be an N–dimensional vector space and let U and W be sub–spaces of V, with dimensions n and N–n respectively such that $V = U \oplus W$. Let $G^n(v)$ be the set of all n–dimensional sub–spaces of V. It has been shown in Doolin and Martin, [2], that $G^n(v)$ is a compact and hausdorff manifold with an atlas of charts $\Gamma(W)$. Where each $\Gamma(W)$ is the set of n–dimensional sub–spaces of the form

$$W_B = \left\{ \begin{pmatrix} u \\ Bu \end{pmatrix} : u \in U and B \in L(U, W) \right\}$$

Let Gl(N) be the set of all $N \times N$ invertible matrices. Also, let β be a one–parameter sub–group of Gl(N). $G^n(V)$ can be regarded as the Grassmann manifold. The action of β on $G^n(V)$ is given as:

$$Gl(N) \times G^n(V) \to G^n(V),$$

where for $\beta \in Gl(N)$ and $w \in G^n(V)$, we have

$$(\beta, w) \to \beta(w).$$

Now we shall describe a two–point boundary value problem of the type, $X'(t) = AX(t)$, for $t \in [0,T]$, subject to $X_1(0) = X_0$ and $X_2(T) = X_T$. We assume that the decomposition is given with respect to two complementary subspaces, U and W with $\dim U = n$ and $\dim W = N - n$. The set of all possible initial conditions for the above problem is described by the following linear subvariety of $\mathbb{R}^N$;

$$\mathcal{V}_0 = X_0 + W$$

and the set of all possible terminal conditions is given by

$$\mathcal{V}_T = X_T + U.$$

If we could determine a particular point of W then we could simple integrate the differential equation. However the entire point of two-point boundary value problems is that the other half of the boundary conditions are not known. Thus we integrate the entire set of boundary conditions or in more geometric language we let the one parameter group, $\exp(At)$ act on $\mathcal{V}_0$. Now the set

$$\exp(At)\mathcal{V}_0$$

is a linear variety of dimension n for all values of t. At time T we have the following intersection to consider

$$\exp(At)\mathcal{V}_0 \cap \mathcal{V}_T.$$

If this intersection is empty there is no solution and if the intersection consists of a single point there is a unique solution to the original problem. If, on the other hand, the solution consist of a proper variety there are multiple solutions. This represented graphically in Diagram's 1 and 2.

To derive a necessary and sufficient condition for the existence of a unique solution of the given boundary value problem, we make use of the following lemma, form Doolin and Martin, [2]. We leave out the proof and refer the reader to Doolin and Martin, [2].

Lemma 4.3 *An n–dimensional sub–space U of V, can be represented by a unique form, W_B, where*

$$W_B = \left\{ \begin{pmatrix} u \\ Bu \end{pmatrix} : u \in U and B \in L(U,W) \right\}$$

for some matrix B, if and only if $U_0 \cap W = 0$.

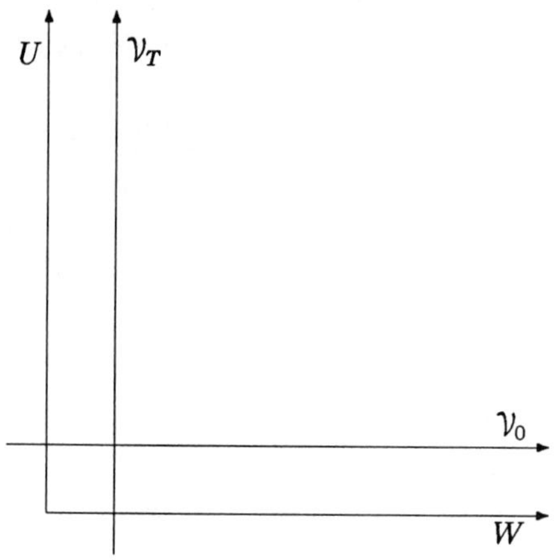

Figure 1: Initial and Terminal Data

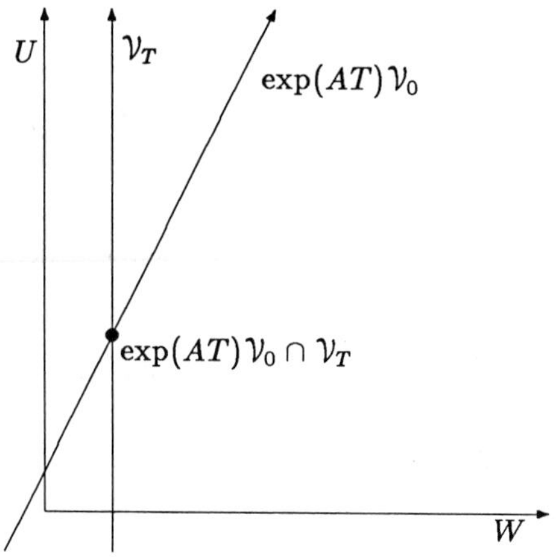

Figure 2: Data at Time T

Now we are in a position to state the necessary and sufficient condition. By the group action,

$$Gl(N) \times G^n(V) \to G^n(V),$$

we have, $\beta(T)W_0 \in G^n(V)$, for $\beta(T) \in Gl(N)$. By lemma 4.3, $\beta(T)W_0$ has a unique decomposition if and only if $\beta(T)W_0 \cap W_T = 0$. Thus $\beta(T)W_0$ has a unique decomposition if and oniy if, $\beta(T)W_0 = 0$. But, $\beta(t)W_0$ gives the solution to the boundary value problem. Thus, we have proved the following theorem.

Theorem 4.1 *A homogeneous boundary value problem of the type,* $X'(t) = AX(t), t \in [0,T]$, *with the boundary conditions* $X_1(0) = 0$ *and* $X_2(T) = 0$; *has a unique solution if and only if* $\beta(T)W_0 \cap W_T = 0$.

Now we can modify this condition and get it in a convenient form as follows.

Theorem 4.2 *The boundary value problem described in theorem 4.1, has a unique solution if and only if* $\beta_1^{-1}(T)$ *exists. Where* $\beta(T)$ *is partitioned as*

$$\beta(T) = \begin{pmatrix} \beta_1(T) & \beta_2(T) \\ \beta_3(T) & \beta_4(T) \end{pmatrix}^{n}_{N-n}$$

Proof: Using theorem 4.1, the given boundary value problem has a unique solution if and only if $\beta(T)W_0 \cap W_T = 0$, where $\beta(T)W_0 \in G^n(V)$. Using the matrix form of $\beta(T)$ we get

$$\beta(T)W_0 = \begin{pmatrix} \beta_1(T)X_1(T) \\ \beta_3(T)X_1(T) \end{pmatrix}$$

But, from lemma 4.3, we have that, $\beta(T)W_T$ has the unique form W_B. Thus, $\beta(T)W_0$ can be written as

$$\beta(T)W_0 = \begin{pmatrix} X \\ \beta_3(T)\beta_1^{-1}(T)X \end{pmatrix}.$$

Where $X = \beta_1(T)X_1(T) \in U$. Such a representation of $\beta(T)$ is possible if and only if $\beta_1^{-1}(T)$ exists. Hence the result.

Theorem 4.2 gives a very important result, for it not only gives conditions for the existence of a unique solution, but it also guarantees the existence of a solution of the corresponding Riccati differential equation. This will be discussed in the next section. In this section, we used some basic facts of differential geometry and derived a necessary and sufficient condition for the existence of a unique solution of a certain homogeneous two–point boundary value problem.

5 THE NECESSARY AND SUFFICIENT CONDITIONS IN TERMS OF THE RICCATI EQUATION

In the previous section, we described a two–point boundary value problem in terms of a flow in the Grassmann manifold and derived a condition for the existence and uniqueness of a solution. In this section, we will continue from there and identify the Riccati equation with a flow generated by a one–parameter group on a Grassmann manifold. For this purpose, at first, we will consider a general problem and derive the Riccati equation, as in the previous setting.

Recall all notations of section 4. To begin with, the action of Gl(N) on $G^n(V)$ is defined by,

$$(\beta, W_B) \to \beta(W_B),$$

for $\beta \in Gl(N)$. That is,

$$\beta(W_B) = \begin{pmatrix} \beta_1(t) + \beta_2(t)Bu \\ \beta_3(t) + \beta_4(t)Bu \end{pmatrix}$$

Now, since $\beta(W_B)(W)$, there exists a matrix A such that

$$\beta(W_B) = \left\{ \begin{pmatrix} z \\ Az \end{pmatrix} : z \in U \right\}$$

This implies that $\beta_1(t) + \beta_2(t)Bu = z$ has a solution for all $z \in U$, and hence, $(\beta_1(t) + \beta_2(t))^{-1}$ exists. Thus,

$$\beta(W_B) = \left\{ \begin{pmatrix} z \\ \{\beta_3(t) + \beta_4(t)B\}\{\beta_1(t) + \beta_2(t)B\}^{-1}z \end{pmatrix} : z \in U \right\}.$$

So, the function

$$\beta : B \to \{\beta_3(t) + \beta_4(t)B\}\{\beta_1(t) + \beta_2(t)B\}^{-1} = A \tag{34}$$

is the generalized fractional transformation that is needed. Already we know that $\beta'(t) = A\beta(t)$, and the induced vector field is $X'(t) = AX(t)$. This is the differential equation for which, we are going to derive the associated Riccati equation. The equation, $\beta'(t) = A\beta(t)$, can be given the following matrix form:

$$\begin{pmatrix} \beta_1' & \beta_2' \\ \beta_3' & \beta_4' \end{pmatrix} = \begin{pmatrix} A_1\beta_1 + A_2\beta_3 & A_1\beta_2 + A_2\beta_4 \\ A_3\beta_1 + A_4\beta_3 & A_3\beta_2 + A_4\beta_4 \end{pmatrix} \tag{35}$$

But, by using the transformation, (34), we get

$$\frac{d}{dt}[\{\beta_3 + \beta_4 B\}\{\beta_1 + \beta_2 B\}^{-1}] = \{\beta_3' + \beta_4' B\}\{\beta_1 + \beta_2 B\}^{-1}$$
$$-\{\beta_3 + \beta_4 B\}\{\beta_1 + \beta_2 B\}^{-1}\{\beta_1' + \beta_2' B\}\{\beta_1 + \beta_2 B\}^{-1}$$

Now, we can use equation (35) to eliminate $\beta_1', \beta_2', \beta_3'$ and β_4' in the above equation and obtain

$$\frac{d}{dt}[\{\beta_3 + \beta_4 B\}\{\beta_1 + \beta_2 B\}^{-1}] = A_3 + A_4\{\beta_3 + \beta_4 B\}\{\beta_1 + \beta_2 B\}^{-1}$$
$$-\{\beta_3 + \beta_4 B\}\{\beta_1 + \beta_2 B\}^{-1}A_1$$
$$-\{\beta_3 + \beta_4 B\}\{\beta_1 + \beta_2 B\}^{-1}A_2\{\beta_3 + \beta_4 B\}\{\beta_1 + \beta_2 B\}^{-1}$$

This implies that the transformation

$$\{\beta_3 + \beta_4 B\}\{\beta_1 + \beta_2 B\}^{-1}$$

satisfies the differential equation,

$$P'(t) = A_3 + A_4 P(t) - P(t)A_1 - P(t)A_2 P(t),$$

with the initial condition, $P(t = 0) = B$. Since $\beta(0) = I_{N\times N}$.

The above differential equation is the matrix Riccati equation and the derivation shows that the Riccati equation is associated with a group action on the Grassmann manifold, $G^n(V)$. Also, we see that it arises as the generator of a linear fractional transformation of the Grassmann manifold.

Now, we shall consider two linear homogeneous boundary value problems and derive the associated Riccati equations in the Grassmannian setting.

Let $X'(t) = AX(t)$, $0 \le t \le T$, subject to the boundary conditions, $X_1(0) = 0$ and $X_2(T) = 0$, be the two–point boundary value problem in consideration. To begin the derivation, first partition X(t) and A as in section 4, such that $X_1(t) \in U$ and $X_2(t \in W$. The boundary conditions can be written as:

$$\begin{pmatrix} X_1(0) \\ X_2(0) \end{pmatrix} = \begin{pmatrix} 0 \\ X_0 \end{pmatrix} = W_0$$

where $X_2(0) = X_0$ and $W_0 = \begin{pmatrix} 0 \\ X_0 \end{pmatrix}$ and

$$\begin{pmatrix} X_1(T) \\ X_2(T) \end{pmatrix} = \begin{pmatrix} X_T \\ 0 \end{pmatrix} = W_T,$$

where $X_1(T) = X_T$ and $W_T = \begin{pmatrix} X_T \\ 0 \end{pmatrix}$. Thus, W_0 forms a chart for $G^n(V)$ and W_T forms a chart for $G^{N-n}(V)$. According to section 4, the given problem has a unique solution if and only if $\beta(T)W_0 \cap W_T = \{0\}$. Now, we will derive the transformation that involves the Riccati equation. This can be done by considering the action of $\beta(t)$ on W_0. That is,

$$\beta(t)W_0 = \begin{pmatrix} \beta_1(t) & \beta_2(t) \\ \beta_3(t) & \beta_4(t) \end{pmatrix} \begin{pmatrix} X_T \\ 0 \end{pmatrix},$$

and this implies that,

$$\beta(W_0) = \begin{pmatrix} \beta_1(t)X_T \\ \beta_3(t)X_T \end{pmatrix}.$$

Since $\beta(W_0) \in G^n(V)$, we have that $\beta_1(t)X_T = z$ has a solution for all $z \in U$. Hence, $\beta_1^{-1}(t)$ exists for all $t \in [0, T]$ and particularly, when $t = T$. But, this is a necessary and sufficient condition for the existence of a unique solution of the given boundary value problem. We proved this in Theorem 4.2, in section 4. This is a very important connection between the solution of the boundary value problem and the solution of the associated Riccati equation. It will be fully realized only after deriving the Riccati equation of the given problem. Therefore, first, we will complete the derivation of the Riccati equation and then discuss about this relationship. $\beta(W_0)$ can be written as

$$\beta(W_0) = \begin{pmatrix} z \\ \beta_3(t)\beta_1^{-1}(t)z \end{pmatrix}.$$

So, the function

$$\beta : 0 \to \beta_3(t)\beta_1^{-1}(t)$$

is the linear fractional transformation that is required. Thus, the Riccati equation is obtained by differentiating $\beta_3(t)\beta_1^{-1}(t)$ with respect to t. That is,

$$\frac{d}{dt}[\beta_3(t)\beta_1^{-1}(t)] = \beta_3'(t)\beta_1^{-1}(t) - \beta_3(t)\beta_1^{-1}(t)\beta_1'(t)\beta_1^{-1}(t).$$

When we substitute the values of $\beta_1'(t)$ and $\beta_3'(t)$ using equation, (35), we get

$$\begin{aligned}\frac{d}{dt}[\beta_3(t)\beta_1^{-1}(t)] &= A_3 + A_4\beta_3(t)\beta_1^{-1}(t) - \beta_3(t)\beta_1^{-1}(t)A_1 - \\ &\quad \beta_3(t)\beta_1^{-1}(t)A_2\beta_3(t)\beta_1^{-1}(t).\end{aligned}$$

Thus, the matrix Riccati equation associated with the given problem is;

$$P'(t) = A_3 + A_4P(t) - P(t)A_1 - P(t)A_2P(t). \tag{36}$$

This equation can be solved with the initial condition $P(t = 0) = 0$.

Here, we like to point out that the same Riccati equation can be obtained by making the Riccati transformation $X(t) = T(t)Y(t)$, where

$$T(t) = \begin{pmatrix} I & 0 \\ P(t) & I \end{pmatrix},$$

to the problem as in section 2.

Now we are in a position to discuss the relationship between the solution of a boundary value problem and the solution of the associated Riccati equation. The Riccati equation, (36), has a well-behaved solution in the interval $[0, T]$, if and only if $\beta_1^{-1}(t)$ exists for all $t \in [0, T]$. If that happens, then $\beta_1^{-1}(T)$ also exists, and this condition guarantees a unique solution to the boundary value problem, due to theorem 4.2 of section 4. Thus, we have proved the main theorem of this paper. That is,

Theorem 5.1 *Consider a two-point boundary value problem of the form, $X'(t) = AX(t)$, $0 \leq t \leq T$, with boundary conditions $X(0) = 0$ and $X(T) = 0$. The Riccati equation (36) associated with the problem has a solution at $t = T$, if and only if the above boundary value problem has a unique solution, over the interval, $[0, T]$.*

Now, we will consider another boundary value problem and derive the corresponding Riccati equation. Let $X'(t) = AX(t), 0 \leq t \leq T$, subject to the boundary conditions $X_1(0) = X_0$ and $X_2(T) = X_T$; be the problem in consideration. As in section 4, we can write the boundary conditions as linear varieties as:

$$\begin{pmatrix} X_1(0) \\ X_2(0) \end{pmatrix} = \begin{pmatrix} X_0 \\ 0 \end{pmatrix} + W_T$$

and

$$\begin{pmatrix} X_1(T) \\ X_2(t) \end{pmatrix} = W_0 + \begin{pmatrix} 0 \\ X_T \end{pmatrix}.$$

The action of $\beta(T)$ on the terminal condition gives,

$$\beta(T)\left\{W_T + \begin{pmatrix} 0 \\ X_T \end{pmatrix}\right\} \cap \left\{\begin{pmatrix} X_0 \\ 0 \end{pmatrix} + W_T\right\} = \beta(T)W_0 \cap W_T. \tag{37}$$

According to the theory, we derived in section 4, the necessary and sufficient condition for the existence of a unique solution to the given boundary value problem is that;

$$\beta(T)\left\{W_0 + \begin{pmatrix} 0 \\ X_T \end{pmatrix}\right\} \cap \left\{\begin{pmatrix} X_0 \\ 0 \end{pmatrix} + W_T\right\} = \{0\}.$$

Thus, from equation (37), the above condition reduces to $\beta(T)W_0 \cap W_T = \{0\}$. This is the exact condition we obtained for the previous problem. Hence, we can state the following Proposition.

Proposition 5.1 *A necessary and sufficient condition for the existence of a unique solution of the boundary value problem,* $X'(t) = AX(t)$ *with the boundary conditions* $X_1(0) = X_0$ *and* $X_2(T) = X_T$; *is the existence of a unique solution of the following boundary value problem:* $X'(t) = AX(t)$, *with the boundary conditions* $X_1(0) = 0$ *and* $X_2(t) = 0$.

Now, we return to the problem and derive the Riccati equation. To get the required linear fractional transformation, we need to know the terminal condition $X_1(T)$. This can be computed by the action of $\beta(T)$ on the terminal condition. That is,

$$\beta(T)\left\{\begin{pmatrix} X_1(T) \\ 0 \end{pmatrix} + \begin{pmatrix} 0 \\ X_T \end{pmatrix}\right\} = \begin{pmatrix} X_0 \\ X_2(0) \end{pmatrix}.$$

So, we have $\beta_1(T)X_1(T) + \beta_2(T)X_T = X_0$. Thus,

$$X_1(T) = \beta_1^{-1}(T)X_0 - \beta_1^{-1}(T)\beta_2(T)X_T$$

provided $\beta_1^{-1}(T)$ exists. To show this, consider the action of $\beta(T)$ on $\begin{pmatrix} X_0 \\ 0 \end{pmatrix}$. That is,

$$\beta(T)\begin{pmatrix} X_0 \\ 0 \end{pmatrix} = \begin{pmatrix} 0 \\ X_T \end{pmatrix}.$$

When $\beta(T)$ is partitioned as before, the above equation gives $\beta_1(T)X_0 = 0$. Thus, $\beta_1^{-1}(T)$ exists. Now the following group action leads to the required transformation. That is,

$$\beta(t)\begin{pmatrix} X_1(T) \\ X_T \end{pmatrix} =$$

$$\begin{pmatrix} \beta_1(t) & \beta_2(t) \\ \beta_3(t) & \beta_4(t) \end{pmatrix}\left\{\begin{pmatrix} \beta_1^{-1}(T)X_0 \\ 0 \end{pmatrix} + \begin{pmatrix} -\beta_1^{-1}(T)\beta_2(T)X_T \\ X_T \end{pmatrix}\right\}.$$

It is sufficient to consider the first term of the above equation, to determine the transformation. That is,

$$\begin{pmatrix} \beta_1(t) & \beta_2(t) \\ \beta_3(t) & \beta_4(t) \end{pmatrix}\begin{pmatrix} \beta_1^{-1}(T)X_0 \\ 0 \end{pmatrix} = \begin{pmatrix} \beta_1(t)\beta_1^{-1}(T)X_0 \\ \beta_3(t)\beta_1^{-1}(T)X_0 \end{pmatrix}. \tag{38}$$

Since $\beta(t)\begin{pmatrix} \beta_1^{-1}(T)X_0 \\ 0 \end{pmatrix} \in G^n(V)$, we can write the eq.(38) as,

$$\beta(T)\begin{pmatrix} \beta_1^{-1}(T)X_0 \\ 0 \end{pmatrix} = \begin{pmatrix} X \\ \beta_3(t)\beta_1^{-1}(t)X \end{pmatrix}.$$

Where, $\beta_1(t)\beta_1^{-1}(T)X_0 = X \in U$ and this equation can be solved for all values of X. Thus, $\beta_1^{-1}(t)$ exists for all $t \in [0, T]$. Therefore, the required transformation is

$$\beta : 0 \to \beta_3(t)\beta_1^{-1}(t).$$

Thus we will get the same Riccati equation as in the previous example. This further strengthens our claim of Proposition 5.1.

In this section, we derived the Riccati equation of a linear homogeneous boundary value problem with different set of boundary conditions. In both cases, we obtain the same Riccati equation. We also found an important relationship between the solution of the boundary value problem and the solution of the associated Riccati equation.

6 CONCLUDING REMARKS

In this paper, we considered mainly a linear homogeneous two–point boundary value problem with separated boundary conditions. At the beginning, in section 2, we used some past results to understand the geometry of a Riccati transformation. Mattheij, in [7], defined such a transformation for a non–homogeneous problem and stated that the transformation is effective when the system matrix is a constant matrix.

In studying some numerical methods to solve boundary value problems, it is important to consider the sensitivity of the problem. In this respect, Mattheij, in [6], introduced the concept of conditioning for boundary value problems. He used condition numbers to decide whether a problem is well–conditioned or ill–conditioned. Also, he related the sensitivity of solutions of linear boundary value problems to perturbations of the boundary conditions.

In section 2, we discussed the method of invariant imbedding. In implementing the method, we derived a Riccati differential equation which is non-linear and most of the time, a solution to this equation does not exist in the given interval. It is because, if the fundamental matrix to the corresponding system is

$$Y(t) = \begin{pmatrix} Y_0(t) & Y_1(t) \\ Y_2(t) & Y_3(t) \end{pmatrix},$$

then the Riccati solution to the system is, $P(t) = Y_1(t)Y_3^{-1}(t)$. But, there is no guarantee that $Y_3^{-1}(t)$ exists in the given interval. However, Taufer, [9], and Keller and Lentini in [3], have shown that by reordering the variables in the problem it is possible to find Riccati solutions bounded in any sub-interval. This amounts to making $Y_3(t)$ non-singular. This was indicated by Lentini, Osborne and Russel, in [4]. Taufer's, [9], method of factorization is explained in section 3. We showed that the method of factorization is closely related to the method of invariant imbedding. Taufer, in [9], has formulated the method by defining a finite number of open sub-intervals to cover the given interval and by giving an appropriate imbedding formulation over each of those sub-intervals. In this way, he made $Y_3(t)$ non-singular.

The Grassmannian manifold is introduced in section 4 to derive the necessary and sufficient condition for the existence of a unique solution to a certain boundary value problem. Here, we made use of the basic properties of tangent spaces, vector fields, Lie groups, one-parameter sub-groups of a Lie group, the group action on a manifold, the Grassmannian manifold and so forth. A nice account of all these is given in Doolin and Martin, [2]. The same Riccati equation which was obtained by the method of invariant imbedding was also identified with a flow generated by a one parameter group on a Grassmann manifold.

In section 5, it was shown that the same condition, namely the existence of $\beta_1^{-1}(t)$, which guarantees the existence of a bounded Riccati solution also guarantees the existence of a unique solution of the boundary value problem.

References

[1] F. Brickell and R. S. Clark, **Differential Manifolds**, Van Nostrand Reinhold Company, Ltd., London, 1972.

[2] B.F.Doolin and C. F.Martin, *Global differential Geometry: an introduction for control engineers*, **NASA Reference Publication 1091**, 1982.

[3] H.B. Keller and M.Lentini, *Invariant embedding, the box scheme and an equivalence between them*, **SIAM J. Numerical Analysis**, Vol. 19, 1982, pp 942–962.

[4] M. Lentini, M.R.Osborne and R.D.Russell, *The close relationship between methods for solving two-point boundary value problems*, **SIAM J. Numerical Analysis**, Vol. 22, 1985, pp 280–309.

[5] C. Martin and J. Smith, *Interpolation, approximation and sampling*, to appear.

[6] R.M.M.Mattheij, *The conditioning of linear boundary value problems*, **SIAM J. Numerical Analysis**, Vol. 19, 1982, pp 963–978.

[7] R.M.M.Mattheij, *Decoupling and stability of algorithms for boundary value problems*, **SIAM Review**, Vol. 27, 1985, pp 1–41.

[8] G. Poyla, *On the mean-value theorem corresponding to a given linear homogeneous differential equation*, **Bull. Amer. Math. Soc.**, Vol. 24, 1922, pp 312-324.

[9] J. Taufer, *On the factorization method*, **Aplikace Matematiky**, Vol. 11, 1966, pp 427–450.

Operator Methods and Singular Control Problems

Stephen L. Campbell
Department of Mathematics
North Carolina State University
Raleigh, North Carolina

INTRODUCTION

Let $L = Ex' + Fx$ where E, F are n×n matrix valued functions and E is singular for all t. The differential equation $Lx = f$ is studied by constructing an mth order differential operator R so that RL is a first order nonsingular differential operator. The construction of R underlies several recent analytic and numerical results on $Lx = f$. Other operator theoretic approaches to analyzing L are discussed and the current status of the approximation problem for L is reviewed.

1. PROBLEM

Linear time varying singular systems of differential equations

$$E(t)x'(t) + F(t)x(t) = f(t) \tag{1.1}$$

arise in a variety of system and control problems. Among these are cheap quadratic control, constrained control, output control, and time varying circuits [1], [2], [3], [12]. In these problems E, F are n×n matrices and E is singular for all t of interest. If E, F are constant, then the structure of (1.1) is well understood, [2], [3]. However, until recently, a general theory for (1.1) has been lacking. Most analysis of (1.1) required making a series of constant rank assumptions [3] and performing repeated time varying reductions. Recently, a general approach has been developed. This note will first explain the operator theoretic basis of this approach. Then we shall examine the current status of the approximation problem.

We are specifically interested in what is sometimes referred to as the higher index case. In the context of the semi-explicit system (1.2)

$$x_1'(t) = A_{11}(t)x_1(t) + A_{12}(t)x_2(t) + f_1(t)$$
$$0 = A_{21}(t)x_1(t) + A_{22}(t)x_2(t) + f_2(t) \qquad (1.2)$$

this means we allow A_{22} to be identically singular. If E in (1.1) has constant rank and $P_1(t)$ is the projection onto the nullspace of E(t), $P_2(t)$ the projection onto the range of E(t), it means we allow rank $((I - P_2)FP_1) <$ rank (P_1) for all t. Such problems arise in some circuit models and in the study of higher order singular arcs in control problems [2], [11].

2. OPERATOR THEORETIC FORMULATION

Assume that E, F are at least 2n-times continuously differentiable on the interval $I = [t_0,t_1]$. Let C^m be the Banach space of m-times differentiable vector valued functions with topology induced by the semi-norms $\|x\|_r = \|x^{(r)}\|_\infty$ for $0 \leq r \leq m$. Let C^∞ be the infinitely differentiable, n-dimensional vector valued functions on I with metric

$$\rho(x,y) = \sum_{i=0}^{\infty} \frac{1}{2^i} \frac{\|x-y\|_i}{1+\|x-y\|_i}$$

Let $\mathcal{D}$ be the operator of differentiation with respect to t. Let L be the differential operator

$$L = E\mathcal{D} + F.$$

so that (1.1) is

$$Lx = f \qquad (2.1)$$

If (1.1) were an explicit differential equation

$$x'(t) - Q(t)x(t) = q(t), \qquad (2.2)$$

then there are a variety of numerical and analytical results available. Let

$$K = \mathcal{D} - Q \qquad (2.3)$$

so that (2.2) is $Kx = q$.

It is known that, in general, the solution of (1.1), and hence any inverse of L, involves not only E, F, f but also their derivatives [9],

[13]. Suppose then that there is a differential operator $\mathcal{R}$ so that

$$\mathcal{R}\mathcal{L} = \mathcal{K} \tag{2.4}$$

on some appropriate domain. Then the solutions of (1.1) are imbedded into those of (2.2) with $q = \mathcal{R}f$. Equation (2.4) is somewhat unusual in that the composition of $\mathcal{L}$ and $\mathcal{R}$ has the same order as $\mathcal{L}$. It is not obvious at first that such an $\mathcal{R}$ need exist, nor that it is computable.

DEFINITION 2.1. Suppose that E, F, f are real analytic on $\mathcal{I}$. Then (1.1) is analytically solvable (a-solvable) if

For every real analytic f there is a real analytic solution to (1.1) (2.5a)

Solutions to (1.1) are defined on all of $\mathcal{I}$ and are real analytic if f is. (2.5b)

Solutions to (1.1) are uniquely determined by their value at $\hat{t} \in \mathcal{I}$. (2.5c)

DEFINITION 2.2. If E, F $\in C^{\infty}$, then (1.1) is ∞-solvable if (2.5a), (2.5b), (2.5c) holds with "infinitely differentiable" replacing real analytic.

DEFINITION 2.3. If E, F $\in C^{2n}$, then (1.1) is solvable if (2.5b), (2.5c) hold and also

Solutions of (1.1) with $f \equiv 0$ are in C^{2n+1}. (2.5d)

and

For every $f \in C^{m}$, $n \leq m \leq 2n$, there is a solution in $C^{(m-n+1)}$. (2.5e)

None of these three definitions imply that E has constant rank and neither solvable nor ∞-solvable imply that (1.1) can be transformed to the usual type of lower triangular canonical forms [4], [8], [13].

We now show how the existence of $\mathcal{R}$ is established and used to numerically solve (1.1).

Suppose that x is a j-times continuously differentiable solution of (1.1). By repeatedly applying $\mathcal{D}$ to (1.1) and using the fact that $\mathcal{D}G = G\mathcal{D} + G'$ for G = E, F we get the system of equations

$$\begin{bmatrix} E & 0 & \cdot & \cdot & \cdot \\ E' + F & E & 0 & \cdot & \cdot \\ E'' + 2F' & 2E' + F & E & 0 & \cdot \\ \cdot & \cdot & \cdot & \cdot & 0 \\ \cdot & \cdot & \cdot & \cdot & E \end{bmatrix} \begin{bmatrix} \mathcal{D}x \\ \mathcal{D}^2 x \\ \mathcal{D}^3 x \\ \cdot \\ \mathcal{D}^j x \end{bmatrix} = \begin{bmatrix} f \\ \mathcal{D}f \\ \mathcal{D}^2 f \\ \cdot \\ \mathcal{D}^{j-1} f \end{bmatrix} - \begin{bmatrix} F \\ F' \\ F'' \\ \cdot \\ F^{(j+1)} \end{bmatrix} x \tag{2.6}$$

Then to get (2.4) it suffices to find

$$\mathcal{R} = \sum_{i=0}^{j-1} R_i(t)\mathcal{D}^i \tag{2.7}$$

such that

$$\begin{bmatrix} R_0 & R_1 & \cdot & R_{j-1} \\ \cdot & \cdot & \cdot & \cdot \\ \cdot & \cdot & \cdot & \cdot \\ \cdot & \cdot & \cdot & \cdot \end{bmatrix} \begin{bmatrix} E & 0 & \cdot & \cdot \\ E' + F & E & \cdot & \cdot \\ \cdot & \cdot & \cdot & 0 \\ \cdot & \cdot & \cdot & E \end{bmatrix} = \begin{bmatrix} I & 0 \\ 0 & H \end{bmatrix} \tag{2.8}$$

where I is n×n.

To show that (2.7) exists and how it is computed, we shall switch to the notation of [8]. Suppose that (1.1) is solvable and that $\hat{t} \in \mathcal{I}$. Let

$$c_i(\hat{t}) = \frac{c^{(i)}(\hat{t})}{i!} \text{ for } c = E, F, f, x;\ \hat{t} \in \mathcal{I} \tag{2.9}$$

so that

$$c(t) = \sum_i c_i \delta^i, \quad \delta = t - \hat{t}. \tag{2.10}$$

(where (2.9) is a Taylor approximation with a remainder if c is not analytic). Substituting these expansions into (1.1) gives that for any $j > 0$, (j less then the smoothness of E, F, x, f), $t \in \mathcal{I}$,

$$\begin{bmatrix} E_0 & 0 & \cdot & \cdot & 0 \\ E_1 + F_0 & 2E_0 & \cdot & \cdot & \cdot \\ E_2 + F_1 & 2E_1 + F_0 & 3E_0 & \cdot & \cdot \\ \cdot & \cdot & & \cdot & \cdot \\ E_{j-1} + F_{j-2} & 2E_{j-2} + F_{j-3} & \cdot & \cdot\ \cdot & jE_0 \end{bmatrix} \begin{bmatrix} x_1 \\ x_2 \\ \cdot \\ \cdot \\ x_j \end{bmatrix} = \begin{bmatrix} f_0 \\ f_1 \\ \cdot \\ \cdot \\ f_{j-1} \end{bmatrix} - \begin{bmatrix} F_0 \\ F_1 \\ \cdot \\ \cdot \\ F_{j-1} \end{bmatrix} x_0$$

or

$$\mathcal{E}_j \underline{x}_j = \mathcal{f}_j - \mathcal{F}_j x_0 = \underline{f}_j \tag{2.11}$$

where all the terms in (2.11) depend on t. The system (2.11) is equivalent to (2.8). Note that $\mathcal{E}_j$ is singular since E_0 is. The system (2.11) is said to be smoothly 1-full on $\mathcal{I}$ if there exists a continuously differentiable nonsingular B(t) on $\mathcal{I}$ such that

$$B(t)E_j(t) = \begin{bmatrix} I & 0 \\ 0 & L(t) \end{bmatrix}$$

where I is n×n. In this case, if $[B_0,\ldots,B_{j-1}]$ is the top n rows of B(t), we have that

$$x_1 = -\sum_{i=0}^{j-1} B_i F_i x_0 + \sum_{i=0}^{j-1} B_i f_i$$

or

$$x'(t) = Q(t)x(t) + q(t) \qquad (2.12)$$

The $B_i(t)$ differ by only a scalar multiple from the $R_i(t)$. Thus the existence of R will follow if E_j is smoothly 1-full.

THEOREM 2.1. [8] The system (1.1) with E, F real analytic is a-solvable if and only if there is an integer $j_0 \in [1, n+1]$ such that ($\mathcal{R}$ denotes range)

$$\text{Range } (E_{j_0}) \text{ is constant on } I. \qquad (2.13)$$

$$E_{j_0} \text{ is 1-full at each } t \in I \qquad (2.14)$$

$$\mathcal{R}(E_{j_0}(t)) + \mathcal{R}(F_{j_0}(t)) = C^{nj} \text{ on } I. \qquad (2.15)$$

THEOREM 2.2. [8] The system (1.1) with E, F infinitely differentiable is ∞-solvable if and only if there exists a $j_0 \geq 1$ such that (2.13), (2.14), (2.15) hold and $j_0 \leq n+1$.

It is easy to show that (2.13), (2.14) imply that E_j is smoothly 1-full and thus R exists. General necessary and sufficient conditions for solvability are unknown if the coefficients are not assumed infinitely differentiable.

THEOREM 2.3. [8] Suppose that (1.1) is solvable on I and E, F are 2n-times continuously differentiable. Then (2.13), (2.14), (2.15) hold for $j_0 = n+1$.

Thus we see that solvability implies that an operator R satisfying (2.4) exists.

3. NUMERICAL IMPLEMENTATION

Given a consistent initial condition [7], the singular system (1.1) can be integrated by integrating (2.2) instead. To do so requires, for methods such as Adams methods or Runge-Kutta methods, the evaluating of

$$x'(\hat{t}) = Q(\hat{t})x(\hat{t}) + q(\hat{t}) \tag{3.1}$$

for a given $\hat{t}$, and estimate of $x(\hat{t})$. This does not require actually computing Q(t), q(t), however. Rather one solves (2.11) for $\underline{x}_j$ and uses x_1 for (3.1) since $x_1 = x'(\hat{t})$. Since E_j has a rank deficiency, but known rank, this may be reliably done using a QR algorithm [6], [17].

One advantage of using (2.11) is that one performs all differentiations directly on the known functions E, F, f. This avoids having to differentiate computed quantities as in most reduction procedures for (1.1). This procedure may also be modified to insure the computed solution lies on the solution manifold of (1.1) and does not drift off it [10].

4. OTHER OPERATOR THEORETIC APPROACHES

There is a large body of literature on, and current activity in studying, the operator L where E, F are again operators. Most of this work has been aimed at examining partial differential equations where E, F are differential operators in the spatial variables. It is natural to wonder if this work sheds light on the problem (1.1) when E, F are n×n matrices.

Most of this work, however, is not applicable to the problem we study. If E, F are partial differential operators in the spatial variables, then $E\mathcal{D} \subset \mathcal{D}E$, which becomes $E\mathcal{D} = \mathcal{D}E$ in our setting implying E is constant. Thus this work, we mention only [18], reduces to the known linear time invariant case when the space is finite dimensional. Other papers utilize range and nullspace assumptions that make their results reduce to the index one case when E, F are matrix valued. An apparent exception is [16]. However, [16] requires that derivatives of the coefficients, when acting as multiplication operators, increase smoothness which is not our case. This last difficulty, incidently, is also why it is so difficult to use the contraction mapping theorem in C^m. Unless the underlying state space is decomposed, and E, F are suitably upper or lower triangular with respect to that decomposition, the coefficients must be able to increase smoothness which is not true for multiplication operators. A more detailed discussion appears in [14]. The situation is somewhat better when dealing with implicit difference equations [15].

5. OPEN PROBLEMS

It is an open problem to establish necessary and sufficient conditions for solvability if the coefficients are less than infinitely differentiable. In the infinitely differentiable case, one can consider K, L, R as everywhere defined operators from C into C. However, in general, R seems to need more smoothness than K or L. It seems possible, however, that by restricting the choice of f or altering slightly the definition of solvability, one could establish a more general version of Theorem 2.2.

One of the major open problems concerns the approximation of solvable operators $L = E\mathcal{D} + F$. It is, of course, relatively easy to tell when two such operators are close. However, we are interested in when the solutions of $Lx = f$, $Lx = f$ are also close. This problem has many important ramifications from the construction of simpler models to the analysis of numerical procedures.

At this stage, it is not yet clear exactly what the correct definitions are. However, let us say $\{L_k\}$ is an <u>approximating sequence</u> for the ∞-solvable operator L if

(i) L_k is ∞-solvable

(ii) $\dim N(L_k) = \dim N(L)$. (N denotes nullspace)

(iii) $L_k \to L$ in the operator norm induced by viewing L_k, L as operators from C^m to C^{m-1} for every $m \geq 1$

(iv) For every $f \in C^\infty$, if a_k is a consistent initial condition for $L_k x = f$ and x_k is the solution of $L_k x_k = f$, $x_k(t_0) = a_k$, and if a is a consistent initial condition of $Lx = f$, $x(t_0) = a$ with solution x, then $a_k \to a$ implies $x_k \to x$ in C^∞.

Using [7], [8] one can show

PROPOSITION 5.1. $\{L_k\}$ is an approximating sequence for L if and only if there exists an integer $j > 0$ such that

$$\operatorname{rank}(E_j^k) \equiv \operatorname{rank}(E_j) \tag{5.1}$$

and

$$(E_k, F_k) \to (E, F) \text{ in } C^\infty \qquad (5.2)$$

where $L_k = E_k\mathcal{D} + F_k$ and (E_j^k, F_j^k) are the matrices E_j, F_j for L_k.

Proof. The assumptions of solvability imply that (5.1) is equivalent to (ii). While (iii) clearly implies (5.2) and conversely. On the other hand, if (5.2) and (5.1) hold, then $K_k \to K$ and (iv) will follow. We omit the remaining details.

PROPOSITION 5.2. If E, F are real analytic and $L = E\mathcal{D} + F$ is solvable, then there is an approximating sequence L_k such that E_k, F_k are polynomials.

Proof. If E, F are real analytic, then there exist real analytic changes of coordinates T_1, T_2 that transfer $Lx = f$ to an upper triangular canonical form [13]. That is, $L = T_1(\hat{E}\mathcal{D} + \hat{F})T_2$ where $\hat{E}$, $\hat{F}$ are upper triangular. Now one takes polynomial approximations of $\hat{E}$, $\hat{F}$, T_1, T_2, to get L_k.

What makes the proof of Proposition 5.2 possible is the canonical form where it is possible to approximate and know we are not altering any of the operator properties. Unfortunately, it is not known if Proposition 5.2 holds with $E, F \in C^\infty$ and the coefficients of L_k chosen from the analytic functions. The difficulty is the lack of a suitably specific canonical form for a general solvable system. Using a weak type of canonical form found in [8] one can reduce this approximation problem to the following one.

PROBLEM 1. Suppose $N \in C^\infty$ and $L = N\mathcal{D} + I$ is solvable. Suppose that $\mathcal{N}(L) = \{0\}$ so that L is an invertible operator of C^∞ onto itself. Does L have an approximating sequence with real analytic coefficients?

In order to appreciate some of the difficulties in Problem 1 we shall give an example.

EXAMPLE 5.1. Let $\{I_i\}$ be a countable collection of disjoint open intervals in [0,1]. Let $\{N_i\}$ be an arbitrary bounded sequence of n×n nilpotent matrices. Let $\phi(t)$ be a C^∞ scalar function such that $1 > \phi > 0$ on $\bigcup_i I_i$ and $\phi \equiv 0$ on $\Sigma = [0,1] \setminus (\bigcup_i I_i)$. Also assume that $\phi^{(i)} \equiv 0$ on Σ for every $i \geq 0$. Define $N(t) = \phi(t)N_i$ if $t \in I_i$, otherwise

$N(t) = 0$. Then L is solvable and one to one. In fact, $(N\mathcal{D})^n = 0$ so that $Lx = f$ has the unique solution

$$x = (I + N\mathcal{D})^{-1}f = \sum_{i=0}^{n-1} (-N\mathcal{D})^i f.$$

Notice that N has variable rank and the projections onto the generalized eigenspaces of N(t) need not be continuous in t.

In general, however, $N\mathcal{D}$ need not even be nilpotent as the example on page 242 of [9] shows. Although, in that example, there is a change of coordinates which puts the operator in the form $\tilde{N}\mathcal{D} + I$ with $\tilde{N}$ nilpotent and $\tilde{N}\mathcal{D} = \mathcal{D}\tilde{N}$.

ACKNOWLEDGEMENT. Research supported in part by the Air Force Office of Scientific Research under Grant No. 84-0240.

REFERENCES

1. K. Brenan, Stability and Convergence of Difference Approximations for Higher Index Differential-Algebraic Systems with Applications in Trajectory Control. Ph.D. Thesis, Mathematics, University of California, Los Angeles, 1983.

2. S. L. Campbell, Singular Systems of Differential Equations, Pitman, Marshfield, Mass., 1980.

3. S. L. Campbell, Singular Systems of Differential Equations II, Pitman, Marshfield, Mass., 1982.

4. S. L. Campbell, One canonical form for higher index linear time varying singular systems of differential equations, Circuits Systems and Signal Processing 2 (1983), 311-326.

5. S. L. Campbell, The numerical solution of higher index linear time varying singular systems of differential equations, SIAM J. Sci. Stat. Comp. 6 (1985), 334-348.

6. S. L. Campbell, Rank deficient least squares and the numerical solution of linear singular implicit systems of differential equations, in Linear Algebra and Its Role in Systems Theory , AMS Cont. Math. Series Vol. 47, 51-64, 1985.

7. S. L. Campbell, Consistent initial conditions for linear time varying singular systems, in Frequency Domain and State Space Methods for Linear Systems, North Holland, 313-318, 1986.

8. S. L. Campbell, A general form for solvable linear time varying singular systems of differential equations, SIAM J. Math. Anal. (to appear).

9. S. L. Campbell, Index two linear time varying singular systems of differential equations, SIAM J. Alg. & Disc. Methods 4 (1983), 237-243.

10. S. L. Campbell, The numerical solution of higher index linear time varying singular systems of differential equations II, report, 1986.

11. S. L. Campbell and K. Clark, Order and the index of singular time invariant linear systems, System and Control Letters 1 (1981), 119-122.

12. S. L. Campbell and C. D. Meyer, Jr., Generalized Inverses of Linear Transformations, Pitman, Marshfield, Mass., 1979.

13. S. L. Campbell and L. Petzold, Canonical forms and solvable singular systems of differential equations, SIAM J. Alg. & Disc. Methods 4 (1983), 517-521.

14. S. L. Campbell and J. Rodriguez, Nonlinear systems and contraction mappings, Proc. Amer. Control Conf. 1984, 1513-1519.

15. S. L. Campbell and J. Rodriguez, Bounded solutions of discrete singular systems on infinite time intervals, IEEE Trans. Aut. Control AC-30 (1985), 165-168.

16. A. Favini, A note on singular and degenerate abstract equations, Academia Nazionale dei Lincei, CCCLXXIX (1983), 128-132.

17. G. H. Golub and C. Van Loan, Matrix Computations, John Hopkins University Press, Baltimore, 1983.

18. J. Lagnese, Singular differential equations in Hilbert space, SIAM J. Math. Anal. 4 (1973), 623-637.

The Euler-Bernoulli Beam Equation with Boundary Energy Dissipation

G. Chen, S.G. Krantz*, D.W. Ma, C.E. Wayne
Department of Mathematics
The Pennsylvania State University
University Park, Pennsylvania

H.H. West
Department of Civil Engineering
Pennsylvania State University
University Park, Pennsylvania

INTRODUCTION

Many problems in structural dynamics involve stabilizing the elastic energy of partial differential equations such as the Euler-Bernoulli beam equation by boundary conditions. Exponential stability is a very desirable property for such elastic systems. The energy multiplier method [1], [2], [7] has been successfully applied by several people to establish exponential stability for various PDEs and boundary conditions. However, it has also been found [2] that for certain boundary conditions the energy multiplier method is not effective in proving the exponential stability property.

A recent theorem of F.L. Huang [4] introduces a frequency domain method to study such exponential decay problems. In this paper, we derive estimates of the resolvent operator on the imaginary axis and apply Huang's theorem to establish an exponential decay result for an Euler-Bernoulli beam with rate control of the bending moment only. We also derive asymptotic limits of eigenfrequencies, which was also done earlier by P. Rideau [8]. Finally, we indicate the realizability of these boundary feedback stabilization schemes by illustrating some mechanical designs of passive damping devices.

*Current affiliation: Washington University, St. Louis, Missouri

Supported in part by NSF Grants DMS 84-01297A01, 85-01306, 84-03664 and AFOSR Grant 85-0253.

§1. BACKGROUND AND MOTIVATION

In this paper, we consider the following uniform Euler-Bernoulli beam equation with dissipative boundary conditions

$$\left.\begin{array}{ll} my_{tt}(x,t)+EIy_{xxxx}(x,t) = 0, & 0 < x < 1, \\ y(0,t) = 0, & \\ y_x(0,t) = 0, & \\ -EIy_{xxx}(1,t) = -k_1^2 y_t(1,t), & k_1 \in \mathbb{R}, \\ -EIy_{xx}(1,t) = k_2^2 y_{xt}(1,t), & k_2 \in \mathbb{R}, \\ (y(x,0),y_t(x,0)) = (y_0(x),y_1(x)), & 0 \le x \le 1. \end{array}\right\} \quad (1.1)$$

where m denotes the mass density per unit length, EI is the flexural rigidity coefficient, and the following variables have engineering meanings:

y = vertical displacement, y_t = velocity

y_x = rotation, y_{xt} = angular velocity

$-EIy_{xx}$ = bending moment

$-EIy_{xxx}$ = shear

at a point x, at time t.

From now on, when we write equation (1.1.j), for example, we mean the jth equation in (1.1).

The above equation and conditions are intended to serve as a simple mathematical model for the mast control system in NASA's COFS (Control of Flexible Structures) Program. See Figure 1. A long flexible mast 60 meters in length is clamped at its base on a space shuttle. The mast is formed with 54 bays but can be idealized as a continuous uniform beam. At the very end of the mast, a CMG (control moment gyro) is placed which can apply bending and torsion rate control to the mast according to sensor feedback.

Boundary conditions (1.1.2) and (1.1.3) signify that the beam is clamped, at the left end, $x = 0$, while boundary conditions (1.1.4) and (1.1.5) at the right end, $x = 1$, respectively, signify

$$\begin{cases} \text{shear } (-EIy_{xxx}) \text{ is proportional to velocity } (y_t) \\ \text{bending moment } (-EIy_{xx}) \text{ is negatively proportional to angular velocity } (y_{xt}) \end{cases}$$

Thus the rate feedback laws (1.1.4) and (1.1.5) reflect some basic features of the CMG mast control system in COFS.

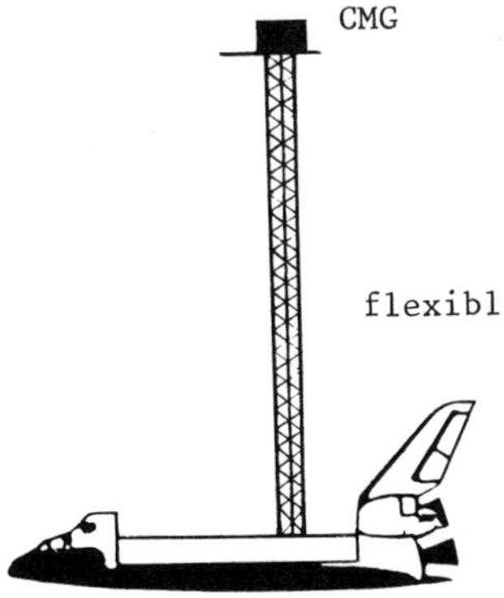

Figure 1 Spacecraft mast control experiment

The elastic energy of vibration, $E(t)$, at time t, for system (1.1) is given by

$$E(t) \equiv \frac{1}{2} \int_0^1 [my_t^2(x,t) + EIy_{xx}^2(x,t)]dx.$$

Note that in (1.1) we have already normalized the beam length to 1.

The qualitative behavior of (1.1) has been studied in an earlier paper [2]. There it is shown that if $k_1^2 > 0$, $k_2^2 \geq 0$ in (1.1.4) and (1.1.5), respectively, then the energy of vibration of the beam decays uniformly exponentially:

$$E(t) \leq Ke^{-\mu t}E(0) \tag{1.2}$$

for some $K, \mu > 0$, uniformly for all initial conditions $(y_0(x), y_1(x))$. Therefore the flexible mast system can be controlled and stabilized.

The proof of the above in [2] was accomplished by the use of energy multipliers and the construction of a Liapounov functional. Nevertheless, a major mathematical question remained unresolved in [2]:

[Q] "Does the uniform exponential decay property (1.2) hold under the assumption of $k_1^2 = 0$, $k_2^2 > 0$?"

This question is of considerable mathematical interest because the feedback scheme using bending moment only is simple and attractive.

For a long time, we have conjectured that the answer to [Q] is affirmative, as asymptotic eigenfrequency estimates obtained in [8] (and §3) have so suggested: Let A denote the infinitesimal generator of the C_0-semigroup corresponding to (1.1) with $k_1 = 0$ and $k_2^2 > 0$, and let $\sigma(A)$ denote the spectrum of A. Then there exists $\beta > 0$ such that

$$\text{Re}\ \lambda \leq -\beta < 0 \quad \text{for all} \quad \lambda \in \sigma(A). \tag{1.3}$$

Nevertheless, it is well known [4] that the following "theorem" is false.

"Let A generate a C_0-semigroup and

$$\sup\{\text{Re}\ \lambda | \lambda \in \sigma(A)\} \leq -\beta < 0 \tag{1.4}$$

for some $\beta > 0$. Then the C_0-semigroup is exponentially stable:

$$\|\exp(tA)\| \leq Me^{-\mu t} \quad \text{for some} \quad M \geq 1, \quad \mu > 0\text{''}. \tag{1.5}$$

Therefore, knowing (1.4) alone is not sufficient to confirm (1.5). This statement remains false even if we assume additionally that A has a compact resolvent.

We have repeatedly tried to refine the energy multiplier technique used in [2] to establish (1.5) without much success, no matter how many different and elaborate multipliers were constructed. There always are boundary terms which cannot be absorbed by terms in the dissipative boundary condition.

A recent theorem by F.L. Huang offers an important direct method for proving exponential stability:

<u>THEOREM 1 (F.L. Huang [4])</u>

Let $\exp(tA)$ be a C_0-semigroup in a Hilbert space satisfying

$$\|\exp(tA)\| \leq B_0, \quad t \geq 0, \quad \text{for some} \quad B_0 > 0. \tag{1.6}$$

Then $\exp(tA)$ is exponentially stable if and only if

$$\{i\omega | \omega \in \mathbb{R}\} \subset \rho(A), \text{ the resolvent set of } A; \text{ and} \tag{1.7}$$

$$B_1 \equiv \sup\{\|(i\omega - A)^{-1}\| | \omega \in \mathbb{R}\} < \infty \tag{1.8}$$

are satisfied. □

Huang's theorem effects a frequency domain method to proving exponential decay properties. As mentioned earlier, the energy multiplier method, which corresponds to a time domain method, has not been successful for the case $k_1^2 = 0$, $k_2^2 > 0$.

Therefore the work is to obtain bounds on the resolvent operator $(i\omega - A)^{-1}$. Here we accomplish this by carrying out a careful analysis on the eigenfunctions and eigenfrequencies of the operator A. This is done in §2.

Associated with [Q] is the question of the asymptotic distribution pattern of eigenfrequencies, as numerical study in [2] suggests that a "structural damping" phenomenon is present at low frequencies. Does it also appear at high frequencies? This is answered in §3. (We must state that the work and numerical verification was done ahead of us by P. Rideau in his recent thesis [8]).

In §4, we present mechanical designs of devices satisfying damping boundary conditions (1.1.3) and (1.1.4) to indicate the realizability of the feedback stabilization scheme using passive dampers.

Notations: We use $\| \ \|$ to denote the $\mathfrak{L}^2(0,1)$ norm. We define the Sobolev space

$$H^k \equiv H^k(0,1) = \{f:[0,1]\to\mathbb{R} \mid \|f\|^2_{H^k(0,1)} \equiv \sum_{j=0}^{k} \int_0^1 |f^{(j)}(x)|^2 dx < \infty\},\ k \in \mathbb{N}.$$

Also, we let

$$H_0^2 \equiv H_0^2(0,1) \equiv \{f \mid f \in H^2(0,1),\ f(0) = f'(0) = 0\}.$$

The underlying Hilbert space $\mathcal{H}$ for the PDE (1.1) is

$$\mathcal{H} = H_0^2(0,1) \times \mathfrak{L}^2(0,1) = \{(f,g) \mid \|(f,g)\|^2_{\mathcal{H}} \equiv \int_0^1 [EI|f''(x)|^2 + m|g(x)|^2] dx < \infty\},$$

whose norm square is the elastic energy.

The unbounded linear operator A associated with (1.1) is given by

$$A = \begin{bmatrix} 0 & I \\ -\alpha^4 \partial_x^4 & 0 \end{bmatrix}, \quad \alpha^4 \equiv \frac{EI}{m}$$

with domain

$$D(A) = \{(f,g) \in H^4\times H^2 \mid -EIf'''(1)=-k_1^2 g(1), -EIf''(1)=k_2^2 g'(1), f(0)=f'(0) = 0\}.$$

§2. ESTIMATION OF THE RESOLVENT OPERATOR ON THE IMAGINARY AXIS. EXPONENTIAL DECAY OF SOLUTIONS.

Consider the resolvent equation: Given $(f,g) \in \mathcal{H}$ and $\lambda \in \mathbb{C}$, find $(w_\lambda, v_\lambda) \in D(A)$ such that

$$(A-\lambda I)\begin{bmatrix} w_\lambda \\ v_\lambda \end{bmatrix} = \left(\begin{bmatrix} 0 & 1 \\ -\left(\alpha\frac{\partial}{\partial x}\right)^4 & 0 \end{bmatrix} - \lambda I_2\right)\begin{bmatrix} w_\lambda \\ v_\lambda \end{bmatrix} = \begin{bmatrix} f \\ g \end{bmatrix} \tag{2.1}$$

This amounts to solving the following boundary value problem for w_λ:

$$\left.\begin{aligned} &\alpha^4 w_\lambda^{(4)}(x) + \lambda^2 w_\lambda(x) = -[\lambda f(x)+g(x)], \quad x \in (0,1) \\ &w_\lambda(0) = 0 \\ &w_\lambda'(0) = 0 \\ &w_\lambda'''(1) - \lambda\tilde{k}_1^2 w_\lambda(1) = \tilde{k}_1^2 f(1) \\ &w_\lambda''(1) + \lambda\tilde{k}_2^2 w_\lambda'(1) = -\tilde{k}_2^2 f'(1) \end{aligned}\right\} \tag{2.2}$$

where

$$\tilde{k}_1^2 \equiv k_1^2(EI)^{-1}, \quad \tilde{k}_2^2 \equiv k_2^2(EI)^{-1}. \tag{2.3}$$

Once w_λ is found we obtain

$$v_\lambda(x) = \lambda w_\lambda(x) + f(x) \tag{2.4}$$

To simplify notation, from now on, unless otherwise specifically mentioned, we set $\alpha^4 = 1$ in (2.2.1) and write k_1, k_2 for $\tilde{k}_1$, $\tilde{k}_2$, respectively.

The main work in this section is to prove estimate (1.8), i.e., to show the existence of some $B_1 > 0$ such that

$$\int_0^1[|w_\lambda''(x)|^2+|v_\lambda(x)|^2]dx \le B_1\int_0^1[|f''(x)|^2+|g(x)|^2]dx \qquad (2.5)$$

for all $\lambda = i\omega$, $\omega \in \mathbb{R}$ and all $(f,g) \in \mathcal{H}$.

<u>LEMMA 2</u> A^{-1} exists and is a compact operator on $\mathcal{H}$. Furthermore, $\sigma(A)$ consists entirely of isolated eigenvalues.

<u>Proof</u>: Let $\lambda = 0$ in (2.2), We see that w_0 in (2.2) is obtained by integrating four times:

$$\begin{aligned} w_0(x) = & -\int_0^x \int_0^{\xi_3} \int_0^{\xi_4} \int_0^{\xi_2} g(\xi_1)d\xi_1 d\xi_2 d\xi_3 d\xi_4 \\ & + \frac{x^2}{2}\left[\int_0^1 \int_0^{\xi_2} g(\xi_1)d\xi_1 d\xi_2 - k_2^2 f'(1)\right] \\ & + \left(\frac{x^3}{6} - \frac{x^2}{2}\right)\left[\int_0^1 g(\xi)d\xi + k_1^2 f(1)\right] \end{aligned}$$

and

$$v_0(x) = f(x).$$

Thus A^{-1} exists and maps $\mathcal{H}$ into $H^4(0,1) \times H_0^2(0,1)$. Therefore A^{-1} is compact. The rest of the lemma follows from Theorem 6.29 in [6, Chapter 3]. □

<u>LEMMA 3</u> The resolvent estimate (2.5) holds for $\lambda = i\omega$, $\omega \in \mathbb{R}$, provided that $|\lambda|$ is sufficiently large.

<u>Proof</u>: For simplicity, let us write (w,v) for (w_λ, v_λ) when no ambiguities will occur.

Let

$$\lambda = i\omega = i\eta^2, \quad \eta \neq 0.$$

We need only consider $\omega = \eta^2 > 0$. The estimates for $\omega < 0$ are similar. First, we find a particular solution $w_p(x)$ of (2.2.1).

$$w_p(x) \equiv -\frac{1}{2}\int_0^x \eta^{-3}[\sinh \eta(x-\xi) - \sin \eta(x-\xi)][i\eta^2 f(\xi) + g(\xi)]d\xi \tag{2.6}$$

Then $w_p(x)$ satisfies

$$\left.\begin{aligned} &w_p^{(4)}(x) - \eta^4 w_p(x) = -[i\eta^2 f(x) + g(x)], \quad x \in (0,1) \\ &w_p(0) = 0 \\ &w_p'(0) = 0. \end{aligned}\right\} \tag{2.7}$$

Consider the solution $\tilde{w}(x)$ of

$$\left.\begin{aligned} &\tilde{w}^{(4)}(x) - \eta^4 \tilde{w}(x) = 0 \\ &\tilde{w}(0) = 0 \\ &\tilde{w}'(0) = 0 \\ &\tilde{w}'''(1) - i\eta^2 k_1^2 \tilde{w}(1) = h_1, \quad h_1 \equiv -w_p'''(1) + i\eta^2 k_1^2 w_p(1) + k_1^2 f(1) \\ &\tilde{w}''(1) + i\eta^2 k_2^2 \tilde{w}'(1) = h_2, \quad h_2 \equiv -w_p''(1) - i\eta^2 k_2^2 w_p'(1) - k_2^2 f'(1) \end{aligned}\right\} \tag{2.8}$$

If we can find $\tilde{w}(x)$, then the solution $w(x)$ of (2.2) is obtainable:

$$w(x) = w_p(x) + \tilde{w}(x). \tag{2.9}$$

But we can solve for $\tilde{w}(x)$ as follows. Since $\lambda^2 = -\eta^4 \neq 0$, we have

$$\tilde{w}(x) = A_1 e^{\eta x} + A_2 e^{i\eta x} + A_3 e^{-\eta x} + A_4 e^{-i\eta x}. \tag{2.10}$$

The coefficients A_i, $1 \le i \le 4$, satisfy

$$M\begin{bmatrix} A_1 \\ A_2 \\ A_3 \\ A_4 \end{bmatrix} = \begin{bmatrix} 0 \\ 0 \\ h_1 \\ h_2 \end{bmatrix} \tag{2.11}$$

where

$$M = \begin{bmatrix} 1 & 1 & 1 & 1 \\ \eta & i\eta & -\eta & -i\eta \\ (\eta^3 - i\eta^2 k_1^2)e^{\eta} & -(i\eta^3 + i\eta^2 k_1^2)e^{i\eta} & -(\eta^3 + i\eta^2 k_1^2)e^{-\eta} & (i\eta^3 - i\eta k_1^2)e^{-i\eta} \\ (\eta^2 + i\eta^3 k_2^2)e^{\eta} & (-\eta^2 - \eta^3 k_2^2)e^{i\eta} & (\eta^2 - i\eta^3 k_2^2)e^{-\eta} & (-\eta^2 + \eta^3 k_2^2)e^{-i\eta} \end{bmatrix} \tag{2.12}$$

If M^{-1} exists for η sufficiently large, then

$$\begin{bmatrix} A_1 \\ A_2 \\ A_3 \\ A_4 \end{bmatrix} = M^{-1} \begin{bmatrix} 0 \\ 0 \\ h_1 \\ h_2 \end{bmatrix}. \tag{2.13}$$

We now begin the estimation of $\int_0^1 |w''(x)|^2 dx$. The work below may seem tedious, but the idea is rather simple. The main observation is that the dominant terms in $w_p(x)$ and $\tilde{w}(x)$ do not satisfy the bounds

$$\int_0^1 |w_p''(x)|^2 dx \le C\int_0^1 [|f''(x)|^2 + |g(x)|^2]dx$$

$$\int_0^1 |\tilde{w}''(x)|^2 dx \le C\int_0^1 [|f''(x)|^2 + |g(x)|^2]dx$$

for $|\lambda|$ large. However, in (2.9), <u>those dominant terms cancel</u>, leaving $w(x)$ with smaller terms which are bounded by $\mathcal{O}(\|f''\| + \|g\|)$.

<u>1st Step</u> <u>Estimation of $w_p''(x)$.</u>

From (2.6),

$$w_p''(x) = -\frac{1}{2}\int_0^x \eta^{-1}[\sinh \eta(x-\xi) + \sin \eta(x-\xi)][i\eta^2 f(\xi) + g(\xi)]d\xi \tag{2.14}$$

$$= -\frac{1}{2}\int_0^x \eta^{-1}[\sinh \eta(x-\xi) + \sin \eta(x-\xi)]g(\xi)d\xi \qquad \text{(integration by parts)}$$

$$-\frac{1}{2}\int_0^x \eta^{-1}[\sinh \eta(x-\xi) - \sin \eta(x-\xi)]if''(\xi)d\xi$$

$$= -\frac{1}{2}\eta^{-1}\int_0^x \sinh \eta(x-\xi)[if''(\xi)+g(\xi)]d\xi + \mathcal{O}(\eta^{-1}[\|f''\|+\|g\|)$$

$$= -\frac{1}{4}\eta^{-1}\int_0^1 e^{\eta(x-\xi)}[if''(\xi)+g(\xi)]d\xi + \mathcal{O}(\eta^{-1}[\|f''\|+\|g\|])$$

$$= -\frac{1}{4}\eta^{-1}\, e^{\eta x}\int_0^1 e^{-\eta\xi}[if''(\xi)+g(\xi)]d\xi + \mathcal{O}(\eta^{-1}[\|f''\|+\|g\|])$$

<u>2nd Step</u> <u>Estimation of h_1 and h_2.</u>

From (2.6), (2.8) and (2.14),

$$h_1 = \int_0^1 K_{1\eta}(\xi)[i\eta^2 f(\xi)+g(\xi)]d\xi + k_1^2 f(1),$$

where

$$K_{1\eta}(\xi) = \frac{1}{2}[\cosh \eta(1-\xi)+\cos \eta(1-\xi)] - \frac{i}{2}k_1^2\eta^{-1}[\sinh \eta(1-\xi)-\sin \eta(1-\xi)]$$

Integration by parts twice for f yields

$$\int_0^1 K_{1\eta}(\xi)i\eta^2 f(\xi)d\xi = \int_0^1 \tilde{K}_{1\eta}(\xi)if''(\xi)d\xi - k_1^2 f(1),$$

where

$$\tilde{K}_{1\eta}(\xi) = \frac{1}{2}[\cosh \eta(1-\xi)-\cos \eta(1-\xi)] - \frac{i}{2}k_1^2\eta^{-1}[\sinh \eta(1-\xi)+\sin \eta(1-\xi)].$$

We get

$$K_{1\eta}(\xi) = \frac{1}{4}\eta^{-1}e^{\eta}(\eta-ik_1^2)e^{-\eta\xi} + \mathcal{O}(1),$$

$$\tilde{K}_{1\eta}(\xi) = \frac{1}{4}\eta^{-1}e^{\eta}(\eta-ik_1^2)e^{-\eta\xi} + \mathcal{O}(1).$$

Therefore

$$h_1 = \frac{1}{4}\eta^{-1}e^{\eta}(\eta-ik_1^2)\int_0^1 e^{-\eta\xi}[if''(\xi)+g(\xi)]d\xi + \mathcal{O}(\|f''\|+\|g\|). \quad (2.15)$$

Similarly, from (2.6) and (2.8),

$$h_2 = \int_0^1 K_{2\eta}(\xi)[i\eta^2 f(\xi)+g(\xi)]d\xi - k_2^2 f'(1)$$

where

$$K_{2\eta}(\xi) = \tfrac{1}{2}\eta^{-1}[\sinh \eta(1-\xi)+\sin \eta(1-\xi)] + \tfrac{i}{2}k_2^2[\cosh \eta(1-\xi)-\cos \eta(1-\xi)].$$

Repeating this same integration by parts procedure twice more for f, we get

$$\int_0^1 K_{2\eta}(\xi) i\eta^2 f(\xi)d\xi = \int_0^1 \tilde{K}_{2\eta}(\xi) i f''(\xi)d\xi + k_2^2 f'(1).$$

As

$$K_{2\eta}(\xi) = \tfrac{1}{4}\eta^{-1}e^{\eta}(1+ik_2^2\eta)e^{-\eta\xi} + O(1)$$

$$\tilde{K}_{2\eta}(\xi) = \tfrac{1}{4}\eta^{-1}e^{\eta}(1+ik_2^2\eta)e^{-\eta\xi} + O(1),$$

we get

$$h_2 = \tfrac{1}{4}\eta^{-1}e^{\eta}(1+ik_2^2\eta)\int_0^1 e^{-\eta\xi}[if''(\xi)+g(\xi)]d\xi + O(\|f''\|+\|g\|). \qquad (2.16)$$

<u>3rd Step</u> <u>Estimation of A_1, A_2, A_3 and A_4.</u>

We first write

$$M = \begin{bmatrix} 1 & & & 0 \\ & \eta & & \\ & & \eta^2 & \\ 0 & & & \eta^2 \end{bmatrix} M_1$$

where

$$M = \begin{bmatrix} 1 & 1 & 1 & 1 \\ 1 & i & -1 & -i \\ (\eta-ik_1^2)e^{\eta} & (-i\eta-ik_1^2)e^{i\eta} & (-\eta-ik_1^2)e^{-\eta} & (i\eta-ik_1^2)e^{-i\eta} \\ (1+i\eta k_2^2)e^{\eta} & (-1-\eta k_2^2)e^{i\eta} & (1-i\eta k_2^2)e^{-\eta} & (-1+\eta k_2^2)e^{-i\eta} \end{bmatrix}.$$

So

$$\begin{bmatrix} A_1 \\ A_2 \\ A_3 \\ A_4 \end{bmatrix} = M_1^{-1} \begin{bmatrix} 0 \\ 0 \\ \eta^{-2}h_1 \\ \eta^{-2}h_2 \end{bmatrix}.$$

Further, write

$$M_1^{-1} = (\det M_1)^{-1} \begin{bmatrix} * & * & \mu_{11} & \mu_{12} \\ * & * & \mu_{21} & \mu_{22} \\ * & * & \mu_{31} & \mu_{32} \\ * & * & \mu_{41} & \mu_{42} \end{bmatrix}.$$

From the evaluation of cofactors,

$$\mu_{11} = (1+i)[(1-\eta k_2^2)e^{-i\eta} + i(1+\eta k_2^2)e^{i\eta} + (1+i)(1-i\eta k_2^2)e^{-\eta}],$$
$$-\mu_{12} = (1-i)[(\eta-k_1^2)e^{-i\eta} - i(\eta+k_1^2)e^{i\eta} + (1-i)(\eta+ik_1^2)e^{-\eta}].$$

Let

$$\begin{aligned} \mathfrak{L}(\eta) &\equiv (\eta-ik_1^2)\mu_{11} + (1+i\eta k_2^2)\mu_{12} \\ &= (1+i)\eta\{[-2k_2^2\eta+(1+i)(1+k_1^2k_2^2)-2ik_1^2\eta^{-1}]e^{-i\eta} \\ &\quad +[2ik_2^2\eta+(1+i)(1+k_1^2k_2^2)+2k_1^2\eta^{-1}]e^{i\eta}\} \\ &\quad + \mathcal{O}(\eta^2e^{-\eta}) \end{aligned}$$

The term in braces above satisfies

$$|\{\ \ \}| \geq |\text{Bracket } 2| - |\text{Bracket } 1|$$

$$\to 2(1+k_1^2k_2^2), \quad \text{as} \quad \eta \to \infty$$

Thus

$$|\{\quad\}| \geq \sqrt{2} \quad \text{for } \eta \text{ sufficiently large,}$$

hence

$$|\mathfrak{L}(\eta)| \geq 2\eta, \quad \mathfrak{L}(\eta) = \mathcal{O}(\eta^2). \tag{2.17}$$

Also

$$D_1 \equiv \det M_1 = (\eta - ik_1^2)e^{\eta}\mu_{11} + (1+i\eta k_2^2)e^{\eta}\mu_{12} + \mathcal{O}(\eta^2)$$

$$= \mathfrak{L}(\eta)e^{\eta} + \mathcal{O}(\eta^2), \quad \text{therefore}$$

$$D_1^{-1} = [\mathfrak{L}(\eta)]^{-1}e^{-\eta} + \mathcal{O}(e^{-2\eta}) \tag{2.18}$$

and by (2.17)

$$D_1^{-1} = \mathcal{O}(\eta^{-1}e^{-\eta}). \tag{2.19}$$

From (2.15), (2.16) and (2.19),

$$A_1\eta^2 = D_1^{-1}(\mu_{11}h_1 + \mu_{12}h_2)$$
$$= (D_1^{-1}) \cdot \tfrac{1}{4}\eta^{-1}e^{\eta}\mathfrak{L}(\eta)\int_0^1 e^{-\eta\xi}[if''(\xi)+g(\xi)]d\xi + \mathcal{O}(D_1^{-1}\eta[\|f''\|+\|g\|])$$

as μ_{11}, μ_{12} are $\mathcal{O}(\eta)$. Continuing from the above:

$$= [\mathfrak{L}^{-1}(\eta)e^{-\eta} + \mathcal{O}(e^{-2\eta})] \cdot \tfrac{1}{4}\eta^{-1}e^{\eta}\mathfrak{L}(\eta)\int_0^1 e^{-\eta\xi}[if''(\xi)+g(\xi)]d\xi$$
$$+ \mathcal{O}(e^{-\eta}[\|f''\|+\|g\|]) \qquad \text{(by (2.17), (2.18))}$$

$$= \tfrac{1}{4}\eta^{-1}\int_0^1 e^{-\eta\xi}[if''(\xi)+g(\xi)]d\xi + \mathcal{O}\left(\eta e^{-\eta}\left|\int_0^1 e^{-\eta\xi}[if''(\xi)+g(\xi)]d\xi\right|\right)$$
$$+ \mathcal{O}(e^{-\eta}[\|f''\|+\|g\|]).$$

But

$$\left|\int_0^1 e^{-\eta\xi}[if''(\xi)+g(\xi)]d\xi\right| = \mathcal{O}(\eta^{-\frac{1}{2}}[\|f''\|+\|g\|]).$$

Thus

$$A_1\eta^2 = \frac{1}{4}\eta^{-1}\int_0^1 e^{-\eta\xi}[if''(\xi)+g(\xi)]d\xi + \mathcal{O}(\eta^{\frac{1}{2}}e^{-\eta}[\|f''\|+\|g\|]). \quad (2.20)$$

As for A_2, we have

$$A_2\eta^2 = D_1^{-1}(\mu_{21}h_1+\mu_{22}h_2) \quad (2.21)$$

where

$$-\mu_{21} = (1-i)(1+i\eta k_2^2)e^{\eta} + 2(1-\eta k_2^2)e^{-i\eta} + (1+i)(1-i\eta k_2^2)e^{-\eta} \quad (2.22)$$

$$\mu_{22} = (1-i)(\eta-ik_1^2)e^{\eta} - 2i(\eta-k_1^2)e^{-i\eta} - (1+i)(\eta-ik_1^2)e^{-\eta} \quad (2.23)$$

By (2.15), (2.16), (2.22) and (2.23), the dominant terms in $\mu_{21}h_1$ and $\mu_{22}h_2$ are $\mathcal{O}(\eta e^{2\eta})$. But their coefficients in $\mu_{21}h_1 + \mu_{22}h_2$ are such that they cancel out. We get

$$A_2\eta^2 = D_1^{-1} \cdot \mathcal{O}(\eta e^{\eta}[\|f''\|+\|g\|]) = \mathcal{O}(\|f''\|+\|g\|) \quad (2.24)$$

Similarly, we can show that

$$A_3\eta^2 = \mathcal{O}(\|f''\|+\|g\|) \quad (2.25)$$

$$A_4\eta^2 = \mathcal{O}(\|f''\|+\|g\|) \quad (2.26)$$

Final Step Estimation of $\|w''\|+\|v\|$.

By (2.14) and (2.20), we have

$$
\begin{aligned}
w''(x) &= w_p''(x) + \tilde{w}''(x) \\
&= \{-\tfrac{1}{4}\eta^{-1}e^{\eta x}\int_0^1 e^{-\eta\xi}[if''(\xi)+g(\xi)]d\xi + \mathcal{O}(\eta^{-1}[\|f''\|+\|g\|])\} \\
&\quad + A_1\eta^2 e^{\eta x} + A_2(-\eta^2)e^{i\eta x} + A_3(-\eta)^2 e^{-\eta x} + A_4 \cdot (-\eta^2)e^{-\eta x} \\
&= \{\mathcal{O}(\eta^{-1}[\|f''\|+\|g\|]) + e^{\eta x} \cdot \mathcal{O}(\eta^{\frac{1}{2}}e^{-\eta}[\|f''\|+\|g\|])\} \\
&\quad + \{-A_2\eta^2 e^{i\eta x} + A_3\eta^2 e^{-\eta x} - A_4\eta^2 e^{-i\eta x}\}.
\end{aligned}
$$

In the first parenthesized term, the $\mathfrak{L}^2$-norm of $e^{\eta x}$ is of order of magnitude

$$
[\int_0^1 (e^{\eta x})^2 dx]^{1/2} \equiv [\tfrac{1}{2\eta}(e^{2\eta}-1)]^{1/2} \equiv \mathcal{O}(\eta^{-1/2}e^{\eta}),
$$

hence

$$
[\int_0^1 |\text{first parenthesized term}|^2 dx]^{1/2} = \mathcal{O}(\|f''\|+\|g\|).
$$

The second parenthesized term also satisfies the above bound, because of (2.24), (2.25), and (2.26).

Hence

$$
\|w''\| \le C(\|f''\|+\|g\|) = \mathcal{O}(\|f''\|+\|g\|), \quad \text{for } \eta \text{ large.} \tag{2.27}
$$

For v, by (2.4) we have

$$
\|v\| \le |\lambda|\,\|w\| + \|f\|. \tag{2.28}
$$

We want to show that

$$
|\lambda|^2\|w\|^2 \le C(\|w''\|^2+\|f''\|^2+\|f\|^2+\|g\|^2), \tag{2.29}
$$

for some constant $C > 0$ independent of λ.

Consider (2.2.1), with $\alpha^4 = 1$ and $\lambda = i\eta^2$, and use k_1^2, k_2^2 for $\tilde{k}_1^2$ and $\tilde{k}_2^2$. Multiply (2.2.1) by $\overline{w}(x)$ and integrate by parts twice. We get

$$w'''(1)\bar{w}(1)-w''(1)\bar{w}'(1) + \|w''\|^2 - \eta^4\|w\|^2 = -\langle\lambda f+g,w\rangle$$
$$= -\langle f,\lambda w\rangle-\langle g,w\rangle.$$

From (2.2.3) and (2.2.4), we get

$$i\eta^2k_1^2|w(1)|^2+k_1^2f(1)\bar{w}(1)+i\eta k_2^2|w'(1)|^2+k_2^2f'(1)\bar{w}'(1)+\|w''\|^2-\eta^4\|w\|^2$$

$$= -\langle f,\lambda w\rangle-\langle g,w\rangle.$$

Hence

$$\eta^4\|w\|^2 = \text{Re}\{\|w''\|^2+\langle f,\lambda w\rangle+\langle g,w\rangle+k_1^2f(1)\bar{w}(1)+k_2^2f'(1)\bar{w}'(1)\}$$

$$\leq \|w''\|^2 + \frac{|\eta|^4}{4}\|w\|^2+2\|f\|^2+\frac{1}{2}\|g\|^2 + \frac{1}{2}\|w\|^2 + C[\|f''\|^2+\|w''\|^2],$$

where we have applied the Poincaré inequality and the trace theorem:

$$|f(1)|^2 + |f'(1)|^2 \leq C\|f''\|^2$$

$$|w(1)|^2 + |w'(1)|^2 \leq C\|w''\|^2.$$

Therefore (2.29) follows for η sufficiently large.

By (2.27) and (2.29), we have

$$\|v\| \leq C(\|f''\|+\|f\|+\|w''\|+\|g\|) \tag{2.30}$$

$$\leq C(\|f''\|+\|g\|)$$

as

$$\|f\| \leq C\|f''\|$$

and (2.27) holds.

Combining (2.27) and (2.30), we have proved (2.5) for $|\lambda|$ sufficiently large, $\lambda = i\omega$, $\omega \in \mathbb{R}$. So Lemma 3 has been proved. □

<u>THEOREM 4</u>. Let $k_1^2 \geq 0$ and $k_2^2 > 0$ in (1.1). Then the uniform exponential decay of energy (1.2) holds.

Proof: In order to apply Theorem 1, we need to verify that assumptions (1.6), (1.7) and (1.8) are satisfied.

We note that (1.6) is satisfied, because A is dissipative and

$$\|\exp(tA)\| \leq 1.$$

The verification of (1.7) and (1.8) is accomplished if we can verify merely (1.7):

$$(\lambda I-A)^{-1} \text{ exists for all } \lambda = i\omega, \quad \omega \in \mathbb{R}, \tag{2.31}$$

because by (2.31) and Lemma 3,

$$\|w_\lambda''\| + \|v_\lambda\| \leq C'(\|f''\|+\|g\|), \quad \forall\lambda = i\omega, \quad \omega \in \mathbb{R},$$

where

$$C' = \max(C,C''), \quad C \text{ as in (2.27) and (2.30)}$$

$$C'' \equiv \max_{|\lambda|\leq B_2} \|(\lambda I-A)^{-1}\|, \quad \text{for some } B_2 \text{ sufficiently large},$$

$$\lambda = i\omega, \quad \omega \in \mathbb{R}.$$

To show (2.31), we assume the contrary that $\sigma(A) \cap \{i\omega | \omega \in \mathbb{R}\} \neq \emptyset$. By Lemma 2, $\sigma(A)$ consists solely of isolated nonzero eigenvalues. Without loss of generality, let

$$\lambda_0 \in \sigma(A), \quad \lambda_0 = i\eta_0^2, \quad \eta_0 \in \mathbb{R}, \quad \eta_0 \neq 0.$$

Then

$$(\lambda_0 I-A) \begin{bmatrix} w_0 \\ v_0 \end{bmatrix} = 0$$

has a nontrivial solution $(w_0,v_0) \in D(A)$. Explicitly, (w_0,v_0) satisfies

$$\begin{cases} i\eta_0^2 w_0 - v_0 = 0 & \text{on } [0,1] \\ w_0^{(4)} + i\eta_0^2 v_0 = 0 & \text{on } [0,1] \\ w_0(0) = 0 \\ w_0'(0) = 0 \\ w_0'''(1) - k_1^2 v_0(1) = 0 \\ w_0''(1) + k_2^2 v_0'(1) = 0 \end{cases}$$

Letting

$$w(x,t) = e^{i\eta_0^2 t} w_0(x),$$

we easily check that w satisfies

$$w_{tt} + w_{xxxx} = 0.$$

Also, the energy

$$\int_0^1 [|w_{xx}(x,t)|^2 + |w_t(x,t)|^2]dx$$

is constant, thus

$$\frac{d}{dt}\int_0^1 [|w_{xx}(x,t)|^2 + |w_t(x,t)|^2]dx$$

$$= 0$$

$$= 2\mathrm{Re}[w_{xx}(x,t)\bar{w}_{xt}(x,t) - w_{xxx}(x,t)\bar{w}_t(x,t)]\Big|_{x=0}^{x=1}$$

$$= -2[k_1^2|w_t(1,t)|^2 + k_2^2|w_{xt}(1,t)|^2].$$

Because $k_2^2 > 0$ and $k_1^2 \geq 0$, we deduce

$$|w_{xt}(1,t)| = \eta_0^2|w_0'(1)| = 0,$$

$$\begin{cases} |w_{xxx}(1,t)| = |w_0'''(1)| = 0, & \text{if } k_1^2 = 0; \\ |w_{xxx}(1,t)| = |w_0'''(1)| = 0, \ |w_t(1,t)| = \eta_0^2 |w_0(1)| = 0, & \text{if } k_1^2 > 0. \end{cases}$$

Thus $w_0(x)$ is a solution to the boundary value problem

$$\left.\begin{aligned} & w_0^{(4)} - \eta_0^4 w_0 = 0 \qquad \text{on } [0,1] \\ & w_0(0) = 0 \\ & w_0'(0) = 0 \\ & w_0'(1) = 0 \\ & w_0''(1) = 0 \\ & w_0'''(1) = 0 \end{aligned}\right\} \tag{2.32}$$

Write

$$w_0(x) = A_{01}\cos \eta_0(x_0-1)+A_{02}\sin \eta_0(x-1)+A_{03}\cosh \eta_0(x-1)+A_{04}\sinh \eta_0(x-1)$$

Then the five boundary conditions in (2.32) require that

$$\begin{bmatrix} \cos \eta_0 & -\sin \eta_0 & \cosh \eta_0 & -\sinh \eta_0 \\ \sin \eta_0 & \cos \eta_0 & -\sinh \eta_0 & \cosh \eta_0 \\ 0 & 1 & 0 & 1 \\ -1 & 0 & 1 & 0 \\ 0 & -1 & 0 & 1 \end{bmatrix} \begin{bmatrix} A_{01} \\ A_{02} \\ A_{03} \\ A_{04} \end{bmatrix} = 0 \tag{2.33}$$

has a nontrivial solution $(A_{01}, A_{02}, A_{03}, A_{04})$. However, it is easy to check that the matrix in (2.33) has rank 4 for any $\eta_0 \in \mathbb{R}$, $\eta_0 \neq 0$, a contradiction.

Therefore the proof of Theorem 4 is complete. □

3. ASYMPTOTIC ESTIMATION OF EIGENFREQUENCIES

From the graphs in [2], one notices that at low frequencies eigenvalues of the damped operator A seem to exhibit a "structural damping" [3] pattern. Does the structural damping pattern continue into high frequencies, or is it only a low frequency phenomenon, for beams with boundary dissipation? To answer this one must examine the asymptotics of eigenfrequencies.

The work of asymptotic analysis was first done by P. Rideau in his thesis [8] (cf. the acknowledgement at the end of the paper). Unaware of his results, we had carried out the analysis independently. We feel that it is of significant interest to include the work here as it will make the study in this paper more complete, and only a minor effort is required.

Let $(\phi(x), \psi(x))$ be an eigenfunction of A belonging to the eigenvalue $\lambda(\neq 0)$. Then by (2.2), setting $f(x) = g(x) \equiv 0$ and $w_\lambda = \phi$, we see that ϕ satisfies

$$\left.\begin{aligned} &\alpha^4\phi^{(4)}(x) + \lambda^2\phi(x) = 0 \\ &\phi(0) = \phi'(0) = 0 \\ &\phi'''(1) - \lambda\tilde{k}_1^2\phi(1) = 0 \\ &\phi''(1) + \lambda\tilde{k}_2^2\phi'(1) = 0 \end{aligned}\right\} \tag{3.1}$$

To simplify notations, we consider the following eigenvalue problem

$$\left.\begin{aligned} &\phi^{(4)}(x) + \lambda^2\phi(x) = 0 \\ &\phi(0) = 0 \\ &\phi'(0) = 0 \\ &\phi'''(1) - \lambda k_1^2\phi(1) = 0 \\ &\phi''(1) + \lambda k_2^2\phi'(1) = 0 \end{aligned}\right\} \tag{3.2}$$

Noting that the following correspondence

$$\left.\begin{matrix} \dfrac{\lambda}{\alpha^2} \\ \alpha^2\tilde{k}_1^2 \\ \alpha^2\tilde{k}_2 \end{matrix}\right\} \text{ in (3.1)} \longleftrightarrow \left\{\begin{matrix} \lambda \\ k_1^2 \\ k_2^2 \end{matrix}\right. \text{ in (3.2)} \tag{3.3}$$

is in effect.

The boundary value problem (3.1) has a nontrivial solution if and only if

$$\det\begin{bmatrix} 1 & 1 & 1 & 1 \\ \mu\sqrt{\lambda} & \mu^3\sqrt{\lambda} & \mu^5\sqrt{\lambda} & \mu^7\sqrt{\lambda} \\ (\mu^3\lambda^{\frac{3}{2}}-k_1^2\lambda)\cdot e^{\mu\sqrt{\lambda}} & (\mu^9\lambda^{\frac{3}{2}}-\lambda k_1^2)\cdot e^{\mu^3\sqrt{\lambda}} & (\mu^{15}\lambda^{\frac{3}{2}}-k_1^2\lambda)\cdot e^{\mu^5\sqrt{\lambda}} & (\mu^{21}\lambda^{\frac{3}{2}}-k_1^2\lambda)\cdot e^{\mu^7\sqrt{\lambda}} \\ (\mu^2\lambda+k_2^2\mu\lambda^{\frac{3}{2}})\cdot e^{\mu\sqrt{\lambda}} & (\mu^6\lambda+\mu^3k_2^2\lambda^{\frac{3}{2}})\cdot e^{\mu^3\sqrt{\lambda}} & (\mu^{10}\lambda+\mu^5k_2^2\lambda^{\frac{3}{2}})\cdot e^{\mu^5\sqrt{\lambda}} & (\mu^{14}\lambda+\mu^7k_2^2\lambda^{\frac{3}{2}})\cdot e^{\mu^7\sqrt{\lambda}} \end{bmatrix} \tag{3.4}$$

$$= 0,$$

where μ is the eighth root of unity, $\exp(i\pi/4)$. The derivation of the above is identical to (2.10)-(2.12).

Evaluating this determinant yields the transcendental equation

$$\begin{aligned} &2\sqrt{2}k_2^2\lambda\left\{ie^{-i\sqrt{2\lambda}}-e^{-\sqrt{2\lambda}}-ie^{i\sqrt{2\lambda}}+e^{\sqrt{2\lambda}}\right\} \\ &+\sqrt{\lambda}\left\{8(1-k_1^2k_2^2)+2(1+k_1^2k_2^2)e^{-\sqrt{2\lambda}}+2(1+k_1^2k_2^2)e^{-i\sqrt{2\lambda}}\right. \\ &\qquad\left.+2(1+k_1^2k_2^2)e^{\sqrt{2\lambda}}+2(1+k_1^2k_2^2)e^{i\sqrt{2\lambda}}\right\} \\ &+2\sqrt{2}k_1^2\left\{-ie^{-i\sqrt{2\lambda}}-e^{-\sqrt{2\lambda}}+ie^{i\sqrt{2\lambda}}+e^{\sqrt{2\lambda}}\right\} = 0. \end{aligned} \tag{3.5}$$

Now, write

$$\lambda = |\lambda|e^{i\theta}. \tag{3.6}$$

As the closed right half plane does not contain any eigenvalues, and because in (3.5), λ is symmetric with respect to the real axis, we need only consider $\frac{\pi}{2} < \theta \le \pi$ in (3.6). We will actually first consider

$$\frac{\pi}{2} < \theta \le \pi - \delta, \quad \text{for any } \delta > 0 \text{ sufficiently small.} \tag{3.7}$$

The case of $\theta \to \pi$ will be considered in (3.11)-(3.12).

Since

$$\sqrt{\lambda} = |\lambda|^{1/2}\exp(i\theta/2) = |\lambda|^{1/2}[\cos(\theta/2)+i\sin(\theta/2)],$$

we see that

$$e^{-\sqrt{2\lambda}} = e^{-\sqrt{2|\lambda|}\cos(\theta/2)} e^{-i\sqrt{2|\lambda|}\sin(\theta/2)}$$

$$e^{i\sqrt{2\lambda}} = e^{-\sqrt{2|\lambda|}\sin(\theta/2)} e^{i\sqrt{2|\lambda|}\cos(\theta/2)}$$

are $\mathcal{O}(e^{-\gamma\sqrt{|\lambda|}})$ for some $\gamma > 0$. Thus, from (3.5)

$$8\sqrt{\lambda}(1-k_1^2k_2^2) + e^{-i\sqrt{2\lambda}}[2i\sqrt{2}k_2^2\lambda+2(1+k_1^2k_2^2)\sqrt{\lambda} - i2\sqrt{2}k_1^2]$$

$$+ e^{\sqrt{2\lambda}}[2\sqrt{2}k_2^2\lambda+2\sqrt{\lambda}(1+k_1^2k_2^2)+2\sqrt{2}k_1^2] = \mathcal{O}(|\lambda|\exp(-\gamma\sqrt{|\lambda|})),$$

which implies

$$e^{\sqrt{2|\lambda|}}[-\sin(\theta/2)+\cos(\theta/2)] = -e^{-i\sqrt{2|\lambda|}}[\cos(\theta/2)+\sin(\theta/2)] \times$$

$$\times \left\{\frac{2i\sqrt{2}k_2^2\lambda+2(1+k_1^2k_2^2)\sqrt{\lambda}-i2\sqrt{2}k_1^2}{2\sqrt{2}k_2^2\lambda+2(1+k_1^2k_2^2)\sqrt{\lambda}+2\sqrt{2}k_1^2}\right\} + \mathcal{O}(|\lambda|\exp(-\gamma\sqrt{|\lambda|}))$$

But (assuming $k_2^2 > 0$) the term in braces equals

$$i + \frac{(1+i)}{\sqrt{2}} \frac{(1+k_1^2k_2^2)}{k_2^2\sqrt{|\lambda|}} e^{-i\theta/2} + \mathcal{O}(|\lambda|^{-1}),$$

Thus we seek λ's satisfying

$$e^{\sqrt{2|\lambda|}}[-\sin(\theta/2)+\cos(\theta/2)] \tag{3.8}$$

$$= \left[-i - \frac{(1+i)}{\sqrt{2}} \frac{(1+k_1^2k_2^2)}{k_2^2\sqrt{|\lambda|}} e^{-i\theta/2}\right] e^{-i\sqrt{2|\lambda|}}[\cos(\theta/2)+\sin(\theta/2)]$$

$$+ \mathcal{O}(|\lambda|^{-1}).$$

We observe immediately that $\theta \to \pi/2$ as $|\lambda| \to \infty$, since the LHS of this equation would decrease to zero otherwise. Furthermore, the equality can be satisfied (up to higher order terms) only when the first term on the RHS is a positive real number. Thus

$$e^{-i\sqrt{2|\lambda|}\,[\cos(\theta/2)+\sin(\theta/2)]} \approx e^{-i\sqrt{2|\lambda|}} \cdot \sqrt{2} \approx i$$

$$2\sqrt{|\lambda|} \approx (2n - \tfrac{1}{2})\pi,$$

or

$$|\lambda| \approx \left[\frac{(2n - \frac{1}{2})}{2}\,\pi\right]^2, \quad n \text{ is a large positive integer.} \qquad (3.9)$$

The above gap of $\mathcal{O}(n^2)$ for eigenvalues is common for Euler-Bernoulli beams with energy conserving boundary conditions. Now we see that Euler-Bernoulli beams with boundary energy dissipation also have this property.

One checks that when the RHS of (3.8) is real, its modulus is

$$1 - \frac{(1+k_1^2k_2^2)[\cos(\theta/2)+\sin(\theta/2)]}{\sqrt{2}k_2^2\sqrt{|\lambda|}} + \mathcal{O}(|\lambda|^{-1}).$$

This in turn implies that the exponent on the LHS of (3.8) must satisfy

$$\sqrt{2|\lambda|}\left[-\sin(\tfrac{\theta}{2})+\cos(\tfrac{\theta}{2})\right] \approx - \frac{(1+k_1^2k_2^2)\left[\cos(\frac{\theta}{2})+\sin(\frac{\theta}{2})\right]}{\sqrt{2}k_2^2\sqrt{|\lambda|}} .$$

If we now write $\theta = \frac{\pi}{2} + \varepsilon$, $\varepsilon > 0$, and expand to lowest order in ε, we have

$$\varepsilon \approx \frac{(1+k_1^2k_2^2)}{k_2^2|\lambda|} .$$

Now suppose $\lambda = \xi + i\eta$. Then $\theta = \tan^{-1}(\eta/\xi)$ and $|\lambda| = (\xi^2+\eta^2)^{1/2}$. Expanding $\tan^{-1}$ about $\eta/\xi = \infty$, we have

$$\varepsilon \approx -\frac{\xi}{\eta} ,$$

or

$$-\frac{\xi}{\eta} \approx \frac{(1+k_1^2k_2^2)}{k_2^2(\xi^2+\eta^2)^{1/2}} .$$

Hence

$$\xi \approx -\frac{1}{k_2^2}(1+k_1^2k_2^2), \quad \text{as} \quad |\lambda| \longrightarrow \infty. \tag{3.10}$$

By (3.7), the only remaining case to be considered is when $\theta \to \pi$, i.e. when λ approaches the negative real axis. We write

$$\sqrt{\lambda} = |\lambda|^{1/2}(\cos(\theta/2)+i\sin(\theta/2))$$

as before, but this time was assume $|\pi-\theta| < \delta$, with δ small -- say $0 \le \delta < \pi/8$. Then one easily checks that

$$\begin{aligned} &|e^{-\sqrt{2\lambda}}| \le c, \\ &|e^{i\sqrt{2\lambda}}| \le c, \end{aligned} \tag{3.11}$$

for some $c > 0$ so that (3.5) can be rewritten as:

$$\begin{aligned} &\{2\sqrt{2}\ ik_2^2\lambda+2\sqrt{2}(1+k_1^2k_2^2)-2\sqrt{2}ik_1^2\}e^{\sqrt{2|\lambda|}(\sin(\theta/2)-i\cos(\theta/2))} \\ &= \{2\sqrt{2}k_2^2\lambda+2\sqrt{\lambda}(1+k_1^2k_2^2)+2\sqrt{2}k_1^2\}e^{\sqrt{2|\lambda|}(\cos(\theta/2)+i\sin(\theta/2))} \\ &\qquad + \mathcal{O}(\lambda). \end{aligned} \tag{3.12}$$

However, for θ in the range of interest, $\sin(\theta/2) > 2\cos(\theta/2)$ and in particular, $\sin(\theta/2) > 0.5$. Thus, the modulus of the L.H.S. of (3.12) will be much larger than that of the R.H.S. (for $|\lambda|$ large) so this equation has no solutions if $|\lambda|$ is large.

<u>THEOREM 5</u> Let $\lambda = \xi + i\eta$ be an eigenfrequency of vibration of the beam equation (1.1). Then for $|\lambda|$ large,

$$|\lambda| \sim \left[\frac{(2n-\frac{1}{2})\pi}{2}\right]^2\left(\frac{EI}{m}\right)^{1/2}, \quad n\text{'s are large positive integers,}$$

$$\xi \to -\frac{[m(EI)]^{1/2}\left\{1+k_1^2k_2^2[m(EI)]^{-1}\right\}}{k_2^2} \quad \text{as} \quad |\lambda| \to \infty \tag{3.13}$$

Proof: Just use (2.3), (3.3), (3.9) and (3.10).

By (3.13), the eigenvalues will be distributed nearly parallel to the imaginary axis at high frequencies. Therefore there is no"structural damping" when the frequencies are high. This has also been confirmed numerically in Rideau's thesis [8].

We note that when $k_2^2 = 0$ and $k_1^2 > 0$, asymptotic limits can be obtained in the similar way.

4. DESIGN OF PASSIVE DAMPING MECHANISMS

The following is a (more or less exhaustive) list of combinations of dissipative boundary conditions for an Euler-Bernoulli beam:

$$\left.\begin{aligned} -EIy_{xxx}(1,t) &= -k_1^2 y_t(1,t), \quad k_1^2 > 0 \\ -EIy_{xx}(1,t) &= 0 \end{aligned}\right\} \tag{4.1}$$

$$\left.\begin{aligned} -EIy_{xxx}(1,t) &= 0 \\ -EIy_{xx}(1,t) &= k_2^2 y_{xt}(1,t), \quad k_2^2 > 0 \end{aligned}\right\} \tag{4.2}$$

$$\left.\begin{aligned} y_x(1,t) &= 0 \\ -EIy_{xxx}(1,t) &= -k_1^2 y_t(1,t), \quad k_1^2 > 0 \end{aligned}\right\} \tag{4.3}$$

$$\left.\begin{aligned} y(1,t) &= 0 \\ -EIy_{xx}(1,t) &= k_2^2 y_{xt}(1,t), \quad k_2^2 > 0 \end{aligned}\right\} \tag{4.4}$$

$$\left.\begin{aligned} -EIy_{xxx}(1,t) &= -k_1^2 y_t(1,t) + c_1 y_{xt}(1,t), \quad k_1^2 > 0, \\ -EIy_{xx}(1,t) &= k_2^2 y_{xt}(1,t) + c_2 y_t(1,t), \quad k_2^2 > 0. \end{aligned}\right\} \tag{4.5}$$

where in (4.5), c_1 and c_2 are real constants satisfying

$$(c_1-c_2)\alpha\beta - k_1^2\alpha^2 - k_2^2\beta^2 \le 0 \quad \forall\ \alpha,\beta \in \mathbb{R}. \tag{4.6}$$

Note that the boundary conditions (1.1.4) and (1.1.5) correspond to $c_1 = c_2 = 0$ in (4.5). Obviously, (4.6) is satisfied in this case.

We want to show that all stabilization schemes (4.1)-(4.5) can be realized in practice, at least by designing passive dampers.

As (4.5) seems to represent the most complicated case among (4.1)-(4.5), we treat it here, at least for certain special values of c_1 and c_2 (cf. (4.7) later). The other cases can be studied similarly.

The following damper arrangement gives a design which effects the coupling of shear (resp. bending moment) with velocity and angular velocity:

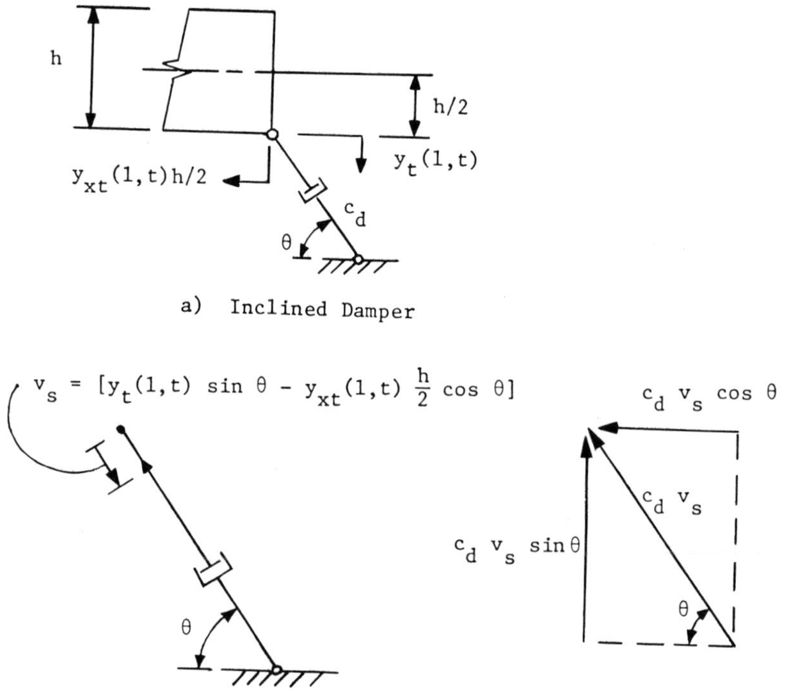

b) Shortening Velocity of Damper

c) Damper Forces on Beam

$$\text{Shear}(1,t) = -c_d\, v_s \sin\theta$$

$$\text{Moment}(1,t) = c_d\, v_s \cos\theta\, h/2$$

Figure 2 Damper arrangement for (4.5)

A single damper (cf. Figure 2a) is attached to the lower end of the beam at an inclination angle θ with respect to the horizontal. Using the velocity at the end of the damper, v_s, and the associated forces shown in Figure 2b and 2c, we obtain

$$\text{Shear}(1,t) = -c_d v_s \sin\theta$$
$$\text{Moment}(1,t) = c_d v_s (\cos\theta)\frac{h}{2},$$

where c_d represents the damping coefficient associated with the damper in use. As

$$v_s = y_t(1,t)\sin\theta - y_{xt}(1,t)\frac{h}{2}\cos\theta,$$

we get

$$\text{Shear}(1,t) = -EIy_{xxx}(1,t) = -(c_d\sin^2\theta)y_t + (\frac{h}{2}\cdot c_d\sin\theta\cos\theta)y_{xt}$$
$$\text{Moment}(1,t) = -EIy_{xx}(1,t) = (\frac{h^2}{4}c_d\cos^2\theta)y_{xt} + (-\frac{h}{2}c_d\sin\theta\cos\theta)y_t.$$

A comparison of the above with (4.5) shows that

$$k_1^2 = c_d\sin^2\theta;\ k_2^2 = \frac{h^2}{4}c_d\cos^2\theta$$
$$c_1 = -c_2 = \frac{h}{2}c_d\sin\theta\cos\theta, \tag{4.7}$$

thus

$$(c_1-c_2)\alpha\beta - k_1^2\alpha^2 - k_2^2\beta^2 = -(k_1\alpha-k_2\beta)^2 \le 0 \quad \forall\ \alpha,\beta\in\mathbb{R}$$

so (4.6) is satisfied and the boundary conditions (4.5) are dissipative.

It is noted that when $\theta = \pi/2$ (vertical damper), the above gain constants reduce to $k_1^2 = c_{d,1}$ and $k_2^2 = c_1 = c_2 = 0$, cf. Figure 3a. Similarly, for $\theta = 0$ (horizontal damper), the gain constants become $k_2^2 = c_{d,2}h^2/4$ and $k_1^2 = c_1 = c_2 = 0$, as shown in Figure 3b. Consequently, the boundary conditions (1.1.4)-(1.1.5) can be realized as in Figure 3c, with $k_1^2 = c_{d,1}$ and $k_2^2 = c_{d,2}h^2/4$.

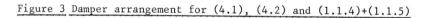

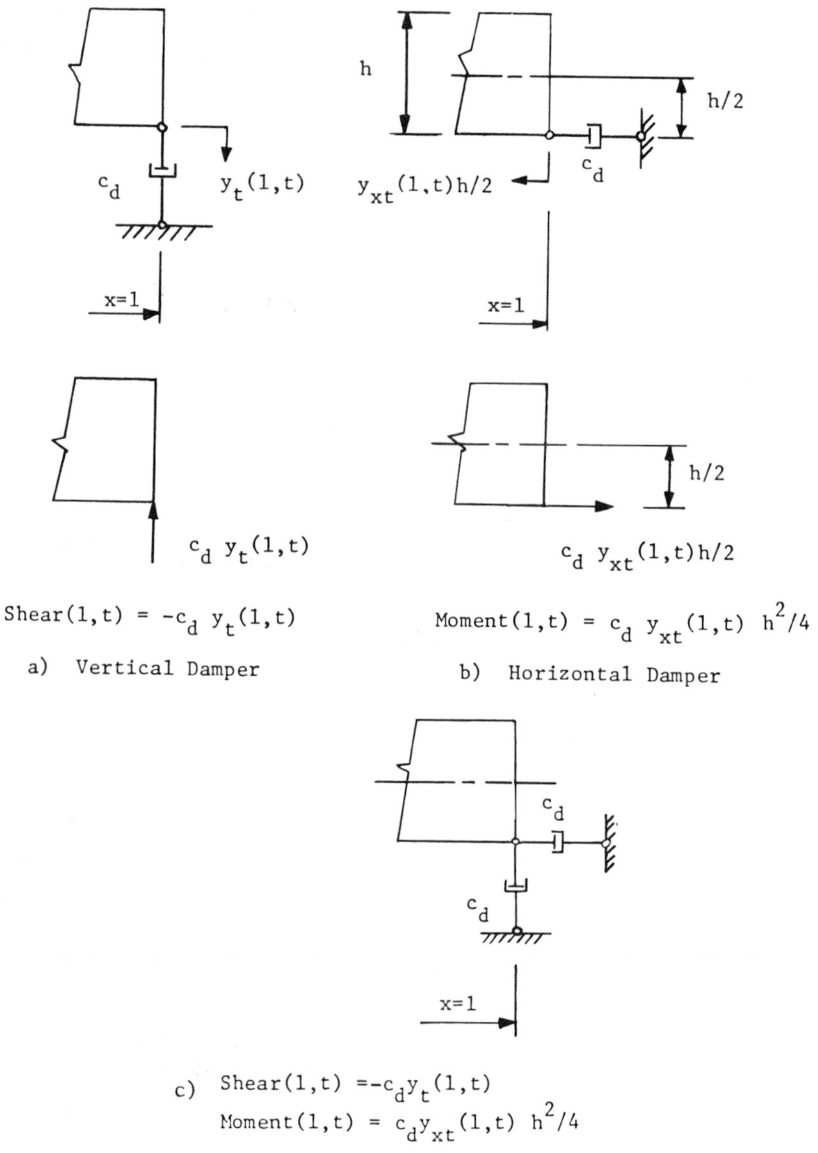

Figure 3 Damper arrangement for (4.1), (4.2) and (1.1.4)+(1.1.5)

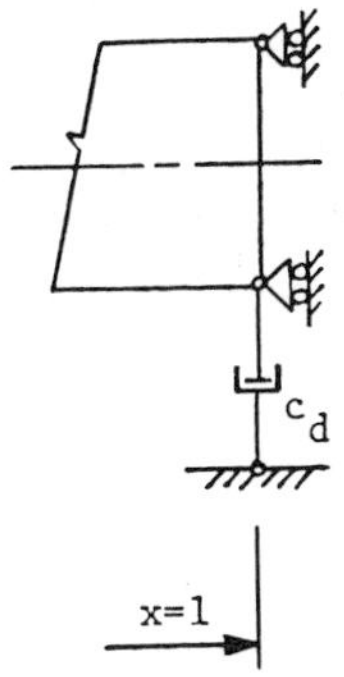

$$\text{Shear}(1,t) = -c_d y_t(1,t)$$
$$\text{Slope}(1,t) = y_x(1,t) = 0$$

Figure 4 Damper arrangement for (4.3)

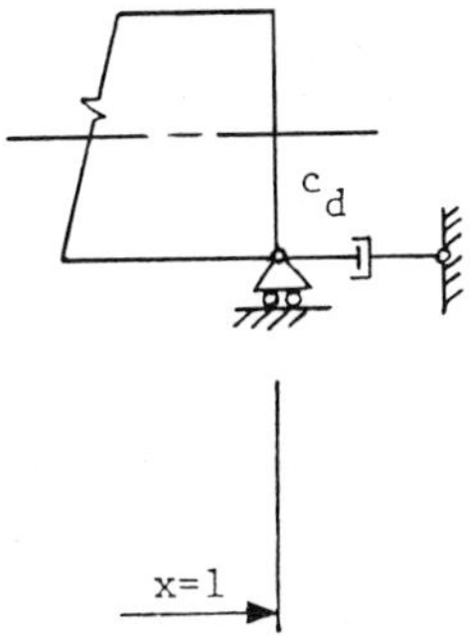

$$\text{Displ.}(1,t) = y(1,t) = 0$$
$$\text{Moment}(1,t) = c_d y_{xt}(1,t)\ h^2/4$$

Figure 5 Damper arrangement for (4.4)

The other boundary conditions (4.1)-(4.4) can be realized and designed, respectively, as in Figures 3a, 3b, 4 and 5.

The method of estimation which we have developed in this paper and Huang's theorem (Thm. 1) can be applied to study exponential stability for all of these boundary stabilization schemes.

ACKNOWLEDGEMENT We wish to thank M.C. Delfour for pointing out the thesis of P. Rideau [8] to us, where Rideau has already carried out the work of asymptotic analysis of eigenfrequencies. The priority should go to him. Recently, we have also found that D.L. Russell [9] and his Ph.D. student S. Hansen [5] have independently obtained similar asymptotic results.

REFERENCES

[1] G. Chen, Energy decay estimates and exact boundary value controlability for the wave equation in a bounded domain, J. Math. Pures Appl. 58(1979), pp. 249-273.

[2] G. Chen, M.C. Delfour, A.M. Krall and G. Payre, Modeling, stabilization and control of serially connected beams, SIAM J. Cont. Opt., to appear.

[3] G. Chen and D.L. Russell, A mathematical model for linear elastic systems with structural damping, Quart. Appl. Math. 39(1981-82), pp. 433-454.

[4] F.L. Huang, Characteristic conditions for exponential stability of linear dynamical systems in Hilbert spaces, Ann. Diff. Eqs. 1(1), 1985, pp. 43-53.

[5] S. Hansen, Private communications.

[6] T. Kato, Perturbation Theory for Linear Operators, Springer-Verlag, New York, 1966.

[7] J. Lagnese, Decay of solutions of wave equations in a bounded region with boundary dissipation, J. Diff. Eq. 50(1983), pp. 163-182.

[8] P. Rideau, Controle d'un assemblage de poutres flexibles par des capteurs-actionneurs ponctuels: étude du spectre du système, Thèse, L'Ecole Nationale Superieure des Mines de Paris, Sophia-Antipolis, France, November, 1985.

[9] D.L. Russell, On mathematical models for the elastic beam with frequency-proportional damping, to appear.

The Time-Optimal Control of Nonlinear Delay Equations

Ethelbert N. Chukwu

Department of Mathematics
University of Tennessee
Knoxville, Tennessee

INTRODUCTION

The paper solves the problem of reaching in minimum time a continuously moving target by a trajectory of the control system described by the nonlinear functional delay differential equation,

$$\dot{x}(t) = f(t, x_t, u(t)) \qquad t \geq 0$$

$$x(t) = \phi(t) \qquad t \in [-h, 0]$$

The state space is either E^n the Euclidean n-dimensional vector space, the space $C = C[-h, 0], E^n)$, of continuous functions from $[-h, 0]$ into E^n with the sup norm or the Sobolev space $W_2^{(1)} = W_2^{(1)}([-h, 0), E^n)$ of functions $[-h, 0] \to E^n$ whose derivatives are L_2 integrable. Admissible controls are measurable functions $u: [0, \infty) \to E^m$ whose values $u(t) \in U$ a compact convex set. We establish the existence of optimal controls and answer the corresponding constrained controllability questions. The necessary condition for time optimal control, the maximum principal, is stated. As a by-product, a detailed treatment of the time optimal control of linear systems is presented.

1. PRELIMINARIES

We shall consider the problem of reaching (in minimum time) a continuously moving target by a trajectory of the nonlinear system described by

$$\begin{aligned} \dot{x}(t) &= f(t, x_t, u(t)) \qquad t \geq 0 \\ x(t) &= \phi(t) \qquad t \in [-h, 0) \end{aligned} \tag{1.1}$$

The existence of an optimal control for (1.1) will be treated in the space $C = C([-h, 0], E^n)$ of continuous functions $x: [-h, 0] \to E^n$, $h \geq 0$ with the sup norm. This is done in the first section. A similar treatment of our specific time optimal problem does not seem to be available, although the basic ideas are implicit in several researches [3, 9, 16]. In section 2 we establish some preliminary results which will help prove, in section 3, global constrained controllability theorems which are sufficient for the existence of time optimal controls. The results here are new and tie up neatly with the treatment of the sufficient condition of section 4. In section 4, we extract from Banks [25] the necessary condition for time optimal control of nonlinear systems with target in E^n. We derive a special form of optimal control for linear systems with target in E^n. Because the special form of optimal control is not available in other function spaces, we derive them for the space C and $W_2^{(1)}([-h, 0], E^n)$, the Sobolev space of absolutely continuous functions whose derivatives are square integrable in $[-h, 0]$. We also obtain criteria for uniqueness of optimal controls. Our treatment opens up the possibility of constructing an explicit optimal feedback control for linear systems.

The vectors x and f will be n-dimensional and are elements of E^n, the Euclidean n-dimensional space with norm denoted by $|\cdot|$. In (1.1) x is an element of $C([\sigma - h, T], E^n)$, $T > \sigma$, the set of continuous functions mapping $[\sigma - h, T]$ into E^n. For each $t \in [\sigma, T]$, we define an element of the set C by the equation

$$x_t(s) = x(t + s), \quad -h \leq s \leq 0$$

The Banach spaces $C([\sigma - h, T], E^n) = C_T$, $C([-h, 0], E^n) = C_0 = C$, have norms given by $||x|| = \sup\{|x(s)|: s \in [\phi - h, T]\}$ and $||\phi|| = \sup\{|\phi(s)|: s \in [-h, 0]\}$. Let U be a set-valued map $U: E \times C([-h, 0], E^n) \to 2^{E^m}$. An admissible control $u: [\sigma, T] \to E^n$ is a measurable function with values at time t in a nonempty compact set $U(t, x_t)$. The problem of this research is to determine a control u^* subject to its constraints, which is such that the solution $x(\sigma, \phi, u^*)$ of (1.1) reaches a continuously moving target $z_t \in C([-h, 0], E^n)$, in minimum time $t^* \geq 0$. Such a control is called time optimal. The existence of an optimal control is first investigated in C.

We assume that

$$f: E \times C \times E^m \to E^n$$

is continuous; for each $t \in E$, the mapping $(\phi, u) \to f(t, \phi, u)$ is continuously differentiable. The mapping $\overline{D}f$, the partial derivative of f with respect to the second and third variables is locally integrably bounded, and $f(t, \phi, u)$ is Lipschitzian in ϕ, u. The mapping $t \to \overline{D}f(t,\phi,u)$ is assumed measurable. We assume enough smoothness conditions on f to guarantee global existence and uniqueness of a solution passing through each $(\sigma, \phi) \in E^+ \times C$ for each $u \in L_2([\sigma, \infty), E^m)$. We also assume that f satisfies the inequality,

$$x^T(t)\, f(t, x_t, u(t)) \leq k(1 + ||x_t||^2) \tag{1.2}$$

for some constant $k > 0$ all $t \in E$, $x_t \in C$, $u(t) \in U(t, x_t)$. Also for any bounded set $N \subseteq C$ there exists an almost-everywehre bounded measurable function $\xi_N: E \to E^n$ such that

$$|f(t, x_t, u(t))| \leq |\xi_N(t)| \qquad \text{a.e. in } t \tag{1.3}$$

for all $x_t \in N$, $u(t) \in U(t, x_t)$.

In order to study the existence of optimal controls, we consider (1.1) in terms of the orientor field, and introduce the set valued function,

$$F(t, \phi) = \{f(t, \phi, u):\ u \in U(t, \phi)\}$$

Because f is continuous and $U(t, \phi)$ nonempty and compact, $F(t, \phi)$ is nonempty and compact. We assume that $U(t, \phi)$ is upper semicontinuous with respect to inclusion in t, ϕ, in the following sense.

Definition 1.1. The set map, $F(t, x_t): E \times C \to 2^{E^n}$, is called semicontinuous with respect to inclusion in t, x_t, if for any $\delta > 0$, there is a $\delta > 0$, such that $F(t_1, x_{t_1}) \subseteq F_\varepsilon(t, x_t)$, whenever $|t_1 - t| \leq \delta$, $||x_{t_1} - x_t|| \leq \delta$. Here F_ε denotes the ε-neighborhood of F in E^n.

We now consider the problem,

$$\dot{x}(t) \in F(t, x_t), \quad \text{a.e. in } [\sigma, T] \quad T > \sigma, \; x_\sigma = \phi \tag{1.4}$$

A solution $x(t)$ of (1.4) is an element $x \in C([\sigma - h, T], E^n)$, such that

(i) x is absolutely continuous on $[\sigma, T]$.

(ii) $x_\sigma = \phi$, and

(iii) $\dot{x}(t) \in F(t, x_t)$.

That every solution of (1.4) can be viewed as a trajectory of the control system (1.1) is deduced from McShane and Warfield generalization [6] of the result of Filippov [7]. It is stated in the next Lemma. We assume throughout that $F(t, \phi)$ is convex for each $t, \phi \in E \times C$.

Lemma 1.1. A function x is a solution of (1.4), if and only if x is a solution of (1.1) for some admissible control u.

Definition 1.2. The attainable set function A is a subset of C given by

$$A(t) = \{x_t\colon\ x \text{ is a solution of (1.1) for some } u(t) \in U(t, x_t)\} .$$

By Lemma 1.1, this is the same as

$$A(t) = \{x_t\colon \dot{x}(t) \in F(t, x_t)\ ,\ t \geq \sigma\ ,\ x_\sigma = \phi\} .$$

In order to prove the existence of an optimal control for the time optimal problem we will need the fact that $A(t)$ is closed in C.

Theorem 1.1. Suppose

(i) $f\colon E \times C \times E^m \to E^n$ is continuous, and is continuously differentiable in the last two arguments;

(ii) f satisfies the inequality (1.2) and (1.3);

(iii) $F(t, x_t)$ is convex for each t, x_t;

(iv) $U(t, x_t)$ is upper semicontinuous with respect to inclusion in t, x_t and is compact.

Then for any $t \geq \sigma$, $A(t)$ is compact.

Proof. It follows from the given properties of $f(t, x_t, u)$ and the semicontinuity property of $U(t, x_t)$ that $F(t, x_t)$ is upper semicontinuous with respect to inclusion, and closed.

Let $\{x_T^k\}$ be a sequence of points in $A(T)$. Then

$$\dot{x}^k(T) \in F(T, x_T)\ ,\ T \geq \sigma \qquad\qquad x_\sigma^k = \phi$$

x^k is a solution of (1.4). We claim that $\{x^k\}$ is equicontinuous and equibounded so that x_T^k is equicontinuous and equibounded. Because f

satisfies (1.2) and (1.3), $(\dot{x}^k(t), \dot{x}^k(t)) \leq K(1 + ||x_t||^2)$, so that $||x^k(t)||^2 \leq (1 + ||x_\sigma^k||^2) \exp[2k(t - \sigma)] - 1$ for $t \in [\sigma, T]$, and $||x^k(t)||^2 \leq (1 + ||\phi||^2) \exp[2k(t - \sigma)] - 1$. The function $\{x^k\}$ is therefore equibounded. Because $\{x^k\}$ is equibounded (1.3) ensures that there exists an almost everywhere bounded function ξ such that $|\dot{x}^k(t)| \leq |\xi(t)| \leq M$ a.e. t . Hence $|\dot{x}^k(t)|$ are uniformly bounded almost everywhere. This implies that $\{x^k\}$ is equicontinuous and therefore compact. Apply the Ascoli theorem and extract a subsequence (we take it to be the original sequence) $\{x^k\}$ converging $(k \to \infty)$, uniformly on $[\sigma, T]$ to some x .

Note that $x_\sigma = \phi$; $|\dot{x}(t)| \leq M$, so that x is absolutely continuous. Our aim is to show that x is a solution of (1.4). Let t_0 be a point in $[\sigma, T]$ for which $\dot{x}(t_0)$ exists. We show $\dot{x}(t_0) \in F(t_0, x_{t_0})$. Now

$$\frac{x(t) - x(t_0)}{t - t_0} = \lim_{k\to\infty} \frac{x^k(t) - x^k(t_0)}{t - t_0}$$

$$= \lim_{k\to\infty} \frac{1}{t - t_0} \int_{t_0}^{t} \dot{x}^k(\tau)\, d\tau = \lim_{k\to\infty} \int_0^1 \dot{x}^k(t - t_0)s)\, ds .$$

For a given $\varepsilon > 0$ choose $\delta > 0$ such that

$$\left|\frac{x(t) - x(t_0)}{t - t_0} - \dot{x}(t_0)\right| \leq \varepsilon \tag{1.5}$$

$$F(t, x_t) \subseteq F_\varepsilon(t_0, x_{t_0}) \tag{1.6}$$

whenever $|t - t_0| < \delta$, $||x_t - x_{t_0}|| < \delta$. The inclusion in (1.6) is satisfied because $F(t, x_t)$ is assumed upper semicontinuous. For almost all $\tau \in [\sigma, t]$, $\dot{x}^k(\tau) \in F(\tau, x^k)$. But x^k converges uniformly to x so that for $N \geq N_0$ (say), $||x_t - x_t^k|| \leq \delta/2$. Therefore $\dot{x}^k(\tau) \in F_\varepsilon(t_0, x_{t_0})$, for almost all $\tau \in [t_0, t]$. Thus $\dot{x}^k(t_0 + (t - t_0)s) \in F_\varepsilon(t_0, x_{t_0})$ for almost all $s \in [0, 1]$. Since $F_\varepsilon(t_0, x_{t_0})$ is convex, the mean value theorem yields

$$\int_0^1 x^k(t_0 + (t - t_0)s)ds \in F_\varepsilon(t_0, x_{t_0}) \quad \text{for } N \geq N_0 \text{ (say).}$$

Thus $\dfrac{x(t) - x(t_0)}{t - t_0} \in F_\varepsilon(t_0, x_{t_0})$ Therefore

$$\dot{x}(t_0) \in F_{2\varepsilon}(t_0, x_{t_0})$$

for arbitrary $\varepsilon > 0$. Since the sets $F(t, x_t)$ are closed,

$$\dot{x}(t_0) \in F(t_0, x_{t_0}) = \bigcap_{\varepsilon>0} F_{2\varepsilon}(t_0, x_{t_0})$$

Since this holds almost everywhere on $[\sigma, T]$, we have shown that x satisfies (1.4). Thus $x_T \in A(T)$, and $A(T)$ is closed. We have shown that $\{x_T^k\}$ is equibounded and equicontinuous and has a sequence which converges to a point of $A(t)$. Therefore $A(T)$ is compact.

Definition 1.3. Let $z_T \in C$ be a target function which is an element of the function space C, and suppose $z_T \in A(T)$, $T \geq \sigma$ then the system (1) is controllable to the target. Thus for each $\phi \in C$, there exists a $T \geq \sigma$ an admissible control $u(t) \in U(t, x_t)$, $t \in [\sigma, T]$ such that the solution of (1.1) satisfies $x_\sigma(\sigma, \phi, u) = \phi$, $x_T(\sigma, \phi, u) = z_T$. In this case we also say that (1.1) is function space z-controllable on $[\sigma, T]$ with constraints.

Definition 1.4. The system (1.1) is null-controllable with constraints if for each $\phi \in C$, there exists a $t_1 < \infty$, and an admissible control $u(t) \in U(t, x_t)$ $t \in [\sigma, t_1]$ such that the solution of (1.1) satisfies

$$x_\sigma(\sigma, \phi, u) = \phi, \quad x_{t_1}(\sigma, \phi, u) = 0 .$$

Definition 1.5. The system (1.1) is locally null-controllable with constraints if there exists an open ball $\mathcal{O}$ of the origin in C with the following property: For each $\phi \in \mathcal{O}$, there exists a $t_1 < \infty$ and an admissible control $u(t) \in U(t, x_t)$ $t \in [\sigma, t_1]$ such that the solution of (1.1) satisfies $x_\sigma(\sigma, \phi, u) = \phi$, $x_{t_1}(\sigma, \phi, u) = 0$. For controllability with constraints we initially assume $U(t, x_t)$ is a compact set. This assumption prevails in Theorem 1.2 below. It can be relaxed to only closedness of $U(t, x_t)$ and upper semicontinuity together with the so-called normality condition of Cesari [8]. If we work in the space W_∞^1 as the state space with controls L_∞, we can take U to be the fixed unit cube in E^m . Controllability concepts can also be defined without assuming constraints.

Definition 1.6. The system (1.1) is controllable in $[\sigma, t]$, $t_1 > \sigma + h$ if for each $\phi, \psi \in C$, there exists a control $u \in L_\infty([\sigma, t_1], E^m)$ such that the solution $x(\sigma, \phi, u)$ of (1.1) satisfies $x_\sigma(\sigma, \phi, u) = \phi$, $x_{t_1}(\sigma, \phi, u) = \psi$. It is null controllable on $[\sigma, t_1]$, $t_1 > \sigma + h$ if for each $\phi \in C$ there exists a control $u \in L_\infty([\sigma, t_1], E^m)$ such that the solution of (1.1) satisfies $x_\sigma(\sigma, \phi, u) = \phi$, $x_{t_1}(\sigma, \phi, u) = 0$. We shall drop the qualify phrase "on the interval $[\sigma, t_1]$ in the definition if they hold on every interval, with $t_1 > h + \sigma$.

In the above definitions, we could replace C by W_p^1 and consider controls on $L_p([\sigma, b], E^m)$. When $p = 2$ this will tie up with the investigations of [9]. For the general nonlinear system (1.1) we limit our investigation of the existence of optimal control to the state space C . In Definition 1.6, if ψ is replaced by $x_1 \in E^n$, we have Euclidean controllability.

Theorem 1.2. (Existence of Time Optimal Control).
(i) Suppose conditions (i)-(iv) of Theorem 1.1 hold.
(ii) For some $T \geq \sigma$ the system (1.1) is function space z-controllable on $[\sigma, T]$ with constraints.
Then there exists an optimal control.

Proof. Let

$$t^* = \inf\{t: T \geq t \geq \sigma \ , \ z_t \in A(T)\} .$$

Since there is at least one T such that $z_T \in A(T)$, t^* is well defined. We now show $z_{t^*} \in A(t^*)$. Let $\{t_n\}$ be a sequence of times which converge to t^* such that $z_{t_n} \in A(t_n)$. For each n , let x^n be a solution of (1.4) with $z_{t_n}^n = x_{t_n}^n$. Then

$$\|z_{t^*}^n - z_{t^*}\| \leq \|x_{t^*}^n - x_{t_n}^n\| + \|x_{t_n}^n - z_{t^*}\| \leq$$
$$\leq \|x_{t^*}^n - x_{t_n}^n\| + \|z_{t_n} - z_{t^*}\| \leq \|z_{t_n} - z_{t^*}\| + \int_{t^*}^{t_n} \xi(\tau)\, d\tau ,$$

where, by condition (1.3), ξ is given almost bounded. Because z_t is continuous in t $z_{t_n} \to z_{t^*}$ as $t_n \to t^*$. Because ξ is bounded a.e. and measurable $\int_{t^*}^{t_n} \xi(\tau)\, d\tau \to 0$ as $t_n \to t^*$. Hence $x_{t^*}^n \to z_{t^*}$ as $n \to \infty$. Since $x_{t^*}^n \in A(t^*)$ and $A(t^*)$ is closed, $z_{t^*} \in A(t^*)$. This completes the proof.

Remark 1.1. Theorem 1.2 is fundamental in time optimal control theory of functional differential equations. It asserts that one must make the controllability assumption and have conditions that guarantee it before the optimality questions can be resolved. Controllability results with controls constrained to lie in a compact subset U of E^m are therefore very useful in the resolution of time optimal control problems of delay equations. In section 3 we treat the problem of constrained controllability which are necessary for the problems of time optimality.

Remark 1.2. The idea of the proof of Theorem 1.2 is available in the literature in more general setting than the problem of Time Optimal Control (See Angell [9]). It is perhaps new in its present formulation.

2. PRELIMINARY CONTROLLABILITY RESULTS

In this section we establish some preliminary results which will be helpful in studying global controllability questions for the system,

$$\dot{x}(t) = f(t, x_t, u(t)), \quad t \geq \sigma,$$
$$x_\sigma = \phi \in C, \tag{2.1}$$

where f satisfies enough smoothness conditions to guarantee global existence and uniqueness of a solution passing through each $(\sigma, \phi) \in E \times C$ for each $u \in L_\infty([\sigma, \infty), E^m)$. Thus for each $(\sigma, \phi) \in E \times C$ and for every $u \in L_\infty^{loc}([\sigma, \phi), E^m)$ there exists a unique response $x(\sigma, \phi, u)$: $E^+ = [\sigma, \infty) \to C$ with initial data $x_\sigma(\sigma, \phi, u) = \phi$, corresponding to the control u. The mapping $x(\sigma, \phi, u)$: $E^+ \to C$ defined by $t \to x_t(\sigma, \phi, u)$ represents a point of C.

We now study and consider the mapping

$$(t, \sigma, \phi, u) \to x_t(\sigma, \phi, u) \ , \ x\colon E^+ \times E^+ \times C \times L_\infty^{loc}([\sigma, \infty), E^n) \to C$$

Lemma 2.1. For each fixed $(t_0, s_0, \phi_0, u_0) \in E^+ \times E^+ \times C \times L_\infty^{loc}([\sigma, \infty), E^m)$, let $\bar{D}x_{t_0}(\sigma_0, \phi_0, u_0)$ denote the partial derivative of x_{t_0} with respect to its last two variables at the point $(t_0, \sigma_0, \phi_0, u_0)$. Then for every $(\underline{h}, u) \in C \times L_\infty^{loc}([\sigma, \infty), E^m)$, we have $\bar{D}x_{t_0}(\sigma_0, \phi_0, u_0)(\underline{h}, u)$ $= \lambda_{t_0}(\sigma_0, \phi_0, u_0, \underline{h}, u)$, where the mapping $t \to \lambda(t, \sigma_0, \phi_0, u_0, \underline{h}, u)$ of E into E^n is the unique absolutely continuous solution of the linear differential equation

$$\dot{z}(t) = D_2 f(t, x_t(\sigma_0, \phi_0, u_0), u_0(t))z_t + D_3 f(t, x_t(\sigma_0, \phi_0, u_0), u_0(t))v(t) \qquad (2.2)$$

$$z_{\sigma_0} = h \qquad (2.3)$$

Also for $v \in L_\infty^{loc}([\sigma, \infty), E^m)$, we have $D_u x_{t_0}(\sigma_0, \phi_0, u_0)(v)$ $= \psi_{t_0}(\sigma_0, \phi_0, u_0, v)$ where the mapping $t \to \psi(t, \sigma_0, \phi_0, u_0, v)$ is the unique absolutely continuous solution of (2.2) satisfying

$$\psi_{\sigma_0}(\sigma_0, \phi_0, u_0, u) = 0 \qquad (2.4)$$

Proof. Let D_u be the partial derivative of $x(t, \sigma, \phi, u)$ with respect to u, the fourth argument. Then

$$D_u x_\sigma(\sigma, \phi, u) = 0 \quad , \quad \text{in } [-h, 0] .$$

$$D_u x(t, \sigma, \phi, u) = \int_\sigma^t D_2 f(s, x_s(\sigma, \phi, u), u) \cdot D_u x_s(\sigma, \phi, u)ds + \int_\sigma^t D_3 f(s, x_s(\sigma, \phi, u), u(s))\, ds \quad , \quad t \geq \sigma .$$

On differentiating this with respect to t, we have

$$\frac{d}{dt} D_u x(t, \sigma, \phi, u)v = D_2 f(t, x_t(\sigma, \phi, u), u(t))\, D_u x_t(\sigma, \phi, u)v + D_3 f(t, x_t(\sigma, \phi, u), u(t))\, v ;$$

$$D_u x_{\sigma_0}(\sigma_0, \phi_0, u_0)v = D_u \phi_0 v = 0 .$$

This proves the second assertion. To prove the first, we note that

$$Dx(t, \sigma, \phi, u)(h, v) = D_3\, x(t, \sigma, \phi, u)h + D_4\, x(t, \sigma, \phi, u)\, v\ .$$

Now,

$$D_3 x(t, \sigma, \phi, u) = I + \int_\sigma^t D_2\, f(s, x_s(\sigma, \phi, u), u(s))\, D_3\, x_s(\sigma, \phi, u) ds$$

$$D_4\, x(t, \sigma, \phi, u) = \int_\sigma^t D_2 f(s, x_s(\sigma, \phi, u), u(s)) \cdot D_3\, x_s(\sigma, \phi, u)\, ds + \int_\sigma^t D_3\, f(s, x_s(\sigma, \phi, u), u(s))\, ds\ .$$

Therefore

$$Dx(t, \sigma, \phi, u)(\underline{h}, v) = \underline{h} + \int_\sigma^t D_2 f(s, x_s, u(s)) D_3\, x_s \underline{h}\, ds + \int_\sigma^t D_2 f(s, x_s, u(s))\, D_4\, x_s v\, ds + \int_\sigma^t D_3\, f(s, x_s, u(s))\, v\, ds\ .$$

Now take the t derivative of this expression to

$$\frac{d}{dt} Dx(t,\sigma,\phi,u)(h,v) = D_2 f(t,x_t,u(t))[D_3\, x_t \underline{h} + D_4\, x_t v] + D_3 f(t,x_t,u(t))v\ .$$

Note that

$$Dx(\sigma, \sigma, \phi, u)(h, u) = \underline{h}\ ,$$

to conclude the proof.

Lemma 2.2. Given $\phi \in C$, $u \in L_\infty([\sigma, t_1], E^m)$. For $t_1 > \sigma + h$, let $t \in [\sigma, t_1]$. Let $u \to x_{t_1}(\sigma, \phi, u)$ be the mapping

$$H: L_\infty([\sigma, t_1], E^m) \to C([-h, 0], E^n \qquad H(u) = x_{t_1}(\sigma, \phi, u)$$

where $x(\sigma, \phi, u)$ is a solution of (2.1). Then

$DH'(u) = \frac{d}{du} H(u): L_\infty([\sigma, t_1], E^m) \to C$ has a continuous local right inverse (is a surjective linear mapping if, and only if, the variational control system (2.2) of f along the response $t \to x_t(\sigma, \phi, u)$, namely

$$\dot{z}(t) = D_2 f(t, x_t(\sigma, \phi, u); u(t))) z_t + D_3 f(t, x_t(\sigma, \phi, u), u(t))\, v(t)\ ,$$

$$z_\sigma(\sigma, \phi, u, v) = 0 \tag{2.2}$$

is controllable on $[\sigma, t_1]$, $t_1 > \sigma + h$.

Proof. For $t \in [\sigma, t_1]$, $t_1 > \sigma + h$, $u \in L_\infty([\sigma, t_1], E^m)$. Let $t \to z(\sigma, \phi, u)$ denote the solution of (2.2). Obviously the linear variational control system of f along the solution $t \to x_t(\sigma, \phi, u)$ is controllable on $[\sigma, t_1]$ if and only if the mapping $u \to z_t(\sigma, 0, \phi, u, v)$ $t \in [\sigma, t_1]$ is surjective. But by Lemma 2.1

$$z_{t_1}(\sigma, \phi, u, v) = D_4\, x_{t_1}(\sigma, \phi, u)v = DH(u)\, v .$$

Therefore the mapping $v \to z_{t_1}(\sigma, \phi, u, v)$ is surjective if and only if $DH(u)$ is surjective. The proof is complete.

If we restrict our attention to the state space, $W_2^{(1)}$, and use the control space $L_2^{loc}([\sigma, \infty), E^m)$ we have a sharp computational criterion for the surjectivity of the map $DH(u)$. See [10, p. 616] and Theorem 5.3 below.

3. CONSTRAINED CONTROLLABILITY RESULTS

This section states constrained null-controllability results for a very general nonlinear functional differential control system

$$\dot{x}(t) = F(t, x_t, u(t)) , \tag{3.1}$$

where

$$F(t, x_t, u(t)) = f(t, x_t) + g(t, x_t, u(t))$$

It strengthens, improves, and corrects the author's earlier contributions in [11] and in [12]. In those papers, the control set is a fixed n-dimensional unit cube C^m of E^m , that is,

$$C^m = \{u: u \in E^m , |u_j| \leq 1 , j = 1,\dots, m\} .$$

In [12] the state space was $W_2^{(1)}$ and the control space is

$$\mathbb{U} = \{u \text{ square integrable } u(t) \in C^m\} .$$

It was inadvertently erroneously assumed that $\mathbb{U}$ contains an open ball, an error pointed out by Colonius. If $\mathbb{U}$ were replaced by a closed and bounded unit ball of the L_2 space Corollary 3.1 of [12] would be correctly proved by the method therein provided some misprints on pages 202, 206, and 207 are corrected. If the state space were $W^{(1)}$, and the control space L_∞^{loc}, then $\mathbb{U} = L_\infty[\sigma, \infty), C^m)$, would contain an open ball and the proof would be correct modulo corrections of misprints. We would assume that:

$$F: [\sigma, \infty) \times C \times E^m \to E^n$$

has the stated properties of f in section 1, and not the assumption of page 200, line 5 from the bottom in [12].

Our aim here is to treat (3.1) and (3.2) and extend the result in [12] to cover the situation where the base system is nonlinear. In what follows, we consider the state space $W_\infty^{(1)}$ and the control set $L_\infty^{(1)}([\sigma, \infty), C^m)$, in studying the system

$$\dot{x}(t) = f(t, x_t) + g(t, x_t, u(t)) . \tag{3.2}$$

Following Shanholt [13], if

$$\dot{y}(t) = f(t, y_t) , \tag{3.3}$$

with each solution $y(\sigma, \phi)$ of (3.3), we define a linear functional differential equation

$$\dot{z}(t) = D_2 f(t, y_t(\sigma, \phi))z_t , \quad t \geq \sigma .$$

Let $z(\sigma, \phi)$ be a solution of (3.4) with $z_\sigma(\sigma, \phi) = \phi$. Let

$$z_t(\sigma, \phi) = T(t, \sigma; \phi)\phi, \; t \geq \sigma .$$

Then

$$D_2 x_t(\sigma, \phi) = T(t, \sigma; \phi) .$$

The solution of (3.1) is given by the analogue of non-linear Alekseev type variation of parameter formula due to Shanholt, by

$$x_\sigma = \phi$$
$$x_t(\sigma, \phi, u) = y_t(\sigma, \phi) + \int_\sigma^t T(t; s, x_s(\sigma, \phi))X_0 g(s, x_s, u(s))ds \tag{3.5}$$

Throughout what follows, we assume in (3.1) that

$$F(t, 0, 0) = 0$$

We assume that F is smooth enough as given in Section 1 for the existence and uniqueness of solution of (3.1) as well as the continuous dependence of the solution on initial data. In particular, if $\phi \equiv 0$, (3.1) was a unique trivial solution when $u = 0$.

Definition 3.1. The attainable set of (3.1) is the subset of $W_\infty^{(1)}$ given by

$$A(t, \sigma) = \{x_t(\sigma, \phi, u): u \in L_\infty([\sigma, t], E^m)$$
$$x_\sigma = \phi , \quad x \text{ is a solution of } (3.1)\}$$

The reachable set of (3.1) is the subset of $W_\infty^{(1)}$ given by

$$r(t, \sigma) = \{\int_\sigma^t T(t, s; x_s(\sigma,\phi,u))X_0 g(s,x_s,u(s))ds: u \in L_\infty([\sigma,t],E^m)\}$$

The constrained reachable set and the constrained attainable set of (3.1) are given respectively by

$$\underline{r}(t,\sigma) = \{\int_\sigma^t T(t,s; x_s(\sigma,\phi,u)X_0 g(s, x_s,u(s))ds: u \in L_\infty([\sigma, t], C^m)\}$$

$$\mathcal{a}(t, \sigma) = \{x_t(\sigma, \phi, u): u \in L_\infty([\sigma, t], C^m\})$$

$$x \text{ is a solution of } (3.2) \qquad x_\sigma = \phi\}$$

Just as in [12] we introduce the following:

Definition 3.2. The system (3.1) is proper on $[\sigma, t]$, $t > \sigma + h$, if

$$0 \in \mathrm{Int}(\underline{r}(t, \sigma)) \tag{3.6}$$

where $\mathrm{Int}\,\underline{r}$ is the interior relative to $W_\infty^{(1)}$. If (3.6) holds on every interval $[\sigma, t]$, $t > \sigma + h$, we say that (3.2) is proper.

Proposition 3.1. The system (3.1) is proper on $[\sigma, t_1]$, $t_1 > \sigma + h$, when-

$$z(t) = D_2 F(t, 0, 0)z_t + D_3 F(t, 0, 0)\, v(t) \tag{3.7}$$

is controllable on $[\sigma, t_1]$, $t_1 > \sigma + h$.

Proof. For the system (3.1), let $u \in L_\infty([\sigma, t_1], E^m)$ and $x(\sigma, \phi, u)$ be its response. Consider the mapping

$$u \to x_{t_1}(\sigma, 0, u) \quad \text{given by} \quad Hu = x_{t_1}(\sigma, 0, u),$$

where $H: L_\infty([\sigma, t_1], E^m) \to W_\infty^{(1)}$. It follows from the conditions on F and (3.5) that

$$H(L_\infty([\sigma, t_1], C^m) = \underline{r}(t_1, \sigma).$$

Also by Lemma 2.1 and Lemma 2.2,

$$DH(0) = \frac{d}{du} H(u)\Big|_{u=0} = D_3 x_{t_1}(\sigma, 0, u)\Big|_{u=0},$$

is a surjective mapping of $L_\infty([\sigma, t_1], E^m) \to W_\infty^{(1)}$. Therefore [14, Lang 193] H is locally open: There is an open ball $B_\rho \subseteq L_\infty([\sigma, t_1], E^m)$ containing zero and an open ball $B_r \subseteq W_\infty^{(1)}$ containing zero such that $B_r \subseteq H(B_\rho)$. Since $L_\infty([\sigma, t_1], C^m)$ contains an open ball containing zero, $r > 0$, $\rho > 0$ can easily be chosen such that $B_r \subseteq H(B_\rho \cap (L_\infty([\sigma, t_1], C^m)))$. Therefore,

$$B_r \subseteq H(L_\infty([\sigma, t_1], C^m)) = \underline{r}(t_1, \sigma),$$

so that $0 \in \mathrm{Int}\,\underline{r}(t_1, \sigma)$.

Definition 3.3. The domain $\underline{D}$ of null controllability of (3.1) is the set of all initial functions ϕ such that the solution $x(\sigma, \phi, u)$ of (3.1) for some $t < \infty$ and some $u \in L_\infty([\sigma, t], C^m)$ satisfies

$$x_\sigma(\sigma, \phi, u) = \phi \ , \ x_t(\sigma, \phi, u) = 0 \ .$$

Proposition 3.2. Assume that:

(i) $F: E \times C \times E^m \to E^n$ is continuous for each $t \in E$, $(\phi, u) \to F(t, \phi, u)$ is continuously differentiable with a locally integrably bounded derivative. Also the mapping $t \to \bar{D}f(t, 0. u)$ (where $\bar{D}f(t, \phi, u)$ is the partial derivative of F with respect to the second and third variables) is measurable.

(ii) $F(t, 0, 0) = 0$, $t \geq \sigma$;

(iii) The system (3.7) is controllable in $[\sigma, t]$, $t > \sigma + h$. Then the domain of null controllability $\underline{D}$ of (3.1) contains zero in its interior, that is

$$0 \in \text{Int}(D) \ .$$

Proof. Assume that 0 is not contained in the interior of $\underline{D}$. Then there is a sequence $\{\phi_n\}_1^\infty$, $\phi_n: [-h, 0] \to E^n$,

$$\phi_n \to 0 \quad \text{as} \quad n \to \infty \ ,$$

and no ϕ_n is in D. Since the trivial solution is a solution of (3.1), $0 \in D$. Hence $\phi_n \neq 0$ for any n. We also have that

$$0 \neq x_t(\sigma, \phi, u)$$

for any $t > \sigma + h$ and any $u \in L_\infty([\sigma,t], C^m)$. It follows from (3.5) that

$$-y_t(\sigma, \phi_n) \equiv \xi_n \neq \int_\sigma^t T(t, s; x_s) X_0 \, g(s, x_s, u(s)) \, ds$$

for any n, any $t > \sigma + h$, and any $u \in L_\infty([\sigma, t], C^m)$. As a consequence, ξ_n is not contained in $\underline{r}(t, \sigma)$ for any $t > \sigma + h$. From the uniqueness and continuity properties of $y(\sigma, \phi_n)$ we have evidently produced a sequence $\{\xi_n\}_{n=1}^\infty$: $\xi_n \to 0$ as $n \to \infty$, $\xi_n \notin \underline{r}(t, \sigma)$ for any $t > \sigma + h$, $\xi_n \neq 0$ for any n. (Note that $-y_t(\sigma, \phi_n) \to 0$ as $\phi_n \to 0$. Thus $0 \notin \text{Int}\,(\underline{r}(t, \sigma))$ for any $t > \sigma + h$. But 3.7 is controllable by (iii), and Proposition 3.1 shows that

$$0 \in \text{Int}\,(\underline{r}(t, \sigma) \ , \ t > \sigma + h \ .$$

This contradiction shows that indeed $0 \in \text{Int}\,(\underline{D})$.

We now state the main result of this section.

Theorem 3.1. In (3.1) assume that:

(i) $F: E \times C \times E^m \to E^n$ is continuous, for each $t \in E$,

$(\phi, u) \to F(t, \phi, u)$ is continuously differentiable with a locally integrably bounded derivative. Also the mapping $t \to \bar{D}F(t,\phi, u)$ (where $\bar{D}F(t, \phi, u)$ is the partial derivative of F with respect to the second and third variables) is measurable;

(ii) $F(t, 0, 0) = 0$;

(iii) The system (3.7) is controllable on $[\sigma, t]$, $t > \sigma + h$.

(iv) The system

$$\dot{x}(t) = F(t, x_t, 0) = f(t, x_t) + g(t, x_t, 0) \tag{3.8}$$

is uniformly exponentially stable, that is, each solution of (3.8) satisfies

$$||x_t(\sigma, \phi)|| \leq k||\phi|| \exp[-\alpha(t - \sigma)] \quad , t \geq \sigma \tag{3.9}$$

for some $k > 1$, $\alpha > 0$. Then (3.1) is null controllable with constraints. that is with controls in $L_\infty([\sigma, t], C^m)$ for some $t > \sigma + h$.

Proof. By Proposition 3.1, and condition (iii), (3.1) is proper on $[\sigma, t]$, so that $0 \in \text{Int}(\underline{r}(t, \sigma))$, $t > \sigma + h$. This implies that $0 \in \text{Int}(\underline{D})$. Let $\mathcal{O}$ be an open ball contained in $\underline{D}$, guaranteed by Proposition 3.2. Because of (3.9), every solution of (3.1) with $u = 0$, that is, every solution of (3.8), satisfies $x_t(\sigma, \phi, 0) \to 0$ as $t \to \infty$. Thus there exists some $t_0 < \infty$ such that $x_{t_0}(\sigma, \phi, 0) \equiv \psi \in \mathcal{O} \subset \underline{D}$, with t_0 as initial time now and ψ as initial function in $\mathcal{O}$ there exists a control v such that the solution $x(t_0, \psi, v)$ of (3.1) satisfies

$$x_{t_0}(\sigma, \psi, u) = \psi \;, \; x_{t_1}(\sigma, \psi, u) = 0 \;.$$

Using

$$w(s) = \begin{cases} 0 & s \in [\sigma, t_0] \\ v(s) & s \in [t_0, t_1] \end{cases}$$

we have $w \in L_\infty([\sigma, t_1], C^m)$ and the solution $x(\sigma, \phi, w)$ satisfies $x_\sigma(\sigma, \phi, w) = \phi$, $x_{t_1}(\sigma, \phi, w) = 0$. This completes the proof.

4. NECESSARY CONDITIONS FOR TIME OPTIMAL CONTROL OF NONLINEAR SYSTEMS

We have proved in Section 1 the existence of a time optimal control under the assumption of constrained controllability. Conditions that guarantee this are treated in section 3. In this section we give the necessary condition for time optimal control for the nonlinear system

$$\dot{x}(t) = f(t, x_t, u(t)) , \quad t \geq 0 ,$$
$$x_\sigma = \phi \tag{4.1}$$

This necessary condition, the maximum principle is extracted from the general treatment Banks [25, [26] and [27] where the minimization of functionals were explored. Though the time optimal problem is implicit in that treatment, the necessary condition was not explicitly deduced, and it has become interesting and important to include it and tie it up with the treatment in the previous sections.

We recall the basic assumptions of f in Theorem 1.1: $f: E \times C \times E^m \to E^n$ is continuous, and continuously differentiable in the last two arguments with the derivative $\overline{D}f$ locally integrably bounded. Also the mapping $t \to \overline{D}f(t, \phi, u)$ is measurable. These basic assumptions are valid throughout this section. Furthermore, let

$$[D_2f(t, x_t, u(t))](\psi) = \int_{-h}^{t} d_s\eta(t, s)\, \psi(s) \tag{4.2}$$

where D_2f is the derivative of f with respect to the second argument and η satisfies enough conditions for the existence of a unique solution of

$$\dot{z}(t) = D_2f(t, x_t, u(t))z_t \tag{4.3}$$

For this see [15, p. 141]. We work in the target state space E^n, with controls $U = L_\infty([\sigma, t_1], C^m)$, the space of admissible, measurable controls u, $u(t) \in C^m$ where $C^m = \{u \in E^m: |\phi_j| \leq 1 , j = 1,\dots, m\}$. In Banks setting, the problem is: minimize $J(\phi, u, s) = s$, $u \in U$ subject to

$$\dot{x}(t) = f(t, x_t, u(t)) , \qquad t \geq 0$$
$$x_\sigma = \phi \in C ,$$
$$(s, x(s)) \in [0, \infty) \times \{x_1\} , \qquad x_1 \in E^n .$$

This is the problem of reaching a fixed point $x_1 \in E^n$ in least time, starting at time σ from the point $\phi \in C$ using controls $u \in U$. The maximum principle can be stated as follows:

Theorem 4.1. Let (u^*, x^*) be the time optimal control response pair with u^* the optimal control, $x^* = x(\sigma, \phi, u^*)$ its optimal trajectory and t^* the optimal time. Suppose t^* is a Lebesgue point of $f(t, x_t^*, u^*(t))$. Then there exists a non-trivial row vector $y = (y_1,\dots, y_n)$ of bounded variation on $[\sigma, t^*]$ continuous at t^* satisfying

(i) $y(t^*) \neq 0$

$$y(s) + \int_s^{t^*} y(\theta)\, \eta(\theta, s)\, d\theta = y(t^*) \qquad s \in [\sigma, t^*] \tag{4.4}$$

where η is defined in (4.2);

(ii) $y(t)\, f(t, x_t^*, u^*(t)) \geq y(t)\, f(t, x_t^*, u(t))$, (4.5)

for all $u \in U$, for almost all $t \in [\sigma, t^*]$.

For the system

$$\dot{x}(t) = L(t, x_t) + B(t)\, u(t) \quad , \tag{4.6}$$

where

$$L(t, x_t) = \int_{-h}^{0} d_\theta\, \eta(t, \theta)\, x(t + \theta) \quad ,$$

we have from (ii) that

$$y(t)[L(t, x_t^*) + B(t)\, u(t)] \leq y(t)[L(t, x_t^*) + B(t)\, u^*(t)] \ ,$$

for all $u \in U$. Thus we deduce the following necessary condition for optimal control: $y(t)B(t)u(t) \leq y(t)B(t)\, u^*(t)$, for all $u \in U$. This yields,

$$u^*(t) = \operatorname{sgn}[y(t)\, B(t)] \ , \tag{4.7}$$

as the form of optimal control, where y is the nontrivial solution of the adjoint equation (4.4).

We note that the existence of an optimal control response pair is guaranteed by the constrained controllability assumptions of Section 3.

5. SOLUTION OF THE TIME OPTIMAL PROBLEM FOR LINEAR SYSTEMS

The necessary condition of Theorem 4.1 and the form of optimal control in (4.7) for the linear system is for target point of E^n . For delay equations and for target points in function spaces C or $W_2^{(1)}$, the author is not aware of treatment of maximum principles which are applicable to time optimal problems. What seems to be available [18] involve quadratic cost functions which are not easily reduced to time optimal problems. We shall therefore deal with the time optimal control of linear delay equations in the spaces C and $W_2^{(1)}$. First we shall state the bang-bang principle of linear systems in E^n, and its lack in C or $W_2^{(1)}$. A sharp, easily computable criterion for E^n controllability and normality is then presented. The controllability criterion is due to Gabasov and Kirillova [17], the normality condition is new. For targets in function space, the unconstrained controllability criterion is contained in [10]. The maximum principle and the forms of optimal control seem to be new.

We study the linear system

$$\begin{aligned}\dot{x}(t) &= L(t, x_t) + B(t)\, u(t)\ , \quad t \geq \sigma \\ x_\sigma &= \phi\ ,\end{aligned} \tag{5.1}$$

where

$$L(t, x_t) = \sum_{k=0}^{N} A_k(t)\, x(t - w_k) + \int_{-h}^{0} A(t, s)\, x(t + s)\, ds$$

with $0 = w_0 \leq w_1 \leq \cdots \leq w_N \leq h$, A_k is a continuous $n \times n$ matrix function, $B(t)$ is a continuous $n \times m$ function, $A(t, s)$ is an $n \times n$ matrix, continuous in t and integrable in s for each t and there is a function $a(t) \in L_1^{loc}((-\infty, \infty), E)$ such that

$$\left|\int_{-h}^{0} A(t, s)\, x(t + s)\, ds\right| \leq a(t)\ \|x_t\|\ ,$$

for all $t \in E$, $x_t \in C$. We shall often deal with the autonomous version of (5.1), namely

$$\begin{aligned}\dot{x}(t) &= L(x_t) + Bu(t) \qquad t \geq \sigma \\ x_\sigma &= \phi\end{aligned} \tag{5.2}$$

where

$$L(x_t) = \sum_{k=0}^{N} A_k\, x(t - w_k)\ .$$

Our state space is either E^n, $C = C([-h, 0], E^n)$ or $W_2^1([-h, 0], E^n)$, and our target is E^n, C or $W_2^{(1)}$. If the state space is E^n or C, the control set is assumed to be

$$C^m = \{u \text{ measurable, } u \in E^m\ , \ |u_j| \leq 1\ , \ j = 1, \ldots, m\}\ .$$

Thus the control set is a subset of $L^{loc}([\sigma, \infty), E^m)$. On using the norm

$$||u||_\infty = ||(u_1, \ldots, u_m)||_\infty = \max(|u_1|, \ldots, |u_m|)$$

then $||u||_\infty \leq 1$. We note that $0 \in \text{Int}(C^m)$.

The control set for the state space $W_2^{(1)}$ is the set

$$\mathbb{F} = \{u(\cdot): \ ||u||_2 \leq 1\}$$

so that the set of admissible controls is

$$\mathbb{U} = \{u \in L_2([\sigma, t], E^m): \ ||u||_2 \leq 1\} \ .$$

If we designate the solution $x(\sigma, \phi, 0)$ of

$$\begin{aligned} \dot{x}(t) &= L(t, x_t) , \quad t \geq \sigma \\ x_\sigma &= \phi \end{aligned} \tag{5.3}$$

by

$$x_t(\sigma, \phi, 0) = T(t, \sigma)\phi \tag{5.4}$$

where $x_t(\sigma, \phi, 0) \in W_2^{(1)}$, the variation of constant formula given in [15, p. 143] yields

$$\begin{aligned} x_\sigma &= \phi \quad \text{in} \quad [-h, 0] \\ x_t(\sigma, \phi, u) &= T(t, \sigma)(\phi) + \int_\sigma^t T(t, s)\, X_0\, B(s)\, u(s)\, ds \ , \quad t \geq \sigma \end{aligned} \tag{5.5}$$

or in E^n

$$x(\sigma, \phi, u)(t) = T(t, \sigma)\, \phi(0) + \int_\sigma^t T(t, s)\, B(s)\, u(s)\, ds \tag{5.6}$$

$$Y(t, s) = T(t, s)\, B(s)$$

then

$$x(\sigma, \phi, u)(t) = T(t, \sigma)\, \phi(0) + \int_\sigma^t Y(t, s)\, u(s)\, ds \tag{5.7}$$

and we have

Definition 5.1. The constrained reachable set of (5.1) is defined by

$$\mathbf{R}(t) = \{\int_\sigma^t Y(t, s)\, u(s)\, ds: \; u \in C^m\} \tag{5.8}$$

The following properties of $\mathbf{R}(t)$ are proved by standard methods.

Proposition 5.1. $\mathbf{R}(t)$ is convex and compact and satisfies the mononicity relation

$$0 \in \mathbf{R}(t) \qquad \text{for each } t \geq \sigma \tag{5.9}$$

$$T(t, s)\, \mathbf{R}(s) \subset \mathbf{R}(t) \qquad 0 \leq s \leq t \tag{5.10}$$

Proof. Convexity follows trivially from that of C^m because $Y(t, s)$ is integrable and $||u(s)||_\infty \leq 1$ a.e. $\mathbf{R}(t)$ is bounded. We use a weak compactness argument to prove that $\mathbf{R}(t)$ is closed. Because $0 \in C^m$, $0 \in \mathbf{R}(t)$ for each $t \geq \sigma$. It remains to prove (5.10). Let $r \in R(s)$; then for some $u \in C^m$

$$r = \int_\sigma^s T(s, \tau)\, B(\tau)\, u(\tau)\, d\tau .$$

Define

$$u^*(\tau) = \begin{cases} u(t) & \sigma \leq \tau \leq s \\ 0 & s \leq \tau \leq t . \end{cases}$$

Then $u^*(\sigma) \in C^m$. Consider the point

$$\begin{aligned} q = T(t, s)r &= \int_\sigma^s T(t, s)\, T(s, \tau)\, B(\tau)\, u(\tau)\, d\tau \\ &= \int_\sigma^s T(t, \tau)\, B(\tau)\, u(\tau)\, d\tau + \int_s^t T(t, \tau)\, B(\tau)\, 0\, d\tau \\ &= \int_\sigma^t T(t, \tau)\, B(\tau)\, u^*(\tau)\, d\tau \in \mathbb{R}(t) \end{aligned}$$

Hence $T(t, s)\, \mathbf{R}(s) \subset \mathbf{R}(t)$.

Remark 5.1. Because of the special nature of $T(t,\phi)$ in (5.6) the measures involved in $T(t, s)$ are nonatomic. If for more general forms of $L(t, \phi)$, Lebesgue Stieltjes measures were involved, then the corresponding $T(t, s)$, given by $X(t, s) = T(t, s)I$,

$$X(t, s) = \begin{cases} \int_s^t L(u, X_u(\cdot, s)du + I & \text{a.e. in } s,\ t \geq s \\ 0 & s - h \leq t < s \end{cases}$$

where for example,

$$L(u, X_u(\cdot, s)du) = \int_{-h}^{0} [d_\theta \eta(u, \theta)]\, X(u + \theta, s)\, du$$

will be defined with possibly atomic measures. To obtain the convexity of $\mathbf{R}(t)$ property Δ_1, Δ_2 of Banks and Jacobs [16] may have to be invoked. Now, define

$$C^{0m} = \{u \in E^m \text{ u measurable },\ |u_j| = 1,\ j = 1,2,\ldots, m\} .$$

Then the following bang-bang principle is true in E^n.

Theorem 5.1. If

$$\mathbf{R}(t) = \{\int_\sigma^t Y(t, s)\, u(s)\, ds : u \in C^m\}$$

$$\mathbf{R}^0(t) = \{\int_\sigma^t Y(tr, s)\, u^0(s)\, ds: u^0 \in C^m\}$$

then $\mathbf{R}(t) = \mathbf{R}^0(t)$, $t \geq \sigma$.

Proof. Because

$$X(t, s) = T(t, \cdot)\, I \in L_\infty([\sigma, t], E^{n^2}]$$

and

$$B(\cdot) \in L_1([\sigma, t], E^{n\times m}),$$

we have

$$Y(t, s) = T(t, s)\ B(s) \in L_1([\sigma, t], E^{n\times m}) .$$

It follows from Corollary 8.2 of Hermes and LaSalle [2] that

$$\mathbf{R}(t) = \mathbf{R}^0(t) .$$

Remark 5.2. The bang-bang principle of Theorem 4.1 is true only in E^n. Banks and Kent [3] have shown that the principle is invalid in function space, by considering the retarded system

$$\dot{x}(t) = x(t - 1) + u(t)$$

$$U = \{v \in E^1: |v| \leq 1\}$$

Target $\xi(t) = \frac{1}{2}(3 - t)$, $t \in [2, 3]$. Starting from $\phi(t) = -t$, $t \in [-1, 0]$, one can attain ξ by a measurable control u with $u(t) \in U$. But ξ cannot be attained with bang-bang control from any initial ϕ.

From the discussion of Section 4, we have the following.

Theorem 5.2. Suppose (5.1) is E^n controllable so that an optimal control u^* exists with minimum time t^*. Then there exists a function $y: [\sigma, t^*]$ which is of bounded variation and which satisfies the adjoint equation

$$\frac{dy(s)}{ds} = -\sum_{n=1}^{\infty} y(s + w_k)A_k(s + w_k) - \int_{-h}^{0} y(s - \xi)\ A(s - \xi, \xi)ds \qquad (5.11)$$

such that

$$u^*(s) = \operatorname{sgn}[y(s), B(s)] , \quad y(t) \neq 0 \qquad t \in [\sigma, t^*] \qquad (5.12)$$

Remark 5.3. By a theorem of Banks and Jacobs [18, p. 476] the adjoint matrix solution of (5.11) is piecewise analytic on each closed interval $[\sigma, \sigma + T]$ for any $T > 0$ whenever $A(t, s) \equiv 0$ in and $t \to A_k(t)$ $k = 0,1,\ldots, N$ is analytic on $[\sigma, \infty)$. Therefore if $B(t)$ is analytic each component, $g(s) \equiv y(s)\ B(s)$, $s \in [\sigma, \sigma + T]$ is either identically zero or vanishes at most at a finite number of times. Therefore (5.12) either is meaningful for all s, except for a finite (and unimportant) set of values of s or it is undefined for all s in a positive interval. In this last case the maximum principle gives no information. It will not determine the optimal control uniquely if for some $j = 1, \dots, m$, the function

$$t \to y(t)\, B(t) \qquad t \in [\sigma, t^*]$$

is zero on some positive interval. When this pathology does not occur and optimal controls are uniquely given by (5.12) then we have a normal system.

Definition 5.2. The system (5.1) is said to be normal on $[\sigma, t]$ if no component of

$$\{y(t)\, B(t)\} \qquad y \neq 0$$

vanishes on a subinterval of $[\sigma, t]$ of positive length.

We now derive a simple test for the normality of

$$x(t) = A_0\, x(t) + \sum_{j=1}^{N} A_j\, x(t - w_k) + B\, u(t) . \tag{5.13}$$

It is proved by the methods of Gabasov and Kirillova [17]. Let

$$B = [B_1\ B_2, \dots, B_m]$$

where B_j is the jth column of B. Let

$$Q_k(s) = A_0\, Q_{k-1}(s) + \sum_{j=1}^{N} A_j\, Q_{k-1}(s - w_j)$$

$$k = 0,1,2,\dots,\ s = 0,\ w_j,\ 2w_j,\dots,\ j = 1,\dots, N$$

$$Q_0(0) = I\ ,\ Q_0(s) \equiv 0\ ,\ s < 0 .$$

Proposition 5.2. A necessary and sufficient condition for (5.13) to be normal on $[0, T]$ is that for each $r = 1, \dots, m$, the matrix

$$\Pi_r(T) = \{Q_k(s)\, B_r\ ,\ u = 0,1,\dots, n - 1,\ s \in [0, T]\}$$

has rank n. The required E^n-controllability condition for Theorem 5.2 is provided in the following refinement of a result of Gabasov and Kirillova [17] for autonomous systems (5.2).

Proposition 5.2. The system (5.2) is Euclidean controllable (i.e. for each $\phi \in C$, $x_1 \in E^n$ there exists a control $u \in L_\infty^{loc}$ such that the solution $x(\sigma, \phi, u)$ of (5.2) satisfies

$$x_\sigma(\sigma, \phi, u) = \phi \quad , \; x(t_1, \sigma, \phi, u) = x_1,$$

if and only if the matrix

$$\Pi(t_1) = \{Q_k(s)B, \; k = 1,\dots, n-1 \; , \; s \in [\sigma, t_1]\}$$

has rank n, where

$$Q_k(s) = A_0 \, Q_{k-1}(s) + \sum_{j=1}^{N} A_j \, Q_{k-1}(s - w_j) \; ,$$

$$k = 0,1,2,\dots, \; s = 0 \; , \; w_j, \; 2w_j \; \dots \; j = 1,\dots, N \; .$$

$$Q_0(0) = B \; , \; Q_0(s) \equiv 0 \qquad \text{for} \quad s < 0 \; .$$

Our study now turns to the space $W_2^{(1)}$. The required controllability result is due to Banks, Jacobs, and Langenhop [10].

Theorem 5.3. In (4.1) for each $t \in E$, let

$$\overline{\eta}(t, 0) = \begin{cases} 0 & \\ \sum_{k=0}^{N} A_k(t) + A(t + \theta) & \theta < 0 \end{cases}$$

Let $B^+(t)$ be the Moore-Penrose generalized inverse of $B(t)$, $t \in E$ (see [19]). Suppose

$$\text{(i)} \quad B(t) \, B^+(t) \, \overline{\eta}(t, \theta) = \overline{\eta}(t, \theta) \qquad -h \le \theta \le 0 \tag{5.14}$$

(ii) $t \to B^+(t)$, $t \in E$ is essentially bounded on $[t_1 - h, t_1]$ (5.15)

Then (5.1) is controllable on $[\sigma, t_1]$ with $t_1 > \sigma + h$ if and only if

$$\text{rank } B(t) = n \quad \text{on } [t_1 - h, t_1] .$$

If we define

$$A(t, \phi) = A(\sigma, t_1, \phi) = \{x_{t_1}(\cdot, \sigma, \phi, u): \ u \in L_2([\sigma, t_1], E^m)$$
$$x \text{ a solution of } (5.1)\} ,$$

then the following is true.

Theorem 5.4. Suppose (5.14) holds. Then $A(t)$ is closed in $W_2^{(1)}([-h, 0], E^n)$, provided (5.15) is satisfied.

Remark 5.4. The proof of Theorem 4.5 in [10, p. 623] also demonstrates that the subset

$$\underline{r}(t) = \{\int_\sigma^t T(t, s) X_0 B(s) u(s) ds: \ u \in \mathbb{U}\}$$

is a closed subset of $W_2^{(1)}$. It is also clearly convex.

We need the next lemma.

Lemma 5.3. In (5.1), let

$$T(t, \sigma)\phi = x_t(\sigma, \phi, 0) , \ t \geq \sigma .$$

Then

$$T(t, \sigma) W_2^{(1)} = W_2^{(1)} , \ t \geq \sigma + h$$

whenever (5.1) is controllable on $[\sigma, t]$, $t \geq \sigma + h$.

Proof. The solution of (5.1) is given in $W_2^{(1)}$ by

$$x_t(\sigma, \phi, u) = T(t, \sigma)\phi + \int_\sigma^t T(t, s) X_0 B(s) u(s) ds .$$

By definition, $a(t) = a(t,\sigma) = \{x_{t_1}(\sigma,\phi,u) \ \ u \in \mathbb{U} , \ x \text{ a solution of } 5.1\}$

$$\underline{r}(t) = \{\int_{\sigma}^{t} T(t, s)\, X_0\, B(s)u(s)ds: \quad u \in U\} \quad .$$

Because (5.1) is controllable the mapping

$$H: L_2([\sigma, t], E^n) \to W_2^{(1)}$$

given by

$$Hu = x_t(\sigma, \phi, u)$$

is surjective. Because it is a bounded linear map, it is an open map [20, p. 104]. Therefore,

$$H(\mathbb{P}) \subseteq H(\mathbb{U}) = a(t)$$

with $H(\mathbb{P})$ open, where

$$\mathbb{P} = \{u \in L_2([\sigma, t], E^m): \; ||u||_2 < 1\} \; .$$

Hence $a(t)$ has an interior point. Because $W_2^{(1)}$ is a Hilbert space, Fattorini's Lemma [21, p. 170] yields that

$$T(t, \sigma)\, W_2^{(1)} = W_2^{(1)} \quad , \; t \geq \sigma + h \; .$$

We now state without proof the following.

Proposition 5.4. The set valued map $t \to a(t, \phi)$ is continuous with respect to the Hausdorff metric, $t \to \underline{r}(t)$ is also continuous with respect to the Hausdorff metric.

The proof depends the continuity of the fundamental matrix $t \to X(t, \sigma)$ of (5.1).

Definition 5.3. The control $u \in \mathbb{U}$ is an extremal control if the associated solution $x(\sigma, \phi, u)$ corresponding to (σ, ϕ) lies on the boundary of $a(t, \phi)$, that is

$$x_t(\sigma, \phi, u) \in \partial a(t, \phi) \quad , \qquad \sigma \leq t \leq t^* \; .$$

Theorem 5.4.   If (5.13) is controllable and w is a time optimal control
for (5.13) then w is extremal.

Proof. We shall first show that if w is time optimal then at the instant
t_1 of arrival at $0 \in W_2^{(1)}$ then the solution will lie on $\partial a(t_1, \phi)$, the
boundary of $a(t_1, \phi)$. We shall then show that if $x_t(\sigma, \phi)$ lies on
$\partial a(t_*, \phi)$ at any fixed time t_* , then

$$x_t(\sigma, \phi, u) \in \partial a(t, \phi) , \qquad \sigma \leq t \leq t_* . \qquad (5.16)$$

Now assume that w is the optimal control which drives ϕ to 0 in time
t_1 , i.e.

$$x_{t_1}(\sigma, \phi, w) = x_{t_1} = 0 ,$$

and suppose $x_{t_1} = 0$ is not on $\partial a(t_1, \phi)$. Then there is a ball
$\mathbb{B}(0; \rho)$ of radius ρ about 0 such that

$$\mathbb{B}(0; \rho) \subset a(t_1, \phi) .$$

Because $a(t, \phi)$ is a continuous function of t we can preserve this
inclusion for t near t_1 if we reduce the size of $\mathbb{B}(0; \rho)$, i.e., if
there is a $\delta > 0$ such that

$$\mathbb{B}(0, \rho/2) \subset a(t, \phi) , \qquad t_1 - \delta \leq t \leq t_1$$

Thus we can reach 0 at time $t_1 - \delta$, contradicting the optimality of
t_1 . We conclude that

$$0 = x_{t_1}(\sigma, \phi, w) \in \partial a(t_1, \phi)$$

For the second part, suppose

$$x^* = x_{t_*}(\sigma, \phi, w) \in \text{Int } a(t_*, \phi)$$

for some $0 < t_* < t_1$. We claim that

$$x_t(\sigma, \phi, w) \in \text{Int } a(t, \phi)$$

for $t > t_*$. Because $x^* \in \text{Int}\, a(t_*, \phi)$, there is a ball $\mathbb{B} \equiv B(x^*, \delta) \subset a(t_*, \phi)$. Therefore each $\phi_0 \in \mathbb{B}$ can be reached from ϕ at time t_* using control u_0. Introduce the new system

$$\dot{z}(t) = L(t, z_t) + B(t)\, w(t)$$

$$z_{t*} = \phi_0 \qquad t_* < t$$

with fixed w. We observe that for $\phi_0 = x^*$, we have by uniqueness of solutions

$$z_t(\sigma, \phi, w) = x_t(\sigma, \phi, w) \quad , \qquad t \geq t_* .$$

The solution of this equation is

$$z_t(\sigma, \phi_0, w) = T(t, \sigma)\phi_0 + \int_0^t T(t, s)\, X_0\, B(s)\, u(s)\, ds ,$$

that is,

$$z_t(\sigma, \phi_0, w) = T(t, \sigma)\phi_0 + c(t)$$

where

$$c(t) = \int_0^t T(t, s)\, X_0\, B(s)u(s)\, ds .$$

Since the system is controllable and the solution operator $T(t, \sigma)$, defined by the relation $x_t(\sigma, \phi) = T(t, \sigma)\phi$ is, by Lemma 4.3, an open map. Hence the mapping $\phi_0 \to z_t(\sigma, \phi_0, w)$ is an open map and takes open sets into open sets. Thus the image of the open ball $\mathbb{B}$ lies in the interior of $a(t, \phi_0)$. Thus the image of x^* which is just $x_t(\sigma, \phi, w)$ lies in $\text{Int}\, a(t, \phi)$.

Remark. Since extremal controls give solutions which lie on a boundary they are intimately related to bang-bang controls which represent maximum use of admissible power. We now know that if the system is controllable and there exists an optimal control, the optimal control is extremal. The next theorem asserts that optimal control satisfies a certain necessary condition.

Theorem 5.5. Assume that (5.1) is function space controllable and both (5.14) and (5.15) hold. There exists a nontrivial $\psi \in W_2^{(1)*}$ such that the optimal control u_0 of (5.1) satisfies

$$u_0(s) = \text{sgn}[g(t, s, \psi)B(s)] \qquad \sigma \leq s \leq t^* \tag{5.16}$$

where $g(t, s, \cdot): W_2^{(1)*} \to E^n$ is given by

$$g(t, s, \psi) \triangleq \{\psi(-h)[T(t, s)X_0](-h) + \int_{-h}^{0} \dot{\psi}(\theta)\, d[T(t, s)\, X_0](\theta) \tag{5.17}$$

Proof. By definition

$$\underline{r}(t^*) = \{\int_{\sigma}^{t^*} T(t, s)\, X_0\, B(s)\, u(s)\, ds: \quad u \in \mathbb{U}\}$$

Theorem 5.4 states that this set is a closed subset of $W_2^{(1)}([-h, 0], E^n)$. It is also convex. Also, $0 \in \text{Int}\,\underline{r}(t^*)$ since (4.1) is function space controllable. Since u_0 is optimal the corresponding solution $x(\sigma, \phi, u_0)$ is extremal. Thus, $x_{t^*}(\sigma, \phi, u_0) \in \partial\alpha(t^*)$. From this we quickly deduce that $x_{t^*}(\sigma, \phi, u_0) - T(t^*\ \sigma)\phi \in \partial\, \underline{r}(t^*)$. But then $\underline{r}(t^*)$ has a support hyperplane through each relative boundary point. Thus by standard separation theorem [24, p. 418] of closed convex sets there exists a $\psi \in W_2^{(1)*}$ such that $\psi \neq 0$, and

$$\langle\psi, v\rangle \leq \langle\psi, x_{t^*}(\sigma, \phi, u_0) - T(t^*, \sigma)\phi\rangle$$

for every $v \in \underline{r}(t^*)$ where t^* is the optimal time. Here $\langle\cdot, \cdot\rangle$ denotes the inner produce in $W_2^{(1)}$ defined by

$$\langle\psi, v\rangle = (\psi(-h), v(-h)) + \int_{-h}^{0} (\dot{\psi}(\theta)\, \dot{v}(\theta))\, d\theta\ .$$

It now follows from this inner product that for all $u \in U$,

$$\int_{\sigma}^{t^*} \{\psi(-h)[T(t^*, s)X_0](-h) + \int_{-h}^{0} \dot{\psi}(\theta)\, d[T(t^*, s)X_0](\theta)\}B(s)\, u(s)\, ds$$

$$\leq \int_{\sigma}^{t^*} \{\psi(-h)[T(t^*,s)\gamma_0](-h) + \int_{-h}^{0} \dot{\psi}(\theta)\, d[T(t^*,s)X_0](\theta)\}B(s)u_0(s)\, ds\ .$$

Thus

$$\int_{\sigma}^{t^*} g(t^*, s, \psi)\, B(s)\, u(s)ds \leq \int_{\sigma}^{t^*} g(t^*, s, \psi)\, B(s)\, u_0(s)\, ds$$

for all $u \in \mathbb{U}$, where g is defined in (5.7). Thus u_0 satisfies

$$u_0(s) = \operatorname{sgn}[g(t, s, \psi)\, B(s)] \quad , \qquad \sigma \leq s \leq t^* \quad ,$$

as claimed in (5.16). Because of the following property of L_2 we must have a unique optimal control whenever the definition of u_0 in (5.16) makes sense:

If $u, v \in L_2$ $||u|| = ||v|| = 1$, $||1/2(u + v)|| = 1$ then $u = v$. But u_0 given in (5.16) will make sense if (5.1) is normal in the following way:

Definition 5.4. The system (5.1) is function space normal if the set

$$G_j(s) = \{s\colon \{g(t^*, s, \phi)\, B(s)\}_j = 0\} \qquad s \in [\sigma, t^*]$$

has measure zero for each $j = 1, \ldots, m$ and each $\psi \not\equiv 0$.

A characterization of normality in terms of the systems coefficients is interesting but difficult and will appear elsewhere.

In the space $C = C([-h, 0], E^n)$ the optimal control is given by

$$u_0(t) = \operatorname{sgn}\{-y(t, t^*)\, B(t)\} = \operatorname{sgn}\{-[T^*(s, t^*)\psi]\, (0^-)B(s)\} \qquad (5.19)$$

where $y(t, t^*)$ is the solution of the adjoint equation (5.11). We note that $[T^*(s, t^*)\psi]\, (0^-)$ is a point of E^{n^*}, a row n-dimensional vector.

To see this, if u_0 is optimal control then, as before

$$z^* = x_{t^*}(\sigma, \phi, u_0) - T(t^*, \sigma)\phi \in \partial r(t^*) \subseteq C$$

Thus there exists $\psi \in \mathbb{B}_0$, the conjugate space of C, $\psi \not\equiv 0$ such that

$$\langle\psi, \rho\rangle \leq (\psi, z^*) \qquad \text{for all } \rho \in \partial \underline{r}(t^*)$$

where

$$\langle\psi, \rho\rangle = \int_{-h}^{0} [d\psi(\theta)]\, \rho(\theta) \ , \ \psi \in \mathbb{B}_0 \ , \ \rho \in C \ .$$

Thus

$$\langle \psi, \int_{\sigma}^{t^*} T(t, s)\, X_0\, B(s)\, u(s)\, ds \rangle \leq \langle \psi, \int_{\sigma}^{t^*} T(t, s) X_0\, B(s) u_0(s)\, ds \rangle$$

It follows as was done in [15, p. 155], Corollary 4.1 that

$$\int_{\sigma}^{t^*} -y(s, t^*)\, B(s)\, u(s) ds \leq \int_{\sigma}^{t} -y(s, t^*)\, B(s) u_0(s)\, ds ,$$

from which (5.19) follows. Since

$$[T^*(s, t)\psi](0^-) = y(s, t) ,$$

where $y(s, t)$ is the solution of the adjoint equation

$$y(s) + \int_{s}^{t^*} y(\theta)\eta\, (\alpha, s - \theta)\, d\theta = \psi(-h) .$$

Remark. An appropriate definition of normality can be made and its criterion determined.

REFERENCES

[1] N. Minorsky, Self-excited mechanical oscillations, J. Appl. Phys. 19(1948), 332-338.

[2] H. Hermes and J.P. LaSalle, Functional Analysis and Time Optimal Control, Academic Press, New York, 1969.

[3] H.T. Banks and G.A. Kent, Control of Functional Differential Equations of Retarded and Neutral Type to Target sets in Function Space, SIAM J. Control. 10(1972), pp. 567-594.

[4] V. Volterra, Variations and Fluctuations of the number of individuals in animal species living together, in Animal Ecology, by R.C. Chapman, pp. 409-448. McGraw-Hill, New York, 1931. Translated from the original, 1928.

[5] N.K. Gupta and D.R.E. Rink, Optimum Control of Epidemics, Biosci. 18(1978), pp. 383-396.

[6] E.J. McShane and R.B. Warfield, On Filippov's Implicit Function Lemma, Proc. American Math. Society, 18(1967), pp. 41-47.

[7] A.F. Filippov, On certain questions in the theory of Optimal Control, SIAM J. Control, 1(1962), pp. 76-84.

[8] L. Cesari, Existence Theorems for Weak and Usual Optimal Solutions in Lagrange Problems with Unilateral Constraints, Trans. American Math. Soc. 124(1966), pp. 369-412.

[9] T.S. Angell, Existence Theorems for Optimal Control Problems Involving Functional Differential Equations, Journal Optimization Theory and Applications, 7(1971), pp. 149-169.

[10] H.T. Banks, Marc Q. Jacobs, and C.E. Lagenhop, Characterization of the Control States in $W_2^{(1)}$ of Linear Hereditary Systems, SIAM J. Control 13(1975), pp. 611-649.

[11] E.N. Chukwu, On the Null-Controllability of Nonlinear Delay Systems with Restrained Controls, J. Math. Anal. Applications, 76(1980), pp. 283-296.

[12] E.N. Chukwu, Null Controllability in Function space of Nonlinear Retarded Systems with Limited Control, J. Math. Anal. Appl. 103 (1984), pp. 198-210.

[13] G.A. Shanholt, A nonlinear variation of constant formula for functional differential equations, Math. Systems Theory, 6(1973), pp. 343-352.

[14] S. Lang, Analysis II, Addison-Wesley Publ. Co., Reading, Mass, 1969.

[15] J. Hale, Theory of Functional Differential Equations, Springer-Verlag, New York, New York, 1977.

[16] H.T. Banks and M.Q. Jacobs, The Optimization of Trajectories of Linear functional Differential Equations, SIAM J. Control, 8(1970), pp. 461-488.

[17] R. Gabasov and F. Kirillova, The qualitative theory of Optimal processes, Marcel Dekker, New York, 1976.

[18] H.T. Banks and Marc D. Jacobs, An Attainable sets Approach to Optimal Control of Functional Differential Equations with Function Space Terminal Conditions, Journal of Diff. Equa., 13, 127-149 (1973).

[19] D.G. Luenberger, Optimization by vector space methods, John Wiley, New York, 1969.

[20] W. Rudin, Real and Complex Analysis, McGraw-Hill, New York, 1974.

[21] H.O. Fattorini, The Time Optimal Control Problem in Banach Spaces, Applied Mathematics and Optimization, 1(1974), pp. 163-188.

[22] J. Kloch, A necessary and sufficient condition for normality of linear control systems with delay, Annals Polonici, Mathematici, 35(1978), pp. 305-312.

[23] E.B. Lee and L. Markus, Foundations of Optimal Control Theory, John Wiley and Sons, New York, New York, 1967.

[24] N. Dunford and J.T. Schwartz, Linear Operators, Part I, Interscience, New York, 1957.

[25] H.T. Banks, A maximum principle for Optimal Control Problems with Functional Differential Systems, Bull. Amer. Math. Soc. 75(1969),

pp. 158-161.

[26] H.T. Banks, Necessary Conditions for Control Problems with Variable Time Lags, SIAM J. Control, 6(1968), pp. 9-47.

[27] H.T. Banks, Variational Problems involving Functional Differential Equations, SIAM J. Control, 7(1969).

Laplace Transform Techniques in Control and Stabilization Problems

Richard Datko

Mathematical Reviews

Ann Arbor, Michigan

ABSTRACT

We present three applications of Laplace transform technique in problem related to the stability and control of linear systems. The first application shows the usefulness of the finite Laplace transform in the computation of controls which steer an object from an initial state to a given terminal state. The second application concerns the effect of certain stability properties on small perturbations of the delay in multidelay systems. The third application indicates that boundary stabilization for linear hyperbolic partial differential equations is not robust with respect to delays in the feedback.

INTRODUCTION

The use of the Laplace transform in linear control theory might be called pervasive. In an implicit or explicit form it occurs in many areas of linear system theory such as sensitivity analysis, synthesis and design, stability analysis and so on. Thus it needs no justification for its use. In fact one could easily argue that the analytic theory of semigroups a lá Hille and Phillips is one aspect of the Laplace transform extended to infinite dimensions. Thus the purpose of this article is not to promote the use of the Laplace transform in system theory. It is to present what we believe are some unusual aspects of its application to control and stabilization problems.

This presentation has three sections each of which is devoted to a different type of problem for which a Laplace transform approach is suitable. In an endeavor to keep the presentation at an accessible level most of the exposition is via examples, which we think are nontrivial and which convey the spirit of each section.

Section I concerns the finite Laplace transform (F.L.T.) and its use in the computation of controls for three types of linear systems. The basic idea is very simple. It consists in observing two things.

One: The F.L.T. of a function is an entire analytic function.

Two: The F.L.T.s of many linear control systems involve initial and terminal states, and since the expressions for the F.L.T.s formally contain poles, the object is to select the F.L.T.s of the controllers to eliminate these poles. This process then leads to numerical procedures for finding a controller which takes an object from an initial to a given terminal state. The examples involve a linear ordinary differential equation and two partial differential equations.

Section II deals with the problem of the preservation of asymptotic stability for all delays of controlled and uncontrolled linear functional differential equations. Two examples are presented in this section, one with no control and the other controlled. The purpose of these examples is to show that there are "cracks" in some aspects of the general theory for these systems. To wit, that in multiple delay systems small perturbations in certain delay parameters effect their stability properties and hence one must question stability theories for multiple delay systems that depend upon an exact knowledge of their delays.

Section III is, from our point of view, the most interesting section since it indicates that boundary feedback stabilization of hyperbolic partial differential equations may not be robust with respect to small delays in the feedback. Two examples, the one dimensional wave equation and an Euler-Bernoulli beam equation, demonstrate this phenomena. We also briefly indicate at the end of this section why this may be a general difficulty for hyperbolic systems.

I. THE USE OF THE FINITE LAPLACE TRANSFORM IN CONTROL PROBLEM.

Consider a differential vector function $f: [0,T] \to R^n$. Let s be an arbitrary complex number. The finite Laplace transform of $\dot{f} = \frac{df}{ds}$ is

$$\begin{aligned} G(s) &= \int_0^T e^{-st}\dot{f}(t)dt = f(T)e^{-sT} - f(0) + s\int_0^T e^{-st}f(t)dt \\ &= f(T)e^{-sT} - f(0) + sF(s) . \end{aligned} \tag{1.1}$$

The complex valued functions F and G are entire and satisfy the conditions

$$\int_{-\infty}^{\infty} |F(i\omega)|^2 d\omega < \infty , \int_{-\infty}^{\infty} |G(i\omega)|^2 d\omega < \infty \tag{1.2}$$

(see e.g. [5]). Exploiting these two properties and the fact that the initial and terminal states of f occur in the finite Laplace transform of f it is possible to compute controls for a variety of autonomous linear control problems expressed by ordinary or partial differential equations. Briefly what one does in the case of ordinary differential equations is converts the given problem to a finite Laplace transform and isolates the transform of the trajectory, $\bar{x}(s)$, on the left hand side of a certain equation. On the right hand side of the equation the initial and terminal states of the system and the finite Laplace transform of the control function explicitly occur in linear combination multiplied by a holomorphic complex valued linear operator. If the initial and terminal states of the trajectory are given then the constraint that its finite Laplace transform be an entire vector function imposes conditions on the transformed control, $\bar{\mu}(s)$, namely that the numerators of certain expressions have zeros of the same order as the poles of the holomorphic operator. These conditions can then be used to find the finite Laplace transform of the control which guides the system from its initial to its terminal state. A variant of the above technique can be used to solve the quadratic regulator problem over finite time intervals.

Preliminaries. The following are definitions and notational conventions which will be used throughout this section.

1. C will denote the complex plane, s will denote a point in C. Re s and Im s will stand for the real and imaginary parts of a complex number.

2. If $f: [0,T] \to R^n$, $T < \infty$, is an L_1 integrable mapping, then the finite Laplace transform of f, denoted by $\bar{f}$, is given by $\int_0^T e^{-st} f(t)dt$. Throughout this section the finite Laplace transform will be abbreviated to F.L.T., and its dependence on T will not in general be emphasized.

3. The characteristic function of a measurable set E in R will be denoted by $\chi_E(t)$.

We now present the fundamental theorem on which this paper is based (see e.g. [5, pp. 238, 241]).

THEOREM 1.1. Let $\bar{f}: C \to C$ be an entire function of exponential type, i.e. $|\bar{f}(s)| \leqq a\, e^{b|s|}$ for all $s \in C$ and fixed constants a and b. Then there exist nonnegative constants T and T' and a function $f \in L^2(-\infty,+\infty)$ with $f(t) = 0$ if $t \notin [-T',T]$ and $\bar{f}(s) = \int_{-T}^{T} e^{-st} f(t)dt$, if and only if $\int_{-\infty}^{\infty} |\bar{f}(i\omega)|^2 d\omega < \infty$. Moreover the constants T' and T satisfy the relations

$$\begin{aligned} T' &= \overline{\lim_{x\to\infty}} \frac{1}{x} \ln|\bar{f}(x)| , \\ T &= \overline{\lim_{x\to\infty}} \frac{1}{x} \ln|\bar{f}(-x)| . \end{aligned} \tag{1.3}$$

Example 1.1. (See e.g. [1, p. 536-540].) Consider the scalar system

$$\dddot{x} + \ddot{x} = \mu \quad , \ x(0) \ , \ \dot{x}(0) \ , \ \ddot{x}(0) \ . \tag{1.4}$$

Assume $|\mu(t)| \leqq 1$ for all $t \geqq 0$. It is desired to drive the initial values to $x(T) = \dot{x}(T) = \ddot{x}(T) = 0$ in some minimum time T. If this is possible the F.L.T. of (2.10) becomes

$$\bar{x}(s) = \frac{1}{s^2(s+1)}[(s^2+s)x(0) + (s+1)\dot{x}(0)+\ddot{x}(0)+\bar{\mu}(s)] \tag{1.5}$$

where $\bar{\mu}(s)$ is the F.L.T. of some measurable μ from $[0,T] \to [-1,1]$. By Theorem 1.1 this implies that

$$\begin{aligned} x(0) + \dot{x}(0) + \bar{\mu}'(0) &= 0 \ , \\ \dot{x}(0) + \ddot{x}(0) + \bar{\mu}(0) &= 0 \ , \\ \ddot{x}(0) + \bar{\mu}(-1) &= 0 \ . \end{aligned} \tag{1.6}$$

Since we may assume $\mu(t)$ is "bang-bang" ([ ]) and that there are at most two switches (see e.g. [1]) $\bar{\mu}(s)$ must be of the form

$$\bar{\mu}(s) = \varepsilon_0 [\int_0^{t_1} e^{-st}dt - \int_{t_1}^{t_2} e^{-st}dt + \int_{t_2}^{T} e^{-st}dt] \tag{1.7}$$

where $\varepsilon_0 = \pm 1$, $0 \leqq t_1 \leqq t_2 \leqq T$. Thus between (1.6) and (1.7) we are led by some simple calculations to the equations

$$\begin{aligned} \ddot{x}(0) &= -\bar{\mu}(-1) = \varepsilon_0[1-2e^{t_1}+2e^{t_2}-e^{T}] \ , \\ \dot{x}(0) &= \bar{\mu}(-1)-\bar{\mu}(0) = \varepsilon_0[-1+2e^{t_1}-2e^{t_2}-e^{T}-2t_1+2t_2-T] , \\ x(0) &= -\bar{\mu}(-1)+\bar{\mu}(0)-\bar{\mu}'(0)=\varepsilon_0[1-2e^{t_1}+2e^{t_2}-e^{T}+2t_1-2t_2+T+t_1^2-t_2^2+\tfrac{T^2}{2}] , \\ & 0 \leqq t_1 \leqq t_2 \leqq T \ , \ \varepsilon_0 = \pm 1 \ . \end{aligned} \tag{1.8}$$

The smallest value of T for which (1.8) is satisfied is the optimal time of transfer to the origin, the switching times are t_1 and t_2 and ε_0 is the value of $\mu(0)$.

Example 1.2. In this example all functions are assumed to have the necessary integrability conditions.

Let

$$S = \{(x,y): 0 \leqq x \leqq 1 \ , \ 0 \leqq y \leqq 1\} \tag{1.9}$$

and on S consider

$$\begin{aligned} &u_{tt} = u_{xx} + u_{yy} , \\ &u(x,y,0) = \phi(x,y) , \\ &u_t(x,y,0) = \psi(x,y) , \\ &u(0,y,t) = u(1,y,t) = 0 . \end{aligned} \tag{1.10}$$

Let the boundary controls be given by

$$\begin{aligned} &a_0(x,t) = u(x,0,t) , \ t > 0 , \ 0 \leqq x \leqq 1 \\ &a_1(x,t) = u(x,1,t) , \ t > 0 , \ 0 \leqq x \leqq 1 . \end{aligned} \tag{1.11}$$

We shall attempt to find conditions on the a_i , $i = 0,1$, such that after some time $T > 0$

$$u(x,y,T) = u_t(x,y,T) = 0 , \ (x,y) \in S . \tag{1.12}$$

The F.L.T. of (1.10) with respect to the t variable is given by the four equations

$$s^2U(x,y,s) - s\phi(x,y) - \psi(x,y) = U_{xx}(x,y,s) + U_{yy}(x,y,s) .$$

$$\begin{aligned} &A_0(x,s) = \int_0^T u(x,0,t)e^{-st}dt = U(x,0,s) , \\ &A_1(x,s) = \int_0^T u(x,1,t)e^{-st} = U(x,1,s) , \\ &U(0,y,s) = U(1,y,s) = 0 . \end{aligned} \tag{1.13}$$

We may assume that because of the boundary conditions in (1.10) A_0 and A_1 in (1.13) are representable in the forms

$$\begin{aligned} &A_0(x,s) = \sum_{n=1}^{\infty} A_0^n(s)\sin(n\pi x) , \\ &A_1(x,s) = \sum_{n=1}^{\infty} A_1^n(s)\sin(n\pi x) . \end{aligned} \tag{1.14}$$

For convenience we shall let

$$\int_0^1 (\sin n\pi\tau)(s\phi(\tau,\sigma)+\psi(\tau,\sigma))d\tau = q_n(\sigma) , \ n = 1,\ldots . \tag{1.15}$$

Using the notation (1.14) and (1.15) and the method of separation of variables the solution of (1.13) can be written

$$U(x,y,s)=\sum_{n=1}^{\infty}\frac{A_0^n(s)\sin(n\pi x)\sinh\sqrt{s^2+n^2\pi^2}(1-y)}{\sinh\sqrt{s^2+n^2\pi^2}}$$

$$+\sum_{n=1}^{\infty}\frac{A_1^n(s)\sin(n\pi x)\sinh\sqrt{s^2+n^2\pi^2}\,y}{\sinh\sqrt{s^2+n^2\pi^2}} \qquad (1.16)$$

$$+\sum_{n=1}^{\infty}\sin n\pi x\int_0^y\frac{\sinh\sqrt{s^2+n^2\pi^2}(1-y)\sinh\sqrt{s^2+n^2\pi^2}\,\sigma}{\sqrt{s^2+n^2\pi^2}\ \sin\sqrt{s^2+n^2\pi^2}}q_n(\sigma)d\sigma$$

$$+\sum_{n=1}^{\infty}\sin(n\pi x)\int_y^1\frac{\sinh\sqrt{s^2+n^2\pi^2}\ y\ \sinh\sqrt{s^2+n^2\pi^2}(1-\sigma)}{\sqrt{s^2+n^2\pi^2}\ \sinh\sqrt{s^2+n^2\pi^2}}q_n(\sigma)d\sigma\ .$$

A necessary condition for (1.16) to be a F.L.T. in the s variable is that for each integer n the corresponding term in the summation be a finite Laplace transform. Since the zeros of $\sinh\sqrt{s^2+n^2\pi^2}$ occur at the points

$$s = \pm\, i\sqrt{m^2 + n^2}\pi\ , \quad m = 1,2,\dots\ ,$$

this is possible only when

$$(-1)^{m+1}A_0^n(\sqrt{n^2 + m^2}\ \pi i) + A_1^n(\sqrt{n^2 + m^2}\ \pi i) \qquad (1.17)$$

$$+\frac{(-1)^{m+1}}{m\pi}\int_0^1(\sin(m\pi\sigma))q_n(\sigma)d\sigma = 0\ ,$$

$m = 1,2,\dots$. A similar expression holds when we replace $\sqrt{(n^2+m^2)}\pi i$ by $-\sqrt{(n^2+m^2)}\pi i$. Taking note of (1.15) and setting

$$\alpha_{mn} = \int_0^1\int_0^1(\sin(m\pi\sigma))(\sin(n\pi\tau))\phi(\tau,\sigma)d\tau\ d\sigma \qquad (1.18)$$

and

$$\beta_{mn} = \int_0^1\int_0^1\sin(m\pi\sigma)\sin(n\pi\tau)\psi(\tau,\sigma)d\tau\ d\sigma \qquad (1.19)$$

we can rewrite the conditions (1.17) in the form

$$(-1)^{m+1}\ A_0^n(\pm\sqrt{n^2+m^2}\ \pi i) + A_1^n(\pm\sqrt{n^2+m^2}\ \pi i) \qquad (1.20)$$

$$+\frac{(-1)^{m+1}}{m}(\pm i)\sqrt{n^2+m^2}\ \alpha_{mn} + \frac{(-1)^{m+1}}{m}\beta_{mn} = 0\ .$$

A choice for $A_0^n(s)$ will be of the form

$$A_0^n(s) = \sum_{m=1}^{\infty} A_0^{mn}(s) , \tag{1.21}$$

where if $\sqrt{n^2+m^2}$ is an integer

$$A_0^{mn}(s) = \frac{s^2 e^{-2s}(\sinh 2\sqrt{s^2+n^2\pi^2})b_{mn}}{\sqrt{s^2+n^2\pi^2}(s^2+(n^2+m^2)\pi^2)} . \tag{1.22}$$

The coefficient b_{mn} is chosen so that

$$A_0^{mn}(\pm i\sqrt{n^2+m^2}\pi) = -\frac{(m^2+n^2)\pi}{m} b_{mn} = \frac{-\beta_{mn}}{m} . \tag{1.23}$$

Thus

$$b_{mn} = \frac{\beta_{mn}}{(m^2+n^2)\pi} ,$$

if $\sqrt{n^2+m^2}$ is an integer.

If $\sqrt{m^2+n^2}$ is not an integer it must be irrational and there is a natural number ℓ such that

$$\ell - 1 < \sqrt{m^2 + n^2} < \ell .$$

We define

$$\frac{k_m}{2} = \frac{\ell}{\sqrt{m^2 + n^2}} \tag{1.24}$$

and observe that the inequality

$$k_m - \frac{2}{\sqrt{m^2 + n^2}} < 2 < k_m \tag{1.25}$$

is satisfied. We then define

$$A_0^{mn}(s) = \frac{s^2 e^{-k_m s} \sinh(k_m-2)\sqrt{s^2+n^2\pi^2}(\sinh 2\sqrt{s^2+n^2\pi^2})b_{mn}}{(s^2+n^2\pi^2)(s^2+(n^2+m^2)\pi^2)} . \tag{1.26}$$

It is not difficult to verify that (1.26) and (1.22) define finite Laplace transforms over intervals of the form $[0,k_m]$, $2 < k_m \leq 2 + \sqrt{2}$. In the case of (1.26) we observe that because $\sqrt{m^2+n^2}$ is irrational $(k_m)m$ is also. Thus $\sin(k_m\pi) \neq 0$ and the equation

$$A_0(\pm\sqrt{n^2+m}\ \pi i) = \frac{(m^2+n^2)}{m^3\pi}(\sin m\pi k_m)b_{mn} = -\frac{\beta_{mn}}{m}$$

has a solution for b_{mn} which is

$$b_{mn} = -\frac{m^2\pi}{m^2+n^2}\frac{\beta_{mn}}{\sin(m\pi k_m)} . \tag{1.27}$$

Similarly we define

$$A_1^n(s) = \sum_{m=1}^{\infty} A_1^{mn}(s) , \tag{1.28}$$

where if $\sqrt{m^2+n^2}$ is an integer

$$A_1^{mn}(s) = \frac{s\ e^{-2s}\sinh 2\sqrt{s^2+n^2\pi^2}}{\sqrt{s^2+n^2\pi^2}\ (s^2+(n^2+m^2)\pi^2)} a_{mn}$$

with

$$a_{mn} = (-1)^m \pi_m \alpha_{mn} . \tag{1.29}$$

If $\sqrt{m^2+n^2}$ is not an integer we define

$$A_1^{mn}(s) = \frac{s\ e^{-k_m s}\sinh(k_m-2)\sqrt{s^2+n^2\pi^2}(\sinh 2\sqrt{s^2+n^2\pi^2})a_{mn}}{(s^2+n^2\pi^2)(s^2+(n^2+m^2)\pi^2)} , \tag{1.30}$$

where

$$a_{mn} = (-1)^{m+1}\frac{m^2\pi^2}{\sin(k_m\Pi_m)}\alpha_{mn} . \tag{1.31}$$

Notice that when $\sqrt{m^2+n^2}$ is not an integer the inequality (1.26) implies that equations (1.27) and (1.31) tend to the quantities $\pm(m/\sqrt{m^2+n^2})\beta_{mn}$ and $\pm m\pi\alpha_{mn}$ respectively. Also observe that

$$A_j^{mn}(\pm i\sqrt{\ell^2+n^2}) = 0 , \ j = 0,1,$$

if $\ell \neq m$.

As a final remark concerning this example it should be pointed out that the convergence of $A_0^n(s)$ and $A_1^n(s)$ in equations (1.21) and (1.28) has not been discussed. Since this example is meant to demonstrate a technique we shall not concern ourselves with this question. Suffice it to say that if $\phi(x,y)$ and $\psi(x,y)$ have finite Fourier expansions i.e. $\{\alpha_{mn}\}$ and $\{\beta_{mn}\}$ contain only a finite number of nonzero terms, then the above constructions will always yield controls which drive the initial state in (1.10) to the zero state in some time $T \leqq 2 + \sqrt{2}$ (i.e. the maximum possible k_m given by inequality (1.25)).

Example 1.3. Consider the heat equation in one dimension

$$\frac{\partial u}{\partial t} = \frac{\partial^2 u}{\partial t^2}, \quad t \geqq 0, \quad 0 \leqq x \leqq 1. \tag{1.32}$$

Assume

$$\begin{aligned} u(x,0) &= \phi(x), \\ u(0,t) &= a_0(t), \\ u(1,t) &= a_1(t), \end{aligned} \tag{1.33}$$

where ϕ, a_0 and a_1 are integrable over finite intervals.

If $u(x,T) = \psi(x)$ for some $T > 0$ and

$$\int_0^T e^{-st} a_i(t)dt = A_i(s), \quad i = 0,1,$$

then using the techniques of Example 1.2, the F.L.T. of (1.32)-(1.33) is given by

$$U(x,s) = \frac{A_0(s)\sinh\sqrt{s}(1-x) + A_1(s)\sinh\sqrt{s}\,x}{\sinh\sqrt{s}} + \int_0^1 F(x,\sigma,s)\,[\phi(\sigma) - \psi(\sigma)e^{-sT}]d\sigma \tag{1.34}$$

where

$$F(x,\sigma,s) = \frac{1}{\sqrt{s}\sinh\sqrt{s}} \sinh\sqrt{s}\,\sigma \sinh\sqrt{s}(1-x) \text{ if } 0 \leqq \sigma \leqq x$$

and

$$F(x,\sigma,s) = \frac{1}{s \sinh\sqrt{s}} \sinh\sqrt{s}\,x \sinh\sqrt{s}(1-\sigma) \text{ if } \sigma \leqq x \leqq 1. \tag{1.35}$$

Suppose it is desired to drive an initial temperature ϕ to the zero temperature in some time $T > 0$. Since the poles of (1.34) are of order one and occur at the points $s = -n^2\pi^2$, $n = 1,2,\ldots$, the numerator in (1.34) must satisfy the equations

$$(-1)^{n+1} A_0(-n^2\pi^2) + A_1(-n^2\pi^2) + \frac{(-1)^{n+1}}{n\pi} \int_0^1 (\sin(n\pi\sigma))\phi(\sigma)d\sigma = 0. \tag{1.36}$$

Thus it is desired to find entire functions $A_0(s)$ and $A_1(s)$ which are finite Laplace transforms over $[0,T]$ and which also satisfy (1.36). This is not always a practical problem as the following special case shows (see e.g.[2]). Let

$$\phi(x) = \phi_0 = \text{constant} \neq 0 . \tag{1.37}$$

For this value of ϕ the equations (4.5) reduce to

$$A_0(-n^2\pi^2) = A_1(-n^2\pi^2), \text{ n even,} \tag{1.38}$$

and

$$A_0(-n^2\pi^2) + A_1(-n^2\pi^2) = \frac{-2\phi_0}{n^2\pi^2} , \text{ n odd.} \tag{1.39}$$

Assume

$$A_0(s)=A_1(s), \text{ i.e. } a_0(t)=a_1(t), \text{ and } |a_0(t)| \leqq 1 \text{ on } [0,T]. \tag{1.40}$$

If we also assume $a_0(t)$ is piecewise constant with a finite number of switches on $[0,T]$ then it is easily seen that (1.39) can never be satisfied. For if

$$a_0(t)= \sum_{j=1}^{N} \alpha_j \chi_{[t_{j-1},t_j)}(t), 0=t_0<\ldots<t_n = T ,$$

then

$$A_0(s) = \frac{1}{s}\sum_{j=1}^{N} \alpha_j(e^{-st_{j-1}} - e^{-st_j}) . \tag{1.41}$$

Clearly $A_0(s)$ given by (1.41) can never satisfy (1.39).

However if we permit an infinite number of switches then it is possible to bring the temperature to zero in any finite time T. For let

$$a_0(t)=a_1(t)=\sum_{j=1}^{\infty} \alpha_j \chi_{[t_{j-1},t_j)}(t), 0=t_0<t_1<\ldots<t_n \to T .$$

Then

$$A_0(s) = A_1(s) = \frac{1}{s}\sum_{j=1}^{\infty} \alpha_j(e^{-st_{j-1}} - e^{-st_j}) . \tag{1.42}$$

and at the points $-n^2\pi^2$ we have

$$A_0(-n^2\pi^2)=A_1(-n^2\pi^2) = \frac{1}{n^2\pi^2} \sum_{j=1}^{\infty} \alpha_j(e^{n^2\pi^2 t_{j-1}} - e^{n^2\pi^2 t_j}). \tag{1.43}$$

Thus (1.39) reduces to the moment problem (see e.g.[2]) of selecting $\{\alpha_j\}$, $|\alpha_j| \leqq 1$ for $j = 1,2,\dots$, and $0 = t_0<t_1<\dots<t_n \to T$ such that

$$\sum_{j=1}^{\infty} \alpha_j (e^{n^2\pi^2 t_{j-1}} - e^{n^2\pi^2 t_j}) = \phi_0 . \tag{1.44}$$

This is a solvable problem (see e.g.[2]) which can be solved for any $T > 0$.

II. ASYMPTOTIC STABILITY AND STABILIZATION OF LINEAR DELAY SYSTEMS.

In this section we shall indicate some "cracks" in the algebraic theory which concerns itself with stabilization of linear delay systems for all values of the delay. To wit, what we wish to indicate, through two examples, is that:

(i) Delay systems of the form

$$\dot{x}(t) = A_0 x(t) + \sum_{j=1}^{\ell} \sum_{r_j=1}^{r_j=q_j} A_{r_j} x(t-\beta_{r_j} h_j) ,$$

where A_j are real $n \times n$ matrices, $h_j \geq 0$ are real numbers and β_{r_j} are fixed positive constants may be uniformly asymptotically stable for all $h_j \geq 0$ for a fixed set of $\{\beta_{r_j}\}$, but may not possess this property for small perturbations of $\{\beta_{r_j}\}$.

(ii) That control systems of the form

$$\dot{x}(t) = A_0 x(t) + \sum_{j=1}^{\ell} \sum_{r_j=1}^{q_j} A_{r_j} x(t-\beta_{r_j} h_j) + Bu(t) ,$$

where B is an $n \times m$ constant matrix and $u(t)$ is a control, may be stabilizable, with an arbitrarily given decay rate, by a feedback of the form

$$u(t) = \sum_{j=0}^{r} K_j x(t-\tau_j) ,$$

where $\{0 \leq \tau_j\}$ are linear combinations of $\{\beta_{r_j} h_j\}$ and K_j are real constant $m \times n$ matrices, for one set $\{\beta_{r_j}\}$, but not for small perturbations of $\{\beta_{r_j}\}$.

Example 2.1. Consider the parametric family of scalar delay equations

$$\dot{x}(t) = -x(t) - x(t-\alpha) - \frac{1}{2}x(t-\beta\alpha) , \qquad (2.1)$$

where $h \geq 0$ is allowed to vary, but $\beta > 0$ is fixed. For $\alpha = 0$ (2.1) is clearly asymptotically stable. If for some $\alpha > 0$ the system loses this property, then there exists $\omega > 0$ and $\alpha \geq 0$ such that

$$i\omega + 1 + e^{-i\alpha} + \frac{1}{2}e^{-i\alpha\beta} = 0 \qquad (2.2)$$

(see e.g. [4]). For $\beta = 2$ Eq. (2.2) reduces to the two equations

$$0 = \frac{1}{2} + \cos\alpha + \cos 2\alpha$$

and

$$\omega = \sin\alpha + \frac{1}{2}\sin 2\alpha ,$$

which have no solution for α real. Thus (2.1) is asymptotically stable for all $\alpha \geq 0$ if $\beta = 2$.

Now let

$$\beta_n = 2 + \frac{1}{2}\frac{1}{2n+1}$$

and

$$\hat{B}_n = 2 - \frac{3}{2}\frac{1}{2n+1} , \quad n = 1,2,\ldots .$$

Then for these values of β_n (2.1) is asymptotically stable when $\alpha = 0$. However when $\alpha = 2(2n+1)\pi$ and $\omega = \frac{1}{2}$ equation (2.2) is satisfied for β_n and $\hat{\beta}_n$. Thus we have two sequences $\{\beta_n\} \to 2^+$ and $\{\beta_n\} \to 2^-$ such that (2.1) is not asymptotically stable for all $\alpha \geq 0$ while it does possess this property for $\beta = 2$.

Actually more may be shown. For example, if $\beta = 2 + \frac{1}{2}(1/(2n+1))$ there is an $\varepsilon > 0$ such that (2.1) is unstable for all α in the open interval $(2(2n+1)\pi ,$ $2(2n+1)\pi+\varepsilon)$. To see this consider the implicitly defined function $\alpha:[0,\infty) \to C$ (the complex plane) described by the relationship

$$s = 1 - e^{-s\alpha} - \frac{1}{2}e^{-s\alpha(2+\frac{1}{2}(1/(2n+1))} . \qquad (2.3)$$

Using the implicit differentiation we find at the point $s_0 = i/2$ and $\alpha_0 = 2(2n+1)\pi$ that

$$\frac{ds}{d\alpha} = \frac{-\frac{i}{2}v}{1+\alpha v} = \frac{-\frac{i}{2}v - \frac{i}{2}\alpha|v|^2}{|1+\alpha v|^2} ,$$

where

$$v = 1 + \frac{(2+\frac{1}{2}\frac{1}{2n+1})i}{2} .$$

Thus, at $s = i/2$ and $\alpha = 2(2n+1)\pi$, $\mathrm{Re}(ds/d\alpha) > 0$. But this implies that for some $\alpha > 2(2n+1)$ (2.3) has a solution with $\mathrm{Re}\, s > 0$, i.e. (2.1) is unstable for that value of α when $\beta = 2 + \frac{1}{2}\frac{1}{2n+1}$.

Example 2.2. Consider the two-dimensional system

$$\begin{aligned} \dot{x}(t) &= -x(t) - y(t-h_1) - x(t-h_2) \\ \dot{y}(t) &= u(t) , \end{aligned} \tag{2.4}$$

where h_1 and h_2 are fixed positive constants and u is a control. We seek a feedback control of the form

$$u(t) = \sum_{\tau=1}^{\ell} [q_{r_1} x(t-\tau_r) + q_{r_2} y(t-\tau_r)] . \tag{2.5}$$

where $\tau_r \geq 0$ are constants such that the resulting system is asymptotically stable. Moreover, if possible, we would like the resulting system to have as its spectrum two predetermined complex conjugate points in the left half plane. As we shall see this depends on the two fixed delays h_1 and h_2 .

The poles of the Laplace transform of (2.4)-(2.5) determine the spectrum of the system. An easy computation shows them to be of the form

$$\det \begin{vmatrix} s+1+e^{-sh_2} & e^{-sh_1} \\ -K_1(s) & s-K_2(s) \end{vmatrix} = q(s) = 0 \tag{2.6}$$

where

$$K_j(s) = \sum_{r=1}^{\infty} q_{r_j} e^{-\tau_r s} , \ j = 1,2, \text{ and } \tau_r \geq 0 . \tag{2.7}$$

For finite pole placement (i.e. two complex conjugate roots of (2.6)) the equation

$$q(s) = s^2 + a s + b \tag{2.8}$$

must hold, where $a, b > 0$. Equating (2.6) to the right hand side of (2.8) we obtain the two following relations for $K_j(s)$, $j = 1,2$.

$$K_1(s) = e^{sh_1}(1-a+b) + (2-a)e^{s(h_1-h_2)} + e^{s(h_1-2h_2)} \tag{2.9)(a}$$

$$K_2(s) = 1 + e^{-sh_1} - a . \tag{2.9)(b}$$

Since τ_r in (2.7) must be nonnegative, we see that $K_1(s)$ in (2.9)(a) may cause some difficulty. For one thing we must have at the very least

$$a = 1 + b . \tag{2.10}$$

Other than that we have the following three possibilities, which are obtained from (2.9)(a) under the condition that the coefficients of s in its exponential terms must be nonnegative:

(i) If $h_2 < h_1 \leq 2h_2$, then $a = 2$,

$$q(s) = s^2 + 2s + 1$$

and the stabilizing feedback is

$$u(t) = x(t-2h_2+h_1)+y(t-h_2)-y(t) .$$

(ii) If $h_1 \leq h_2$, then

$$q(s) = s^2 + a s + a - 1, \quad a - 1 > 0$$

and the stabilizing feedback is

$$u(t) = (2-a)x(t-h_2+h_1)+x(t-2h_2+h_1) + y(t-h_2)+(-a+1)y(t) .$$

(iii) If $2h_2 < h_1$, there is no finite placement of the spectrum, i.e. (2.8) cannot be satisfied using a feedback of the type (2.5). However the system may still be stabilized by a feedback of the form (2.5), but the spectrum of the resulting system will be infinite. A feedback which stabilizes the system is

$$u(t) = -a\,y(t)\ ,\ a > 0\ .$$

In fact this last feedback stabilizes the system in all three cases.

What this example shows is that certain notions such as finite pole placement, even for simple systems with constant delays, have features which are too fine to be covered entirely by the algebraic methods that are currently so popular, even though these methods are basically rooted in a Laplace transform framework.

III. BOUNDARY FEEDBACK STABILIZATION OF HYPERBOLIC SYSTEMS IS NOT ALWAYS ROBUST WITH RESPECT TO "SMALL" TIME DELAYS.

In this section two examples of hyperbolic partial differential equations are presented in which time delays of arbitrarily small magnitude, when introduced into the feedback control, destabilize a system which has been stabilized by boundary feedback of Neumann type. As we shall indicate after the examples this phenomena is probably not pathological, but one which can occur in many hyperbolic systems, and calls into question the entire concept of boundary feedback stabilization for such systems. An interesting by-product of the presentation in this section is that the Laplace approach appears to be essential.

Example 3.1. Consider the equation

$$u_{tt}-u_{xx}+2au_t+a^2u = 0\ ,\quad 0<x<1\ ,\ t>0\ , \tag{3.1}$$

with boundary conditions

$$u(0,t) = 0\ ,\ t > 0\ , \tag{3.2}$$

$$u_x(1,t) = -ku_t(1,t-\varepsilon)\ ,\ t>0\ , \tag{3.3}$$

where $\varepsilon \geq 0$, $a \geqq 0$ and $k \geqq 0$ are constants.

The initial values of u and u_t are

$$u(x,t) = \phi(x,t)\ , \tag{3.4}$$

$$u_t(x,t) = \psi(x,t) \tag{3.5}$$

for $0 \leq x \leq 1$, $-\varepsilon \leq t \leq 0$, and for simplicity we assume ϕ and ψ are infinitely differentiable in (x,t) .

The Laplace transform of (3.1), (3.2), (3.4) and (3.5) with respect to the variable t results in a one parameter family of function defined over $[0,1]$ by the equation:(3.6)

$$U(x,\omega)=A(\omega)\sinh(\omega+a)x-\frac{1}{\omega+a}\int_0^x \sinh(\omega+a)(x-\sigma)[(\omega+2a)\phi(\sigma)+\psi(\sigma)]\,d\sigma,$$

where ω is a complex parameter, and the function $A(\omega)$ is to be determined by applying the boundary condition (3.3).

Using (3.6) and taking the Laplace transform of (3.3), where ω is the transform variable, leads to the following equation for $A(\omega)$.

$$A(\omega)(\omega+a)\cosh(\omega+a)-\frac{1}{\omega+a}\int_0^1 \cosh(\omega+a)(1-\sigma)[(\omega+2a)\phi(\sigma)+\psi(\sigma)]\,d\sigma$$

$$= k\phi(1,-\varepsilon) - k\omega e^{-\varepsilon\omega}[A(\omega)\sinh(\omega+a) - \tag{3.7}$$

$$\frac{1}{\omega+a}\int_0^1 \sinh[(\omega+a)(1-\sigma)][(\omega+2a)\phi(\sigma)+\psi(\sigma)]\,d\sigma\,]$$

$$-k\omega\int_{-\varepsilon}^0 \phi(1,\sigma)e^{-\omega(\sigma+\varepsilon)}d\sigma\ .$$

Collecting the terms in (3.7) multiplied by $A(\omega)$ we obtain the equation

$$A(\omega)[(\omega+a)\cosh(\omega+a)+k\omega e^{-\varepsilon\omega}\sinh(\omega+a)] = q(\omega)\ , \tag{3.8}$$

where q is an entire function of ω with the property that

$$\lim_{\mathrm{Re}\,\omega\to\infty} |q(\omega)| \leq q_0 < \infty\ .$$

Thus since the term containing the integral in (3.6) is also an entire function the only singularities of $U(\cdot,\omega)$ which can occur are in the term $A(\omega)\sinh(\omega+a)x$. A moments reflection shows that this is equivalent to finding the zeros of an entire function (extended by l'Hôpital's rule at $\omega = -a$) given by the equation:

$$(\omega+a)\cosh(\omega+a) + k\omega e^{-\varepsilon\omega}\sinh(\omega+a)/(\omega+a) = 0\ . \tag{3.8}$$

When $k = 1$ and $\varepsilon = 0$ equation (3.8) has no solution.

What this means is that $U(x,\omega)$ in (3.6) is the finite Laplace transform (F.L.T.) of a function over $[0,2]$. In fact in this instance the system (3.1)-(3.5) is another example of the F.L.T. technique demonstrated in Section I.

If $a = 0$ and $k \neq 1$ (3.8) has an explicit solution which satisfies

$$\mathrm{Re}\ \omega = \frac{1}{2} \log \left|\frac{k-1}{k+1}\right| < 0 .$$

If $a \neq 0$ then (3.8) satisfies the equation

$$e^{2(\omega+a)} = \frac{(k-1)\omega-a}{(k+1)\omega+a} . \qquad (3.9)$$

Assume that $|e^{2(\omega+a)}| \geq 1$, then a simple computation on the right hand side of (3.9) leads to the inequality

$$-|\omega|^2 \geq a\ \mathrm{Re}\ \omega .$$

From which we can deduce that all zeros $\{\omega_n\}$ of (3.9) lie in some half plane

$$\mathrm{Re}\ \omega \leq \alpha < 0 \text{ and satisfy } \mathrm{Im}\ \omega_n \to \infty \text{ as } n \to \infty .$$

Now let us consider the case $\varepsilon > 0$, $k > 0$. Then the following theorem holds.

THEOREM 3.1. Let $K = e^{-2a}$. The system (3.1)-(3.5) has the following stability properties:

(i) If $0 < k < (1-K)/(1+K)$, for each $\varepsilon > 0$ there exists $\beta(\varepsilon) > 0$ such that the spectrum of the system lies in $\mathrm{Re}\ \omega \leq -\beta$.

(ii) If $k = (1-K)/(1+K)$, for each $\varepsilon > 0$ the spectrum lies in $\mathrm{Re}\ \omega < 0$, but there is a countably dense set R in $(0,\infty)$ such that for each ε in R there is a sequence $\{\omega_n\}$ in the spectrum such that

$$\lim_{n\to\infty} \mathrm{Re}\ \omega_n = 0 .$$

(iii) If $k > (1-K)/(1+K)$, there is dense open set D in $(0,\infty)$ such that for each ε in D the system admits exponentially unstable solutions.

To prove the theorem, two lemmas will be needed. The first lemma is a special case of [4, Lemma 2.3].

Lemma 3.1. Let

$$h(\varepsilon,\omega) = \omega[1+K\ e^{-2\omega}+k\ e^{-\varepsilon\omega}(1-K\ e^{-2\omega})]+a(1+K\ e^{-2\omega}) \qquad (3.10)$$

$$= \omega f(\varepsilon,\omega)+a(1+K\ e^{-2\omega}) .$$

If, for fixed $\varepsilon, f(\varepsilon,\omega)$ has a zero at $\omega_0 = \xi_0 + i\eta_0$, then for any $\delta > 0$ the vertical strip $\{\omega:\xi_0-\delta<\text{Re}\,\omega<\xi_0+\delta\}$ has an infinite number of zeros of both $f(\varepsilon,\omega)$ and $h(\varepsilon,\omega)$.

Lemma 3.2. Let $K = e^{-2a}$ and $k > (1-K)/(1+K)$. Then there exists for each such k an open dense set, $\mathcal{D}$, in $(0,\infty)$ such that for every ε in $\mathcal{D}, f(\varepsilon,\omega) = 0$ has at least one solution with $\text{Re}\,\omega > 0$.

Proof of Lemma 3.2. Let K and k satisfying the hypotheses of the lemma be fixed. Consider the mapping from the complex ω-plane into the complex ε-plane defined implicitly by

$$k = \frac{1+K\ e^{-2\omega}}{K\ e^{-2\omega}-1}\ e^{\varepsilon\omega} . \qquad (3.11)$$

For $\omega \neq 0$ solutions of equation (3.11) can be found among the infinite family of meromorphic functions given by the equations

$$\varepsilon = \frac{1}{\omega}\left[\log k + \log\left(\frac{K\ e^{-2\omega}-1}{K\ e^{-2\omega}+1}\right) + 2m\pi i\right] , \qquad (3.12)$$

where log is the principal value of the logarithm and m is a positive integer.

Now let n be a fixed positive integer and

$$\omega = \omega_1 + \frac{(2n+1)}{2}\pi i , \ \omega_1>0 . \qquad (3.13)$$

Then ε in (3.12) has the form

$$\varepsilon = \frac{\log k+\log((K\ e^{-2\omega_1}+1)/(1-K\ e^{-2\omega_1}))+(2m+1)\pi i}{\omega_1 + ((2n+1)/2)\pi i} . \qquad (3.14)$$

When ω_1 satisfies the equation

$$\omega_1 = \frac{2n+1}{2(2m+1)}\left[\log k + \log\left(\frac{K\ e^{-2\omega_1}+1}{1-K\ e^{-2\omega_1}}\right)\right] , \qquad (3.15)$$

equation (3.14) satisfies

$$\varepsilon = \frac{2(2m+1)}{2n+1} . \tag{3.16}$$

Equation (3.15) always has a solution for some $\omega_1 > 0$. To see this, notice that because $k > (1-K)/(1+K)$ the right side of (3.15) is positive for $\omega_1 = 0$ and as ω_1 tends to infinity the right side tends to $(2n+1)/(2(2m+1))\log k$.

Next observe that points of the form (3.16) are dense on $(0,\infty)$. Furthermore because of the open mapping property of meromorphic functions (see e.g. [6, p. 116] about each point of the form (3.16) there is an open interval, with $2(2m+1)/(2n+1)$ as its center which is contained in the image, under the mapping (3.12), of some open ball in $\mathrm{Re}\ \omega > 0$. This completes the proof of the lemma.

Proof of the theorem. The spectrum of (3.1)-(3.5) is determined by equation (3.8). The result is equivalent to

$$h(\varepsilon,\omega) = 0 \tag{3.17}$$

where h is defined by (3.10) and $K = e^{-2a}$.

(i) Suppose $k \leqq (1-K)/(1+K)$ and $\varepsilon > 0$. Since $\omega = 0$ is not a solution of (3.17), that equation can be rewritten as

$$1 = k\ e^{-\varepsilon\omega} \frac{K\ e^{-2\omega}-1}{K\ e^{-2\omega}+1} - \frac{a}{\omega} . \tag{3.18}$$

If (3.18) has a solution $\omega = \xi + i\eta$ with $\xi > 0$, then

$$1 = k\ \mathrm{Re}\left[e^{-\varepsilon\omega} \frac{K\ e^{-2\omega}-1}{K\ e^{-2\omega}+1}\right] - \frac{a\xi}{\xi^2+\eta^2} < k\ \frac{1-K\ e^{-2\xi}}{1-K\ e^{-2\xi}} < k\ \frac{1+K}{1-K} \leqq 1 \tag{3.19}$$

which is a contradiction. Thus (3.17) has no solution with $\mathrm{Re}\ \omega > 0$, for any $\varepsilon > 0$, provided $k \leqq (1-K)/(1+K)$. Also, if $k < (1-K)/(1+K)$ the string of inequalities in (3.19) again leads to a contradiction whenever $\xi = \mathrm{Re}\ \omega \geqq 0$, so that all zeros of (3.17) must satisfy $\mathrm{Re}\ \omega < 0$ when $k < (1-K)/(1+K)$. Moreover, in this case it is not possible for a sequence of such zeros to accumulate at the imaginary axis. For suppose there were a sequence $\omega_n = \xi_n + i\eta_n$ of zeros such that $\lim \xi_n = 0$, $\underline{\lim}|\eta_n| > 0$. From (3.18)

$$1 \leqq \overline{\lim}\ k\left\{\mathrm{Re}\left[e^{-\varepsilon\omega_n}\frac{K\ e^{-2\omega_n}-1}{K\ e^{-2\omega_n}+1}\right] - \frac{a\xi_n}{\xi_n^2+\eta_n^2}\right\}$$

$$\leqq \lim\ k\ \frac{1+K\ e^{-2\xi_n}}{1-K\ e^{-2\xi_n}} = k\ \frac{1+K}{1-K} < 1\ ,$$

a contradiction. Thus for $\varepsilon > 0$ and $k < (1-K)/(1+K)$, the spectrum of (3.1)-(3.5) must lie in a half-plane $\mathrm{Re}\ \omega \leqq -\beta$, $\beta > 0$.

(ii) From the proof of (i), if $k = (1-K)/(1+K)$ and $\varepsilon > 0$ the spectrum lies in $\mathrm{Re}\ \omega \leqq 0$. If (3.17) has a zero $\omega = i\eta$, $\eta \neq 0$, then from (3.18)

$$1 + \frac{a}{i\eta} = k\ e^{-i\varepsilon\eta}\left(\frac{K\ e^{-2i\eta}-1}{K\ e^{-2i\eta}+1}\right) . \tag{3.20}$$

Taking the modulus of each side of (3.20) results in the contradiction

$$1 + \frac{a^2}{\eta^2} = k^2\ \frac{1+K^2-2K\ \cos\ 2\eta}{1+K^2+2K\ \cos\ 2\eta} \leqq k^2\left(\frac{1+K}{1-K}\right)^2 = 1\ .$$

However, it is easily seen that

$$f\left(\frac{2(2m+1)}{2n+1}\ ,\ \frac{(2n+1)\pi i}{2}\right) = 0$$

where f is defined in (3.10) and m,n are arbitrary. Thus, by Lemma 1, given any $\delta > 0$ the vertical strip $\{\omega\colon -\delta < \mathrm{Re}\ \omega < 0\}$ contains an infinite number of points of the spectrum of (3.1)-(3.5).

(iii) Let $k > (1-K)/(1+K)$. By Lemmas 3.1 and 3.2 it follows that for each such k there is an open dense set, D, in $(0,\infty)$ such that (3.17) is satisfied for each ε in D and some ω with $\mathrm{Re}\ \omega > 0$. This completes the proof.

The next example is of an Euler-Bernoulli beam equation which is stabilized by a boundary feedback and then destabilized by small time delays in the feedback. The undelayed system is the example presented by G. Chen et al in these proceedings with certain coefficients normalized to one.

Example 3.2. Consider the system

$$u_{tt} + u_{xxxx} = 0 \ , \ 0 < x < 1 \ , \ t > 0 \ , \tag{3.21}$$

$$u(0,t) = u_x(0,t) = u_{xxx}(1,t) = 0 \tag{3.22}$$

$$u_{xx}(1,t) = -u_{xt}(1,t-\varepsilon) \ , \text{ where } \ \varepsilon \geq 0 \ \text{ is fixed.} \tag{3.23}$$

The initial values of u and u_t are

$$u(x,t) = \phi(x,t) \ , \tag{3.24}$$

$$u_t(x,t) = \psi(x,t) \ , \tag{3.25}$$

for $0 \leq x \leq 1$, $-\varepsilon \leq t \leq 0$, and for simplicity we assume as in Example 3.1 that ϕ and ψ are infinitely differentiable in (x,t) .

The Laplace transform of (3.21) with the initial conditions (3.24) and (3.25) satisfies the two parameter family of ordinary differential equations in x

$$s^2 U(x,\omega,\varepsilon) + \frac{d^4 U}{dx^4}(x,\omega,\varepsilon) = \omega\phi(x,0)+\psi(x,0) \ . \tag{3.26}$$

If we let

$$\sqrt{\omega}\, e^{\frac{\pi}{4}i} = \tau \ , \ \omega = -i\tau^2 \tag{3.27}$$

and apply the first two conditions in (3.22) we can write the solutions of (3.26) in the form

$$U(x,\tau,\varepsilon) = A(\cosh \tau x - \cos \tau x)+B(\sinh \tau x-\sin \tau x) \tag{3.28}$$

$$+ \frac{1}{2\tau^3} \int_0^x [\sinh \tau(x-\sigma)-\sin\tau(x-\sigma)]\,[-i\tau^2\phi(\sigma)+\psi(\sigma)]\,d\sigma$$

$$= A(\cosh \tau x-\cos \tau x)+B(\sinh \tau x-\sin\tau x)+U_p(x,\tau) \ .$$

The following statement is obvious but will be needed below so we state it as a proposition.

Proposition 3.1. (i) The expression $U_p(x,\tau)$ (implicitly defined in (3.28)) is an entire function in τ .

(ii) The term $(\cosh \tau x - \cos \tau x)$ in (3.28) has a zero of order two at $\tau = 0$.
(iii) The term $(\sinh \tau x - \sin \tau x)$ in (3.28) has a zero of order three at $\tau = 0$.

To obtain the coefficients A and B in (3.28) we apply the last boundary condition in (3.22) and the boundary condition (3.23) via their Laplace transforms. The first condition leads to the equation

$$\tau^3[A(\sinh \tau - \sin \tau) + B(\cosh \tau + \cos \tau)] = \tag{3.29}$$

$$-\frac{1}{2}\int_0^1 (\cosh \tau(1-\sigma)+\cos \tau(1-\sigma))(-i\tau^2\phi(\sigma)+\psi(\sigma))d\sigma .$$

The second condition results in the equation

$$\tau^2[A(\cosh \tau + \cos \tau) + B(\sinh \tau + \sin \tau)] + \tag{3.30}$$

$$\frac{1}{2\tau}\int_0^1(\sinh \tau(1-\sigma)+\sin \tau(1-\sigma))(-i\tau^2\phi(\sigma)+\psi(\sigma))d\sigma$$

$$= i\tau^3 + e^{i\varepsilon\tau^2}[A(\sinh \tau + \sin \tau) + B(\cosh \tau - \cos \tau)]$$

$$+ i\tau^2 \frac{dU_p}{dx}(1,\tau) + \phi_x(1,-\varepsilon)+i\tau^2 e^{i\varepsilon\tau^2}\int_{-\varepsilon}^0 \phi_x(\sigma,1)e^{i\tau^2\sigma}d\sigma .$$

Dividing (3.29) by τ^3 and (3.30) by τ^2 and rearranging terms we obtain the following equations for the solution of A and B.

$$A(\sinh \tau - \sin \tau) + B(\cosh \tau + \cos \tau) = \frac{1}{\tau^3} f_1(\tau) , \tag{3.31}$$

$$A[(\cosh \tau + \cos \tau) - i\tau e^{i\varepsilon\tau^2}(\sinh \tau + \sin \tau)] + \tag{3.32}$$

$$B[(\sinh \tau + \sin \tau) - i\tau e^{i\varepsilon\tau^2}(\cosh \tau - \cos \tau)]$$

$$= g(\tau) + \frac{1}{\tau^2} f_2(\tau) ,$$

where f_1, f_2 and g are entire functions of τ. The matrix equation associated with (3.31) and (3.32) has the structure

$$\begin{bmatrix} (\sinh\tau - \sin\tau) & (\cosh\tau + \cos\tau) \\ (\cosh\tau + \cos\tau) + \tau q_1(\tau) & (\sinh\tau + \sin\tau) + \tau q_2(\tau) \end{bmatrix} \begin{pmatrix} A \\ B \end{pmatrix}$$

$$= \begin{pmatrix} \frac{1}{\tau^3} f_1(\tau) \\ g(\tau) + \frac{1}{\tau^2} f_2(\tau) \end{pmatrix}, \quad = R(\tau) \begin{pmatrix} A \\ B \end{pmatrix} \tag{3.33}$$

where $q_1(\tau)$ and $q_2(\tau)$ are entire functions of τ. Notice that at $\tau = 0$ $\det R(0) \neq 0$. The inversion of (3.33) leads to the following equations for A and B.

$$A(\tau) = \frac{(\sinh\tau + \sin\tau + \tau q_2(\tau))}{\det R(\tau)} \frac{f_1(\tau)}{\tau^3} - \frac{(\cosh\tau + \cos\tau)\left(g(\tau) + \frac{f_2(\tau)}{\tau^2}\right)}{\det R(\tau)}; \tag{3.34}$$

$$B(\tau) = \frac{-(\cosh\tau + \cos\tau + \tau q_1(\tau))}{\det R(\tau)} \frac{f_1(\tau)}{\tau^3} + \frac{(\sinh\tau - \sin\tau)}{\det R(\tau)} \left(g(\tau) + \frac{f_2(\tau)}{\tau^2}\right). \tag{3.35}$$

Thus (3.34) and (3.35) show that $A(\tau)$ has a pole of order at most two and $B(\tau)$ a pole of order at most three at $\tau = 0$. Hence by Proposition 3.1 it follows that $U(x,\tau,\varepsilon)$ has no poles at $\tau = 0$, and we can state the following proposition.

Proposition 3.2. The poles of $U(x,\tau,\varepsilon)$ are determined by the zeros of $\det R(\tau)$, i.e. by

$$F(\tau,\varepsilon) = -\frac{1}{2}\det R(\tau) = (1 + \cosh\tau\cos\tau) - i\tau e^{i\varepsilon\tau^2}(\sinh\tau\cos\tau + \cosh\tau\sin\tau) = 0. \tag{3.36}$$

Proof. By the discussion preceding the statement of Proposition 3.2 we know that $\tau = 0$ is not a pole of $U(x,\tau,\varepsilon)$ and by Proposition 3.1 that $U_p(x,\tau)$ is entire.

Thus the only poles of (3.28) i.e. of $U(x,\tau,\varepsilon)$ that can occur are in the coefficients $A(\tau)(\cosh \tau x - \cos \tau x)$ and $B(\tau)(\sinh \tau x - \sin \tau x)$. By (3.34) and (3.35) these occur where $\det R(\tau) = 0$, which proves the proposition.

Since Chen et al have proven that when $\varepsilon = 0$ the system (3.21)-(3.25) is uniformly exponentially stable we know that when we replace τ in (3.36) by $\tau = e^{\frac{\pi}{4}i}\sqrt{\omega}$ and set $\varepsilon = 0$ the resulting equation has all its roots in $\mathrm{Re}\,\omega < -\alpha$, $\alpha > 0$, for some α.

Proposition 3.3. There exist $\omega = -i\omega_n$, $\omega_n \to \infty$ and $\varepsilon_n > 0$, $\varepsilon_n \to 0$, which satisfy equation (3.36) when ω is substituted for τ via the relations (3.27).

Proof. Let $n = 1,2,\ldots$, and consider the interval $I_n = (2n\pi - \frac{\pi}{2}, 2n\pi - \frac{\pi}{4})$. Choose ε_n such that $\tau^2\varepsilon_n = \frac{\pi}{2}$. Then the left hand side of (3.36) is

$$F(\tau,\frac{\pi}{2\tau^2}) = (1+\cosh \tau \cos \tau) + \tau(\sinh \tau \cos \tau + \cosh \tau \sin \tau). \qquad (3.37)$$

Notice that the value of (3.37) at the right hand end point of I_n satisfies the condition

$$1+\frac{\sqrt{2}}{2}\cosh(2n\pi-\frac{\pi}{4})+(2n\pi-\frac{\pi}{4})\frac{\sqrt{2}}{2}(\sinh(2n\pi-\frac{\pi}{4})-\cosh(2n\pi-\frac{\pi}{4})) > 0$$

for n sufficiently large, whereas the left hand endpoint of (3.37) satisfies

$$1 - (2n\pi-\frac{\pi}{2})\cosh(2n\pi-\frac{\pi}{2}) < 0$$

for all n. Thus for n sufficiently large we can find a $\tau_n \in I_n$ for which (3.37) is zero, and when $\varepsilon_n = \frac{\pi}{2\tau_n^2}$ and $\tau = \tau_n$ equation (3.36) is satisfied. The corresponding value of ω is from (3.27) $\omega_n = -i\tau_n^2$ which proves the proposition.

We proved in Proposition 3.3 that there exists

$$\tau_n^2 = -i\omega_n \text{ and } \varepsilon_n = \frac{\pi}{2\tau_n^2} \text{ with } \tau_n \to \infty \qquad (3.38)$$

such that (3.36) is satisfied for these values of $-i\omega_n$ and ε_n. This is equivalent to stating that for these values of ε_n the spectrum of (3.21)-(3.25) has points on the imaginary axis. We shall now show that there exist $\varepsilon_n \to 0$ such that the spectrum of (3.21)-(3.25) has points in the right half plane. First notice that (3.36) is actually a function of ω and ε, because of the relations (3.27), which implicitly define function $\omega(\varepsilon)$ or $\varepsilon(\omega)$. We seek to compute their derivatives at the points which satisfy Proposition 3.3. The partial derivatives of F are:

$$\frac{\partial F}{\partial \omega} = \frac{\partial F}{\partial \tau}\frac{\partial \tau}{\partial \omega} = \frac{i}{2\tau}\frac{\partial F}{\partial \tau} = \frac{i}{2\tau}[(\sinh \tau \cos \tau - \cosh \tau \sin \tau)$$

$$- ie^{i\varepsilon\tau^2}(\sinh \tau \cos \tau + \cosh \tau \sin \tau) +$$

$$2\tau^2 \varepsilon e^{i\varepsilon\tau^2}(\sinh \tau \cos \tau + \cosh \tau \sin \tau)$$

$$-2i\, \tau e^{i\varepsilon\tau^2} \cosh \tau \cos \tau],$$

$$\frac{\partial F}{\partial \varepsilon} = \tau^3 e^{i\tau^2}(\sinh \tau \cos \tau + \cosh \tau \sin \tau) .$$

Evaluating $\frac{\partial F}{\partial \omega}$ and $\frac{\partial F}{\partial \varepsilon}$ at the points obtained in Proposition 3.3 and noting that $e^{i\varepsilon_n\tau_n^2} = i$ we obtain the following expression

$$\left.\frac{\partial F}{\partial \omega} \Big/ \frac{\partial F}{\partial \varepsilon}\right|_{\substack{\varepsilon=\varepsilon_n, \\ \omega=-i\omega_n}} = \frac{1}{\tau_n^4}\,\frac{\sinh\tau_n \cos \tau_n + \tau_n \cosh \tau_n \cos \tau_n}{\sinh\tau_n \cos \tau_n + \cosh \tau_n \sin\tau_n} + i\, q(\tau_n) , \tag{3.39}$$

where $q(\tau_n)$ is real. The denominator in the first term on the right in (3.39) cannot be zero for large values of n, because (3.36) must also be satisfied, this would imply that $|\cos \tau_n| \cong \frac{1}{\sqrt{2}}$ and $1 + \cosh \tau_n \cos \tau_n = 0$ which is impossible for n sufficiently large. Similarly the numerator of the first term on the right in (3.39) cannot be zero, since this would imply $\cos \tau_n = 0$, which is impossible if (3.36) is satisfied.

Since (3.39) has the form $a + ib$, where $a \neq 0$, and since $\frac{-1}{a+ib} = \frac{-(a-ib)}{a^2+b^2}$ this implies that

$$\mathrm{Re}\left(\frac{d\omega}{d\varepsilon}\right)\Bigg|_{\varepsilon=\varepsilon_n\,,\ \omega=-i\omega_n^2} \neq 0 . \tag{3.40}$$

Hence for ε "near" ε_n (3.36) has solutions in $\mathrm{Re}\,\omega > 0$. Thus we can state the following result.

Theorem 3.2. The system (3.21)-(3.25) is exponentially stable for $\varepsilon = 0$. However there exist values $\varepsilon_n \to 0$ for which the resulting system has its spectrum in $\mathrm{Re}\,\omega > 0$, i.e., is unstable.

Proof. The proof follows from Proposition 3.2, which essentially states that the spectrum of (3.21)-(3.25) is determined by (3.36), and the relation (3.40), which states that for some sufficiently small values of $\varepsilon \to 0$ equation (3.36) has solutions in $\mathrm{Re}\,\omega > 0$.

We shall briefly indicate that the two examples in this section are by no means pathological. That in fact they conform to a well established pattern in the feedback stabilization of hyperbolic partial differential equations. Both examples are of the abstract form

$$C\frac{d^2y}{dt^2} + Ay = 0$$

when the boundary conditions are homogeneous, where C and A are positive operators in some Hilbert space H . The stabilizing boundary control results in a system of the abstract form

$$C\frac{d^2y}{dt^2} + B\frac{dy}{dt} + Ay = 0 ,$$

which is uniformly exponentially stable. If we introduce a delay into the feedback, the system assume the abstract form

$$\frac{C d^2 y(t)}{dt^2} + \frac{B dy(t-\varepsilon)}{dt} + Ay(t) = 0 \qquad (3.41)$$

(we need the $y(t)$, $y(t-\varepsilon)$ arguments now). The point spectrum of (3.41) is obtained by solving an equation of the form

$$(\omega^2 C + \omega e^{-\omega\varepsilon} B + A)\,\phi = 0 . \qquad (3.42)$$

We now let $\omega\varepsilon = \frac{i\pi}{2}$, $\varepsilon > 0$ and look for solutions of the equation

$$\left(-\frac{\pi^2}{4}\frac{1}{\varepsilon^2} C + \frac{\pi}{2\varepsilon} B + A\right)\phi_\varepsilon = 0 . \qquad (3.43)$$

Since C,B and A are positive operators and A is usually unbounded (3.43) often has an infinite number of solutions for $\varepsilon > 0$ with $\varepsilon \to 0$. For a simple example of this phenomena see [3 , pg. 160].

REFERENCES

[1] M. Athans and P.L. Falb, Optimal Control, An Introduction to the Theory and Its Applications, McGraw-Hill, New York, 1966.

[2] A.G. Butkovskiy, Distributed Control Systems, American Elsevier, New York, 1969.

[3] R. Datko, Representation of solutions and stability of linear differential-difference equations in a Banach space, J. Differential Equations, 29 (1978), pp. 105-166.

[4] ________, A procedure for determination of the exponential stability of certain differential-difference equations, Quart. Appl. Math., 36 (1978), pp. 279-292.

[5] G. Doetsch, Handbuch der Laplace Transformation, Vol. III, Birkhaüser Verlag, Basel, 1956.

[6] S. Saks and A. Zygmund, Analytic Functions, Elsevier Publishing Company, Amsterdam, 1971.

Optimal Control of Nonlinear Systems: Convergence of Suboptimal Controls, I

H. O. Fattorini [1]

University of California

Department of Mathematics

Los Angeles, California 90024

INTRODUCTION. We consider optimal problems with set targets for general nonlinear nonconvex input-output systems in Hilbert spaces. Using Ekeland's variational principle, we obtain an approximate maximum principle (called here the convergence principle) for suboptimal controls. We treat in detail applications to distributed parameter systems described by quasilinear parabolic equations, where, in certain situations, the convergence principle implies convergence of sequences of suboptimal controls in L^p norms, $1 \leq p < \infty$.

1. SYSTEMS. We denote by E, F arbitrary Banach spaces; U is a subset of F called the *control set*. Given $k \geq 0, T > 0$, the *control space* $W(-k,T; U)$ is

[1] This work was supported in part by the National Science Foundation under grant DMS 82-00645.

the set of all (equivalence classes of) strongly measurable F-valued functions $u = u(\hat{t})$ defined in $-k \leq t \leq T$ such that

$$u(t) \in U \text{ a.e.}$$

(We indicate here and below by $u(\hat{t})$ the function $t \to u(t)$; $u(t)$ is the value at t. The same convention applies to other functions.) The control space $W(-k,T;U)$ is a complete metric space equipped with the distance

$$d(u,v) = \text{meas}\,\{t;\, u(t) \neq v(t)\}\,,$$

which we shall call the *Ekeland distance*.

The *trajectory* or *output* space $C(0,T;E)$ consists of all E-valued continuous functions $y(\hat{t})$ defined in $0 \leq t \leq T$.

A *system* is, by definition, a map

$$X : W(-k,T;U) \to C(0,T;E)$$

that satisfies the three postulates (a), (b) and (c) below:

(a) *Causality*. Let $0 \leq \bar{t} \leq T$. Then the values of the *trajectory*

$$y(t,u) = (Xu)(t) \tag{1.1}$$

in $0 \leq t \leq \bar{t}$ depend only on the values of u in $0 \leq t \leq \bar{t}$.

(b) *Pointwise continuity*. For $\bar{t}$ as in (a), the map

$$u(\hat{t}) \to y(\bar{t},u) \tag{1.2}$$

from $W(-k,\bar{t};U)$ (endowed with the Ekeland distance) into E (endowed with its original norm) is continuous.

(c) *Differentiability* (with respect to spike perturbations). For every $u(\hat{t}) \in W(-k,\bar{t};U)$ there exists a set $e = e(u)$ of full measure in $0 \leq t \leq \bar{t}$ such that, if $s \in e$, the limit

$$\xi(\bar{t},s,u,v,u(s)) = \lim_{\rho\to 0+} \rho^{-1} (y(\bar{t},u_{s,\rho,v}) - y(\bar{t},u)) \tag{1.3}$$

exists; here $u_{s,\rho,v}(\hat{t})$ is the familiar *spike perturbation* of the control $u(\hat{t})$ defined by $u_{s,\rho,v}(t) = v$ in $s - \rho < t \le \rho$, $u_{s,\rho,v}(t) = u(t)$ elsewhere.

In this generality, systems provide a common framework to describe the input-output relationship generated by different types of nonlinear equations (ordinary differential equations, partial differential equations with boundary or distributed control, functional differential equations), thus results on optimal problems for systems can be applied to all these cases. (The constant k in $W(-k,\bar{t}; U)$ accounts for delays in control action; for details and examples see [13]. In the examples in the present paper, where the state equation is purely differential, $k = 0$).

This approach (which is a generalization of that of Ekeland and Clarke [6], [5] for systems described by ordinary differential equations) has been initiated in [11] and continued in [12], [13], under the postulates above, with (c) slightly reinforced to include continuity of ξ. The main result in these papers is a very general version of Pontryagin's maximum principle that yields as particular cases all the different formulations of the maximum principle available for particular problems.

We retake here the study of optimal problems for general systems, with a view to obtaining convergence results for *suboptimal* (i.e. close to optimal) controls. We consider the *set target* problem, that is, the case where trajectories are required to hit (actually, to come close to) a "sufficiently large" set, in a sense to be precised in §2. The main results (which apply to systems in full generality) are the *convergence principles* in Sections 3 and 4, the first for general optimal problems, the second more precise but restricted to the time optimal problem. These convergence principles apply especially well to quasilinear equations where the solution operator of the linear part is *compact*, thus one of the natural applications is to controlled quasilinear parabolic equations. We collect all the necessary results on quasilinear equations in Banach spaces in Sections 5 and 6 and study the implications of the convergence principle in Sections 7 and 8; the specific applications to quasilinear parabolic distributed parameter systems are in §9 and §10. Compactness of solution operators of course excludes systems described by nonlinear hyperbolic equations; these are considered in [14] using somewhat different technical means.

The *point target* problem is somewhat more involved and has been considered in [15].

We note that convergence results for suboptimal controls using Ekeland's principle have been obtained in [18] in the particular case of a controlled quasilinear hyperbolic equation.

We point out finally that approximation schemes for optimal controls in distributed parameter systems have been proved by different methods (see [3], [4] and bibliography therein). The author is grateful to Professor V. Barbu for kindly providing a pre-publication version of [4].

2. OPTIMAL CONTROL PROBLEMS. We consider in this section a system $X : W(-k,T; U) \to C(0,T; E)$ in the sense of the Introduction, and a second system $X^0 : W(-k, T; U) \to C(0,T; \mathbf{R})$ ($\mathbf{R}$ the real numbers) called the *cost functional*. We denote by $\xi^0(t,s,u,v,w)$ the function in (1.3) corresponding to the system X^0. The *augmented system* $\tilde{X}$ is defined by

$$(\tilde{X}u)(t) = ((X^0,X)u)(t) = (y^0(t,u),y(t,u)) , \tag{2.1}$$

where $y^0(t,u) = (X^0u)(t)$. We also use the notation $\tilde{y}(t,u) = (\tilde{X}u)(t)$ for the trajectories of the augmented system in $\mathbf{R} \times E$; in the same fashion, we write $\tilde{\xi}(t,s,u,v,w) = (\xi^0(t,s,u,v,w), \xi(t,s,u,v,w))$.

Let Y be a subset of E (called the *target set*). We consider the *optimal control problem* of identifying the times $\bar{t}$ and the controls $\bar{u}$ in $W(-k, \bar{t}; U)$ such that

$$y(\bar{t},\bar{u}) \in Y , \tag{2.2}$$

and

$$y^0(\bar{t},\bar{u}) = m = \inf y^0(t,u) \, , \tag{2.3}$$

the infimum taken over all times $t > 0$ and all controls $u(\hat{t})$ in $W(-k, t; U)$ such that $y(t,u) \in Y$; we assume that

$$-\infty < m < \infty \, . \tag{2.4}$$

A sufficient condition for the first inequality (2.4) to hold is that the cost functional $y^0(t,u)$ be *bounded below* in $W(-k,T; U)$. The second inequality simply means that there exists a control u such that $y(t,u)$ hits Y at some time.

In case $Y = \{\bar{y}\}$, (2.2) becomes

$$y(\bar{t},\bar{u}) = \bar{y} \tag{2.5}$$

and we speak of the *point target* problem; the general case is the *set target* problem.

In the point target problem, a control $u(\hat{t})$ in $W(-k,\bar{t}; U)$ is called $(\bar{t},\epsilon)$-*suboptimal* if

$$\|y(\bar{t},u) - \bar{y}\| \le \epsilon, \quad y_0(\bar{t},u) \le m + \epsilon \, . \tag{2.6}$$

This definition is generalized in an obvious way in the set target case: a control $u(\hat{t}) \in W(-k,\bar{t}; U)$ is $(\bar{t},\epsilon)$- *suboptimal* if

$$\operatorname{dist}(y(\bar{t},u),Y) \le \epsilon, \quad y_0(\bar{t},u) \le m + \epsilon \tag{2.7}$$

where, as usual, $\operatorname{dist}(y,Z) = \inf\{\|y - z\| \, ; z \in Z\}$.

We shall assume throughout that the target set Y is closed, so that it satisfies the *saturation condition*

$$Y = \bigcap_{\epsilon>0} \{y; \operatorname{dist}(y,Y) \le \epsilon\} \, . \tag{2.8}$$

3. THE CONVERGENCE PRINCIPLE. Let $\{u^n\}$ be a sequence of controls, each in a different space $W(-k,t_n; U)$. Assume that $t_n \to t_0$. We say that $\{u^n\}$ *converges weakly in* $L^p(-k,t_0; F)$ *to* $u \in L^p(-k,t_0; F)$ if u^n, extended to $t > t_n$ (if $t_n < t_0$) by setting $u^n(t) = 0$ there, or chopped off at t_0 (if $t_n > t_0$) converges to u weakly in $L^p(-k,t_0; U)$.

We shall assume from now on that the system X satisfies assumption (d) below:

(d) Let $\{u^n\}$ be a sequence of controls, $u^n \in W(-k,t_n; U)$, such that $t_n \to t_0$. Then there exists a subsequence of $\{u^n\}$ (which we denote by the same symbol) and a $\bar{u}$ in $W(-k,t_0; U)$ such that

$$u^n(\hat{t}) \to \bar{u}(\hat{t}) \text{ weakly in } L^2(-k,t_0; U) , \tag{3.1}$$

$$y(t_n,u^n) \to y(t_0,\bar{u}) \text{ strongly in } E . \tag{3.2}$$

We can replace L^2 by L^p in (3.1) for any p, $1 < p < \infty$. In the examples examined later in this paper (see §7 and §8) where control appears linearly, Assumption (d) essentially requires *convexity* of the control set U. However, the nonconvex case, where (d) may not hold, is also amenable to the theory; see Remark 7.4.

Finally, we require y^0 to be *lower semicontinuous*, that is,

$$y^0(t_0, u) \leq \lim\inf y^0(t_n, u^n)$$

for any numerical sequence $\{t_n\}$ and any sequence of controls $\{u^n\}$ satisfying the assumptions in (d) and (3.1).

THEOREM 3.1. *Let Y be closed and let y^0 be bounded below. Assume a sequence $\{u^n\}$ of (t_n,ϵ_n)- suboptimal controls with $\{t_n\}$ bounded and $\epsilon_n \to 0$ exists. Then there exists an optimal control $\bar{u}$.*

The proof is immediate; selecting a subsequence, we may assume that the t_n are convergent, which allows us to use Assumption (d). Note that we *must* have

$$y^0(t_n, u^n) \to m , \tag{3.3}$$

where m is the infimum in (2.3); otherwise, the optimal control $\bar{u}$ just constructed would satisfy $y_0(\bar{t},\bar{u}) < m$, which is impossible.

Let Y be an arbitrary set in a Hilbert space H. We say that a point $\bar{y} \in Y$ satisfies the *θ-cone condition* $(0 \leq \theta \leq \pi/2)$ if and only if Y contains a cone $\Gamma = \Gamma(\bar{y},z,\theta,b)$ of summit $\bar{y}$, aperture θ and height $b > 0$, defined as the set of all y such that

$$\|y - \bar{y}\| \cos\theta \leq \langle y - \bar{y}, z\rangle \leq b \, .$$

We assume that $\|z\| = 1$ and call z the *generator* of Γ. We say that the set Y satisfies the *fat cone condition* if every $\bar{y}$ *in* Y *satisfies the* θ*-cone condition with* $\theta > \pi/4$.

THEOREM 3.2. *(The convergence principle) Assume that the target set* Y *satisfies the saturation condition (2.8) and the fat cone condition. Let* $\{u^n\}$ *be an arbitrary sequence of* (t_n,ϵ_n)*- suboptimal controls with* $\{t_n\}$ *bounded,* $\epsilon_n \to 0$. *Then there exists a subsequence of* $\{u^n\}$ *(which we denote by the same symbol) such that*

$$t_n \to \bar{t} \, , \tag{3.4}$$

an optimal control $\bar{u} \in W(-k,\bar{t}; U)$ *with*

$$u^n(\hat{t}) \to \bar{u}(\hat{t}) \ \ \textit{weakly in} \ \ L^2(-k,t_0; U) \, , \tag{3.5}$$

$$y(t_n,u^n) \to y(\bar{t},\bar{u}) \in Y \ \ \textit{strongly in} \ \ E \, , \tag{3.6}$$

a second sequence $\{\tilde{u}^n\}$ *of controls with* $\tilde{u}^n \in W(-k,t_n; U)$ *such that*

$$d_n(u^n,\tilde{u}^n) \to 0 \, , \tag{3.7}$$

(d_n *denotes the distance in* $W(-k,t_n; U)$*), a sequence* $\{\tilde{y}^n\} = \{(\mu_n,y^n)\}$ *in* $\mathbf{R} \times E$ *such that* $\|\tilde{y}^n\| = 1$,

$$\tilde{y}^n \to \tilde{y} \neq 0 \ \ \textit{weakly in} \ \ \mathbf{R} \times E \, , \tag{3.8}$$

and a set e *of full measure in* $0 \leq s \leq \bar{t}$ *such that*

$$\mu_n\xi^0(t_n,s,\tilde{u}^n,v,\tilde{u}^n(s)) + \langle y^n,\xi(t_n,s,\tilde{u}^n,v,\tilde{u}^n(s))\rangle =$$

$$\langle \tilde{y}^n,\tilde{\xi}(t_n,s,\tilde{u}^n,v,\tilde{u}^n(s))\rangle \geq -\,\delta_n \to 0 \quad (v \in U, \, s \in e \cap (0,t_n)) \, . \tag{3.9}$$

Proof: Obviously, (3.5) and (3.6) can be obtained from Assumption (d) passing to a subsequence and using Theorem 3.1; if necessary taking again a subsequence, we may also assume that, if

$$\bar{y} = y(\bar{t},\bar{u}) ,$$

then

$$\|y(t_n,u^n) - \bar{y}\| \le \epsilon_n . \tag{3.10}$$

Observe next that, since the target set Y satisfies the fat cone condition, there exists an angle $\theta > \pi/4$, a number $b > 0$, and a vector z, $\|z\| = 1$ such that the cone $\Gamma = \Gamma(\bar{y},z,\theta,b)$ belongs to Y.

Let $\gamma > 0$. Define

$$\kappa_n = \epsilon_n/\gamma . \tag{3.11}$$

Since $\kappa_n \to 0$ we may assume that $\kappa_n \le b/3$. Let

$$\bar{y}_n = \bar{y} + \kappa_n z , \tag{3.12}$$

$$F_n(u) = \left\{(y^0(t_n,u) - (m - \epsilon_n))^2 + \|y(t_n,u) - \bar{y}_n\|^2\right\}^{1/2} = \|\tilde{y}(t_n,u) - \tilde{y}_n\| ,$$

(where $\tilde{y}_n = ((m - \epsilon_n), \bar{y}_n) \in \mathbf{R} \times E$) in the space $W(-k,t_n; U)$. Plainly, we have $F_n(u) > 0$ (otherwise we could hit the target Y with value $m - \epsilon_n$ of the cost functional y^0, which is impossible). Moreover, F_n is continuous. Note that, in view of (3.3), we may assume (passing if necessary to a subsequence) that, if u^n is an element of the sequence of controls in the statement of Theorem 3.2, we have

$$|y^0(t_n,u^n) - m| < \epsilon_n . \tag{3.13}$$

We have

$$F_n(u^n) = \left\{(y^0(t_n,u^n) - (m - \epsilon_n))^2 + \|y(t_n,u^n) - \bar{y}_n\|^2\right\}^{1/2} \le$$

$$|(y^0(t_n,u^n) - (m - \epsilon_n))| + \|y(t_n,u^n) - \bar{y}\| + \|\bar{y} - \bar{y}_n\| \le$$

$$2\epsilon_n + \epsilon_n + \kappa_n = (1 + 3\gamma)\kappa_n , \tag{3.14}$$

by virtue of (3.10). We apply Ekeland's variational principle [6, p. 99] to the function $F_n(u)$ at the approximate minimum u^n and deduce in this way the existence of another control $\tilde{u}^n \in W(-k,t_n; U)$ such that

$$F_n(\tilde{u}^n) = \|\tilde{y}(t_n,\tilde{u}^n) - \tilde{y}_n\| =$$

$$= \left\{(y^0(t_n,\tilde{u}^n) - (m - \epsilon_n))^2 + \|y(t_n,\tilde{u}^n) - \bar{y}_n\|^2\right\}^{1/2} \le$$

$$\le F_n(u^n) \le (1 + 3\gamma)\kappa_n , \tag{3.15}$$

$$d_n(u^n,\tilde{u}^n) \le (1 + 3\gamma)^{1/2}k_n^{1/2} , \tag{3.16}$$

$$F_n(w) \ge F_n(\tilde{u}^n) - (1 + 3\gamma)^{1/2}\kappa_n^{1/2}d_n(w,\tilde{u}^n) \quad (w \in W(-k,t_n; U)) . \tag{3.17}$$

We use (3.17) for needle perturbations $\tilde{u}^n_{s,\rho,y}$ of $\tilde{u}^n(t)$, taking advantage of postulate (c) (see [12], [13] for further details). The result is (3.9), with $\tilde{y}^n = (\mu_n,y^n) = (\lambda_n,x^n)/\|(\lambda_n,x^n)\|$,

$$(\lambda_n,x^n) = \tilde{y}(t_n,\tilde{u}^n) - \tilde{y}_n = \left(y^0(t_n,\tilde{u}^n) - (m - \epsilon_n), y(t_n,\tilde{u}^n) - \bar{y}_n)\right) , \tag{3.18}$$

thus we only have to show (3.8); since $\|\tilde{y}^n\| = 1$, it is enough to prove (after selecting if necessary a subsequence) that $\mu_n \ge \delta > 0$ or that, in case $\mu_n \to 0$, y^n (or its nonnormalized companion x^n) stays in a fixed cone of summit 0 and aperture $\theta_\gamma < \pi/2$. This is handled as follows. For each n, either

(a) $y(t_n,\tilde{u}^n)$ belongs to Γ, or

(b) $y(t_n,\tilde{u}^n)$ does not belong to Γ.

If (a) takes place, $y(t_n,\tilde{u}^n) \in Y$ and

$$y^0(t_n,\bar{u}^n) \geq m$$

(otherwise, Y could be hit with value $< m$ of the cost functional y^0). In view of (3.15) we have

$$\|(\lambda_n,x^n)\| = (\lambda_n + \|x^n\|^2)^{1/2} \leq (1 + 3\gamma)\kappa_n ,$$

so that

$$\mu_n \geq \gamma\kappa_n/(1 + 3\gamma)\kappa_n \to \gamma/(1 + 3\gamma) > 0 .$$

On the other hand, it is geometrically clear that, since $\theta > \pi/4$, for γ sufficiently small, (b) will imply that x^n belongs to a cone $\Gamma(0,z,\theta_\gamma,\infty)$ with $\theta_\gamma < \pi/2$ (precisely, with

$$\theta_\gamma = \pi - \theta \text{ arc sin}((1 + 3\gamma)^{-1/2} \sin \theta)$$

as resolution of triangles in the two dimensional subspace generated by y^n and z shows: see again [12], [13]). Obviously, y^n belongs to the same cone.

If (a) happens for infinitely many n's, we obtain, passing to a subsequence, that $\mu_n \to \mu \neq 0$, so that (3.8) holds. Otherwise, (b) will occur an infinite number of times and, taking a subsequence, either $\|y_n\| \to 0$ (in which case, $\mu_n \to 1$ and we obtain again (3.8)) or $\|y^n\| \geq \eta > 0$; since $y^n \in \Gamma(0,z,\theta_\gamma,\infty)$, no subsequence of $\{y^n\}$ may converge weakly to zero. This ends the proof.

4. THE CONVERGENCE PRINCIPLE FOR THE TIME OPTIMAL PROBLEM. As usual, results for general optimal problems do not necessary yield nontrivial information in the time optimal case: in fact, in this case we have $y^0(t,u) = t$, so that $\xi^0(t,s,u,v,w) = 0$. This renders the convergence principle in §3 empty in the case where $y^n \to 0$ weakly. Thus, a different formulation and a different argument are needed.

We note in passing that Theorem 3.1 holds without changes. Of course, the assumption that $\{t_n\}$ is bounded is automatically satisfied. In particular, existence of the sequence $\{u^n\}$ is assured if there exists a single trajectory $y(t,u)$ that hits Y at any time $t \geq 0$.

THEOREM 4.1 *(The convergence principle for the time optimal problem). Assume the target set Y satisfies the saturation condition (2.8) and the fat cone condition. Let $\{u^n\}$ be an arbitrary sequence of (t_n,ϵ_n)-suboptimal controls with $\epsilon_n \to 0$. Then there exists a subsequence of $\{u^n\}$ (which we denote by the same symbol), a second sequence $\{\tilde{t}_n\}$ such that*

$$\tilde{t}_n < \bar{t}, \quad \tilde{t}_n \to \bar{t}\,, \tag{4.1}$$

($\bar{t}$ the optimal time), an optimal control $\bar{u} \in W(-k,\bar{t},U)$ with

$$u^n(\hat{t}) \to \bar{u}(\hat{t}) \quad \text{weakly in } L^2(-k,\bar{t};\, U)\,, \tag{4.2}$$

$$y(t_n,u^n) \to \bar{y} = y(\bar{t},\bar{u}) \quad \text{strongly in } E\,, \tag{4.3}$$

a second sequence of controls $\{\tilde{u}^n\}$ with $\tilde{u}^n \in W(-k,\tilde{t}_n;\, U)$ such that

$$d_n(u^n,\tilde{u}^n) \to 0\,, \tag{4.4}$$

a sequence $\{y^n\}$ in E such that $\|y^n\| = 1$,

$$y^n \to y \neq 0 \quad \text{weakly in } E\,, \tag{4.5}$$

and a set e of full measure in $0 \leq s \leq \bar{t}$ such that

$$\langle y^n, \xi(\tilde{t}_n,s,\tilde{u}^n,v,\tilde{u}^n(s))\rangle \geq -\,\delta_n \to 0 \quad (v \in U,\, s \in e \cap (0,\tilde{t}_n)\,. \tag{4.6}$$

Proof: Once again, (4.2) and (4.3) are achieved passing to a subsequence (see (3.3)). We check easily that $\bar{t} = \lim t_n$ must be the optimal time. The sequence $\{\tilde{t}_n\}$ is any sequence satisfying (4.1); again, passing to a subsequence, we may assume that (3.10) holds. The cone $\Gamma(\bar{y},z,\theta,b)$ and the sequences $\{\kappa_n\}$ and $\{\bar{y}_n\}$ are chosen exactly as in the proof of Theorem 3.2. The function under consideration this time is

$$F_n(u) = \|y(\bar{t}_n,u) - \bar{y}_n\| ,$$

which is everywhere positive, since we cannot hit Y in time $\bar{t}_n < \bar{t}$. We have

$$F_n(u^n) = \|y(\bar{t}_n,u^n) - \bar{y}_n\| \leq \|y(\bar{t}_n,u^n) - \bar{y}\| + \|\bar{y} - \bar{y}_n\| \leq$$

$$\leq (1 + \gamma)\kappa_n \leq \inf F_n(u) + (1 + \gamma)\kappa_n . \qquad (4.7)$$

Ekeland's variational principle produces a sequence of controls $\{\bar{u}^n\}$ satisfying (4.4) and (4.6) with $y^n = x^n/\|x^n\|$, where

$$x^n = y(\bar{t}_n,\bar{u}^n) - \bar{y}_n . \qquad (4.8)$$

Using the argument following (3.16) we show that x^n (thus y^n) belongs to the cone $\Gamma(0,z,\theta_\gamma,\infty)$, where θ_γ, given by

$$\theta_\gamma = \pi - \theta - \arcsin\ ((1 + \gamma)^{-1} \sin\ \theta)$$

will satisfy $\theta < \pi/2$ for γ sufficiently small. This ends the proof of Theorem 4.1.

5. QUASILINEAR EQUATIONS IN BANACH SPACES. We consider the initial value problem

$$y'(t) = Ay(t) + f(t,y(t),u(t)), \quad (0 \leq t \leq T) \qquad (5.1)$$

$$y(0) = y_0 , \qquad (5.2)$$

where A is the infinitesimal generator of a strongly continuous semigroup $\{S(t); t \geq 0\}$ in the Banach space E. By definition, a *solution* of (5.1)-(5.2) is a continuous solution of the integrated version

$$y(t) = S(t)y_0 + \int_0^t S(t - \sigma)f(\sigma,y(\sigma),u(\sigma))\, d\sigma\ . \quad (0 \leq t \leq T) . \qquad (5.3)$$

We assume that U is bounded, that $f(t,y,u)$ has a Fréchet derivative $\partial_y f(t,y,u)$ with respect to y and that f (resp. $\partial_y f$) is continuous (resp. strongly continuous) and bounded on bounded subsets of $[0,T] \times E \times U$. Under these conditions, (5.3) can be solved by succesive approximations: setting $y_0(t) = y_0$,

$$y_{m+1}(t) = S(t)y_0 + \int_0^t S(t - \sigma)f(s,y_m(\sigma),u(\sigma))\, d\sigma, \quad (0 \le t \le T) \tag{5.4}$$

we obtain a sequence $\{y_m(\hat{t}); n \geq 1\}$ that converges absolutely and uniformly in some interval $0 \leq t \leq T_1 \leq T$ due to the differentiability assumption on f, which implies a local Lipschitz condition in y: this also shows uniqueness of solutions in $0 \leq t \leq T_1$. To insure existence in $0 \leq t \leq T$ we need conditions on $f(t,y,u)$ that guarantee *a priori* boundedness of solutions. An assumption that covers many interesting applications is: (i) $A - \omega I$ is *dissipative* for some ω, that is

$$\langle \bar{y}, Ay \rangle \leq \omega \|y\|^2 \ (y \in D(A), \bar{y} \in \Theta(y)) , \tag{5.5}$$

where $\Theta(y) \subset E^*$ is the duality set of u, consisting of all $\bar{y} \in E^*$ with $\|\bar{y}\|^2 = \|y\|^2 = \langle \bar{y}, y \rangle$; here $\langle \bar{y}, y \rangle$ denotes the value of $\bar{y}$ at y. (ii)

$$\langle \bar{y}, f(t,y,u) \rangle \leq C(1 + \|y\|^2) \ \ (0 \leq t \leq T, y \in E, \bar{y} \in \Theta(y), u \in U) . \tag{5.6}$$

LEMMA 5.1 *Let (5.5) and (5.6) be verified. Then any solution of (5.2) in an interval* $0 \leq t \leq T_1$ *satisfies*

$$\|y(t)\|^2 \leq e^{\beta t} \|y_0\|^2 \tag{5.7}$$

for $\beta = \omega + C(1 + \|y_0\|^{-2})$.

Proof: The estimate (5.7) is not difficult to show for strong solutions of (5.1), (5.2); however, since the solutions under consideration here are just generalized solutions, an approximation argument becomes necessary.

Let $y(t)$ be the solution in the statement, $\{u_m(\hat{t})\}$ a sequence of continuous controls such that $u_m(t) \to u(t)$ a.e. Let $\epsilon > 0$ and denote by $y_m(\hat{t})$ be the solution of (5.3) corresponding to $u_m(\hat{t})$ and by $[0,T_m]$ the maximum subinterval of $[0,T_1]$ such that $y_m(t)$ exists and satisfies

$$\|y_m(t) - y(t)\| \leq \epsilon \quad (0 \leq t \leq T_m) . \tag{5.8}$$

We have

$$y_m(t) - y(t) =$$

$$= \int_0^t S(t-\sigma)\,\{f(\sigma, y_m(\sigma), u_m(\sigma)) - f(\sigma, y(\sigma), u_m(\sigma))\}\,d\sigma\, +$$

$$+ \int_0^t S(t-\sigma)\,\{f(\sigma), y(\sigma), u_m(\sigma)) - f(\sigma, y(\sigma), u(\sigma))\}\,d\sigma$$

$$(0 \le t \le T_m)\,. \tag{5.9}$$

We take norms in (5.9), using in the first term the Lipschitz condition for f (note that, by (5.8), the $\{y_m(\hat{t})\}$ are uniformly bounded) and in the second term the dominated convergence theorem; applying Gronwall's lemma, we deduce that

$$\|y_m(t) - y(t)\| \le C_m \to 0 \tag{5.10}$$

in $0 \le t \le T_m$. If $T_m < T_1$ and $C_m < \epsilon$ we can extend the solution (still satisfying (5.8)) to an interval $[0,\tilde{T}_m]$ with $\tilde{T}_m > T_m$ using the local existence theorem, which contradicts the maximality of $[0,T_m]$. Hence, we shall eventually have $T_m = T_1$; moreover, since ϵ in (5.10) is arbitrary, it is obvious that we only have to show (5.7) for solutions $y(\hat{t})$ corresponding to $u(\hat{t})$ continuous.

For the next step in the proof we take a sequence $\{\mu_m\}$, $\mu_m \to +\infty$ and $P_m = \mu_m R(\mu_m; A)$, so that $\|P_m\| \le C$, $P_m \to I$ strongly: we consider the integral equation

$$y_m(t) = S(t)y_0 + \int_0^t S(t-\sigma)P_m f(\sigma, y_m(\sigma), u(\sigma))\,d\sigma \quad (0 \le t \le T)\,, \tag{5.11}$$

where the control $u(\sigma)$ is continuous. Since $P_m f(\sigma, y_m(\sigma), u(\sigma))$ and $AP_m f(\sigma, y_m(\sigma), u(\sigma))$ are continuous, familiar theorems on the nonhomogeneous equation $y'(t) = Ay(t) + f(t)$ ([17], [9]) imply that the solution of (5.11), wherever it exists, is a strong solution of

$$y_m'(t) = Ay_m(t) + P_m f(t, y_m(t), u(t)) \tag{5.12}$$

$$y_m(0) = y_0\,. \tag{5.13}$$

Let, as before $\epsilon > 0$ and let $[0,T_m]$ the maximum subinterval of $[0,T_1]$ such that $y_m(t)$ exists and satisfies

$$\|y_m(t) - y(t)\| \leq \epsilon \quad (0 \leq t \leq T_m) . \tag{5.14}$$

We argue now as after (5.8): this time, the integral equation is

$$y_m(t) - y(t) =$$

$$= \int_0^t S(t - \sigma)P_m \{f(\sigma,y_m(\sigma),u(\sigma)) - f(\sigma,y(\sigma),u(\sigma))\} \, d\sigma +$$

$$+ \int_0^t S(t - \sigma)(P_m - I)f(\sigma,y(\sigma),u(\sigma)) \, d\sigma . \tag{5.15}$$

We take norms in (5.15). In the first term on the right we use the Lipschitz condition for f; in the second, we note that the set

$$\{f(\sigma,y(\sigma),u(\sigma)); 0 \leq \sigma \leq T\}$$

is compact in E, so that $(P_m - I)f(\sigma,y(\sigma),u(\sigma)) \to 0$ uniformly in $0 \leq \sigma \leq T$. Applying Gronwall's lemma we obtain an estimate similar to (5.10). Arguing in the same way we deduce that $T_m = T_1$ for sufficiently large m. Hence, it is plain that we only have to prove (5.7) for solutions $y_m(t)$ of (5.12)-(5.13) for m large enough.

Let $\gamma > \beta$ and s_m a point in the interval $0 \leq t \leq T_1$ where $e^{-\gamma t} \|y_m(t)\|^2$ reaches its maximum. Consider the function

$$\phi_m(t) = e^{-\gamma t} \langle \bar{y}_m, y_m(t) \rangle$$

in $0 \leq t \leq T_1$, where $\bar{y}_m \in \Theta(y_m(s_m))$; since

$$\phi_m(t) \leq e^{-\gamma t} \|y_m(s_m)\| \, \| y_m(t)\| \leq e^{-\gamma s_m} \|y_m(s_m)\|^2 =$$

$$= e^{-\gamma s_m} \langle \bar{y}_m, y_m(s_m) \rangle , \tag{5.16}$$

$\phi_m(t)$ also reaches its maximum at $t = s_m$. However, its derivative there equals

$$- \gamma e^{-\gamma s_m} \|y_m(s_m)\|^2 + e^{-\gamma s_m} \langle \bar{y}_m, Ay_m(s_m) \rangle +$$

$$+ e^{-\gamma s_m} \langle \bar{y}_m, f(s_m,y_m(s_m),u(s_m)) \rangle +$$

$$+ e^{-\gamma s_m} \langle \tilde{y}_m, (P_m - I) f(s_m, y_m(s_m), u(s_m)) \rangle \le$$

$$\le (\omega - C - \gamma) e^{-\gamma s_m} \|y_m(s_m)\|^2 + e^{-\gamma s_m} C +$$

$$+ \langle \tilde{y}_m, (P_m - I) f(s_m, y_m(s_m), u(s_m)) \rangle . \tag{5.17}$$

Now, passing if necessary to a subsequence we may assume that $s_m \to s$. Then $f(s_m, y_m(s_m), u(s_m)) \to f(s, y(s), u(s))$, Thus we deduce that $(P_m - I) f(s_m, y_m(s_m), u(s_m)) \to 0$ as $m \to \infty$; accordingly, if m is sufficiently large, the derivative on the left side of (5.17) is negative, which shows that $s_m = 0$. This ends the proof of Lemma 5.1.

6. QUASILINEAR EQUATIONS IN BANACH SPACES (CONTINUATION).

The *a priori* bound (5.7) implies that (5.1)-(5.2) can be uniquely solved in $0 \le t \le T$. We define a map $X : W(0,T;U) \to C(0,T;E)$ by

$$(Xu)(t) = y(t,u) = y(t) , \tag{6.1}$$

where $y(t)$ is the solution of (5.1)-(5.2). corresponding to u. The proof that (6.1) is a system in the sense of §1 is carried out by means of standard applications of Gronwall's lemma; for details see [13, Section 4]. We note that the function $\xi(t,s,u,v,w)$ in (1.1) is given by

$$\xi(t,s,u,v,w) = S(t,s;u) \{f(s,y(s,u),v) - f(s,y(s,u),u(s))\} , \tag{6.2}$$

where $S(t,s;u)$ is the solution operator of the *linearized system*

$$z'(t) = Az(t) + \partial_y f(t,y(t,u),u(t)) z(t) \quad (0 \le s \le t \le T) , \tag{6.3}$$

$$z(s) = z_0 . \tag{6.4}$$

The treatment of (6.3)-(6.4) is essentially the same as (but simpler than) that of (5.1)-(5.2): solutions are defined as solutions of the corresponding integrated version

$$z(t) = S(t - \sigma) z_0 + \int_s^t S(t - \sigma) \partial_y f(\sigma, y(\sigma,u), u(\sigma)) z(\sigma)\, d\sigma$$

$$(0 \le s \le t \le T) \tag{6.5}$$

so that $S(t,s;u)$ is the (only) strongly continuous solution of the integral equation

$$S(t,s;u)z = S(t-s)z + \int_s^t S(t-\sigma)\partial_y f(\sigma,y(\sigma,u),u(\sigma))S(\sigma,s;u)z\,d\sigma$$

$$(0 \le s \le t \le T)\,. \tag{6.6}$$

We assume from now on that the operators $S(t)$ are *compact* for $t > 0$.

LEMMA 6.1 *The operator*

$$(\Pi u)(t) = \int_0^t S(t-\sigma)u(\sigma)\,d\sigma$$

from $L^2(0,T;U)$ *into* $C(0,T;E)$ *(endowed with its usual supremum norm) is compact.*

Proof: Let $S_m(t) = S(kT/m)$ for $(k-1)T/m \le t < kT/m$, $k = 1,2,\ldots,m$. Define

$$(\Pi_m u)(t) = \int_0^t S_m(t-\sigma)u(\sigma)\,d\sigma =$$

$$\sum_{1\le k\le mt} S(kT/m)\int_{t-k/m}^{t-(k-1)/m} u(\sigma)\,d\sigma\,.$$

On account of the fact that $S(\hat{t})$ is continuous in the uniform topology of operators in $t > 0$ (see [17]) we show easily that $\Pi_m \to \Pi$ in that topology; accordingly, we only have to show that each Π_m is compact. Let $\{u_n\}$ be a sequence in $L^2(0,T;F)$ weakly convergent to u. If $(\Pi_m u_n)(t)$ does not converge uniformly to $(\Pi_m u)(t)$ then there exists a sequence $\{t_n\}$ such that

$$\|(\Pi_m(u_n - u))(t_n)\| \ge \eta > 0\,. \tag{6.7}$$

Obviously, we may assume that $t_n \to t$. However, weak convergence of the $\{u_n\}$ is easily seen to imply that

$$\int_{t_n-k/m}^{t_n-(k-1)/m} (u_n(\sigma) - u(\sigma))\,d\sigma \to 0\,,$$

weakly in E, which contradicts (6.7). This ends the proof of Lemma 6.1.

We shall assume from now on that the control set U is *closed, bounded* and *convex* and that the nonlinearity $f(t,y,u)$ in (5.1) has the form

$$f(t,y,u) = f(t,y) + Bu \ , \tag{6.8}$$

where $B : F \to E$ is a bounded linear operator and $f(t,y)$ satisfies the corresponding differentiability assumptions.

THEOREM 6.2 *The system defined by the map (6.1) satisfies assumption (d).*

Proof: Let $\{u^n\}$ be the sequence in the statement of (d), and let $T > t_n$; extend the u^n to $L^2(0,T;F)$ by setting $u^n(t) = 0$ in $t > t_n$. We may assume, by passing to a subsequence, that $\{u^n\}$ is weakly convergent, so that (3.1) will hold; since $W(-k,t;U)$ is bounded, closed and convex (hence weakly closed) in $L^2(-k,T;U)$, the limit $\bar{u}$ belongs to $W(-k,T;U)$. We take a look at the approximations $y_{n,m}(t)$, $n = 1,2,\ldots$ (see (5.4)) used to compute $y(t,u^n)$: these are $y_{n,0}(t) = 0$,

$$y_{n,m+1}(t) = S(t)y_0 +$$

$$\int_0^t S(t-\sigma)f(\sigma,y_m(\sigma))\, d\sigma + \int_0^t S(t-\sigma)Bu^n(\sigma)\, d\sigma \ . \tag{6.9}$$

Applying Lemma 6.2 inductively we deduce that

$$y_{n,m}(t) \to y_m(t) \tag{6.10}$$

uniformly in $0 \le t \le T$ as $n \to \infty$, where the $y_m(t)$ are the approximations in (5.4) to $y(t,\bar{u})$. On account that both $\{y_m(t)\}$ and $\{y_{n,m}(t)\}$ converge absolutely and uniformly as $m \to \infty$ (independently of n) in an interval $0 \le t \le T_1$, (6.10) shows that

$$y(t,u^n) \to y(t,\bar{u}) \tag{6.11}$$

uniformly in $0 \le t \le T_1$. Applying the same argument in intervals

$[T_1,T_2],[T_2,T_3],\ldots$ (whose lengths do not tend to zero in view of the estimate (5.7)), we deduce that the convergence in (6.11) is uniform in $0 \le t \le T$. This is, of course, much more than what is necessary to prove (3.2), so that the proof of Theorem (6.2) is complete.

THEOREM 6.3. *Let* $\{s_n\}$, $\{t_n\}$ *be two convergent numerical sequences with*

$$0 \le s_n \le t_n \le T ,$$

$$s_0 = \lim s_n < t_0 = \lim t_n ,$$

and let $\{u^n\}$ *be a sequence of controls,* $u^n \in W(0,t_n; U)$ *such that*

$$u^n \to \bar{u} \in W(0,t_0; U) \text{ weakly in } L^2(0,t_0; U) .$$

Then

$$S(t_n,s_n; u^n) \to S(t,s; \bar{u}) \tag{6.12}$$

in the uniform topology of operators.

The proof is carried out in three steps. In the first, we show that $S(t,s; u)$ is Lipschitz continuous in the uniform operator norm with respect to t in

$$0 \le s + \rho \le t \le T , \tag{6.13}$$

uniformly with respect to s and $u \in W(0,T; U)$. To see this, we note that, if $t \le t'$,

$$S(t',s; u)z - S(t,s; u)z = S(t' - s)z - S(t - s)z +$$

$$\int_t^{t'} S(t' - \sigma)\partial_y f(\sigma,y(\sigma,u))S(\sigma,s; u)z \, d\sigma +$$

$$\int_s^t (S(t' - \sigma) - S(t - \sigma))\partial_y f(\sigma,y(\sigma,u))S(\sigma,s; u)z \, d\sigma ,$$

thus we only have to use the Lipschitz continuity of $S(\hat{t})$ in the operator norm.

The second step is to show (Lipschitz) continuity with respect to s in the same region; if $s < s'$, the pertinent equality is

$$S(t,s'; u)z - S(t,s; u)z = S(t - s')z - S(t - s)z +$$

$$\int_s^{s'} S(t-\sigma)\partial_y f(\sigma,y(\sigma,u))S(\sigma,s;u)z\,d\sigma\ +$$

$$\int_{s'}^t S(t-\sigma)\partial_y f(\sigma,y(\sigma,u))(S(\sigma,s';u) - S(\sigma,s;u))z\,d\sigma$$

and the result follows this time from Gronwall's lemma.

Finally, the third step consists in proving that

$$S(t,s;u^n) \to S(t,s;u) \tag{6.14}$$

in the uniform norm, uniformly for s,t in $0 \le s \le t \le T$. To do this, we note that

$$S(t,s;u^n) - S(t,s;u) =$$

$$\int_s^t S(t-\sigma)\{\partial_y f(\sigma,y(\sigma,u^n)) - \partial_y f(\sigma,y(\sigma,u))\}S(\sigma,s;u^n)\,d\sigma\ +$$

$$\int_s^t S(t-\sigma)\partial_y f(\sigma,y(\sigma,u))((S(\sigma,s;u^n) - S(\sigma,s;u))\,d\sigma\ .$$

We use in the first term on the right-hand side continuity of $\partial_y f$ and Theorem 6.2, and apply Gronwall's lemma.

THEOREM 6.4 *Let* $0 \le s < t \le T$, $u \in W(0,t;U)$. *Then the operator* $S(t,s;u) : E \to E$ *is compact.*

The proof is somewhat similar to that of Lemma 6.1: this time we express $S(t,s;u)$ as the limit of the sequence of operators

$$S_m(t,s;u)z = S(t-s)z\ +$$

$$\sum_{1\le k\le mt} S(kT/m)\int_{t-m/k}^{t-(k-1)/m} \partial_y f(\sigma,y(\sigma,u))S(\sigma,s;u)z\,d\sigma$$

each of which is compact.

7. CONVERGENCE RESULTS. We translate the convergence principles in Sections 3 and 4 into concrete convergence results for suboptimal controls applicable to the controlled differential equations in §5 and §6. All the assumptions there are in force: in particular, $S(t)$ is an analytic semigroup, the control set U is convex, closed and bounded (but see Remark 7.4) and the nonlinearity $f(t,y,u)$ is of the form (6.8); the target set Y satisfies (2.7) and the fat cone condition.

Example 7.1. We consider the time optimal problem. Let $\{u^n(\hat{t})\}$ be a sequence of suboptimal controls. Let $\{\tilde{u}^n(\hat{t})\}$ be the second sequence of suboptimal controls provided by Theorem 4.1 satisfying (4.4) and (4.6): via the calculation of the function ξ in (6.2), this last inequality becomes

$$\langle B^*S(\bar{t}_n,s;\tilde{u}^n)^*y^n, v - \tilde{u}^n(s)\rangle \geq -\delta_n \to 0 \quad (v \in U,\ s \in e \cap (0,\bar{t}_n)) . \tag{7.1}$$

Write

$$S(\bar{t}_n,s;\tilde{u}^n)^*y^n = (S(\bar{t},s;\bar{u})^* - S(\bar{t}_n,s;\tilde{u}^n)^*)y^n + S(\bar{t},s;\bar{u})^*y^n . \tag{7.2}$$

Since each $S(t,s;u)$ is compact for $t > s$ and $S(t,s;u)$ (thus $S(t,s;u)^*$) is continuous in the uniform topology of operators for $t > s$ and u in the weak $L^2(0,T;E)$ topology (see Theorems 6.3 and 6.4), it follows from (7.2) that

$$S(\bar{t}_n,s;\tilde{u}^n)^*y^n \to S(\bar{t},s;\bar{u})^*y \tag{7.3}$$

strongly in $0 \leq s < \bar{t}$. Accordingly, we obtain from (7.1) that

$$\langle B^*S(\bar{t},s;\bar{u})^*y, v - \tilde{u}^n(s)\rangle \geq -\delta_n \to 0 \quad (v \in U,\ s \in e \cap (0,\bar{t}_n)) . \tag{7.4}$$

Let $z \neq 0$, $U(z,\delta)$ the set of all $u \in U$ such that

$$\langle z, v - u\rangle \geq -\delta \quad (v \in U)$$

(see Figure 1). In view of (7.4), we obviously have

THEOREM 7.2. *Assume that the control set* U *is such that*

$$\operatorname{diam} U(z,\delta) \to 0 \quad (\delta \to 0) . \tag{7.5}$$

Then, if the time optimal control $\bar{u}$ *is unique and* d *is the set of all* s *in* $0 \leq s \leq \bar{t}$ *(* $\bar{t}$ *the optimal time) where*

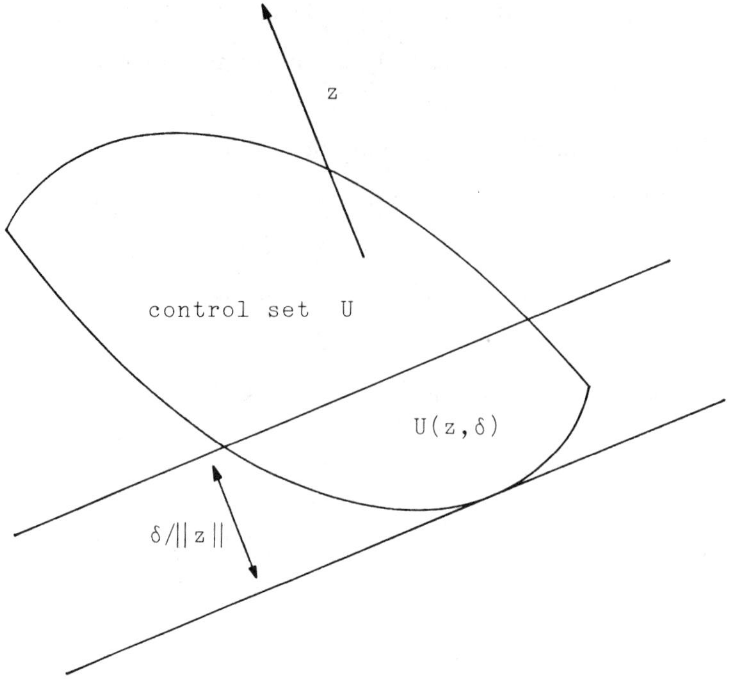

FIGURE 1

($\bar{t}$ the optimal time) where

$$B^*S(\bar{t},s;\bar{u})^*y \neq 0\ , \tag{7.6}$$

any sequence $\{u^n\}$ of (t_n,ϵ_n)- suboptimal controls with $\epsilon_n \to 0$ converges in $L^p(d;F)$, $(1 \leq p < \infty)$ to $\bar{u}$.

The proof has essentially been completed; we only note that the convergence statements follow from (7.4) and (7.5) and thus refer to the controls $\tilde{u}^n$ (rather than the u^n), and to the interval $[0,\bar{t}_n]$, rather than $[0,t_n]$; however, in view of the second relation (4.1) and of (4.4), they will hold for $\{u^n\}$ as well. Of course, L^p convergence follows from pointwise convergence and the dominated convergence theorem. The fact that the entire sequence (not just a subsequence) converges to $\bar{u}$ is a consequence of the postulated uniqueness of $\bar{u}$. This is seen as follows. Assume there exists a subsequence of $\{u^n\}$ that does not converge to $\bar{u}$ in $L^p(0,\bar{t},U)$. Then we can select a subsequence (that we denote with the same symbol) such that

$$\|u^n - \bar{u}\| \geq \rho > 0 .$$

Applying the arguments above, we can select a subsequence converging to a (necessarily different) optimal control, which is a contradiction. Naturally, in case the optimal control is not unique, we can only insure convergence of a subsequence.

To identify the set d where (7.6) holds may not be easy, since neither the optimal control $\bar{u}$ nor the vector y are known. Some information can be obtained in particular cases, however. Note that the solutions of the following nonhomogeneous version of the linearized system

$$z'(t) = \{A + \partial_y f(t,y(t,u))\}\, z(t) + Bv(t) \quad (0 \leq t_0 \leq t \leq T) , \tag{7.7}$$

$$z(t_0) = z_0 , \tag{7.8}$$

can be written in the form

$$z(t) = S(t,t_0; u)z_0 + \int_{t_0}^{t} S(t,\sigma; u)Bv(\sigma)\, d\sigma . \tag{7.9}$$

Let c be a measurable set of positive measure in $[t_0,t]$. We say that (7.7)-(7.8) is *c-null reachable in time* t if all elements of the form (7.9) with $z_0 = 0$ and $v(\hat{\sigma})$ an arbitrary F-valued measurable, bounded function are dense in E. An obvious duality argument shows that (7.7)-(7.8) is c-null reachable in time t if and only if

$$B^*S(t,s; u)^*y = 0 \quad (s \in c) \text{ implies } y = 0 . \tag{7.10}$$

Accordingly, the set d will have full measure in $0 \leq s \leq \bar{t}$ if and only if (7.7)-(7.8) is c-null reachable for any set c in $[0,\bar{t}]$ with positive measure. However, this equivalence does not seem to be of much help in ascertaining (7.10).

Condition (7.10) can of course be established in the *linear* case ($f = 0$), for instance, when A generates an *analytic* semigroup $S(\hat{t})$, for in this case $S(t,s;u)^* = S(t-s)^*$ will be analytic as well in $t \geq 0$.

Some additional information can be had in the case

$$F = E, \quad B = I \,. \tag{7.11}$$

Assuming, for simplicity, that E is *reflexive*, the operator

$$S^*(s,t;u) = S(t,s;u)^* \quad (0 \leq s \leq t \leq T) \tag{7.12}$$

is the solution operator of the *backwards* initial value problem

$$z'(s) = \{-A^* - \partial_y f(s,y(s,u))^*\}z(s) \quad (0 \leq s \leq t)\,, \tag{7.13}$$

$$z(t) = 0\,. \tag{7.14}$$

Accordingly, the statement

$$S(t,s;u)^* y \neq 0 \ \text{ in } \ [0,\bar{t}] \ \text{ if } \ y \neq 0 \tag{7.15}$$

simply means that (7.13)-(7.14) possesses the *forward uniqueness property*: if a solution $y(\hat{s})$ *vanishes at* $s \in [0,\bar{t}]$ *then it vanishes in* $s \leq t \leq \bar{t}$ (of course its counterpart, the backwards uniqueness property is automatically satisfied since the backwards Cauchy problem for (7.13)-(7.14) is well posed).

Condition (7.5) is for instance satisfied if U is the unit sphere of F. In finite dimensional spaces, it is equivalent to all boundary points of U being extremal (I owe this observation to Pascal Thomas).

REMARK 7.3. The requirement that each $S(t)$ be compact can be somewhat weakened at the cost of additional conditions on B. We sketch the necessary modifications to the theory. The alternate assumptions are:

(i) $R(\lambda; A) = (\lambda I - A)^{-1}$ is compact for some λ.

(ii) B^* is compact.

Although (i) does *not* guarantee that assumption (d) holds, it *does* imply the following weakened version thereof:

(d') Let $\{u^n\}$ be a sequence of controls, $u^n \in W(0,t_n; U)$, with $t_n \to t_0$. Then there exists a subsequence of $\{u_n\}$ (which we denote by the same symbol) and a $\bar{u}$ in $W(0,t_0; U)$ such that

$$u^n(\hat{t}) \to \bar{u}(\hat{t}) \quad \text{weakly in } L^2(0,t_0; F) , \tag{7.16}$$

$$y(\hat{t},u^n) \to y(\hat{t},\bar{u}) \quad \text{strongly in } L^2(0,t_0; E) , \tag{7.17}$$

$$y(t_n,u^n) \to y(t_0,\bar{u}) \text{ weakly in } E . \tag{7.18}$$

Under this modified assumption, it is easily seen that Theorem 3.1 survives. Both convergence principles survive as well, but with some modifications. In Theorem (3.2), the sequence $\{t_n\}$ must be replaced by a second sequence $\{\bar{t}_n\}$ with $t^n - \bar{t}_n \to 0$; (3.4), (3.6) and (3.9) hold with respect to this sequence, while the conditions on $\{y^n\}$ are unchanged. In the statement of Theorem 4.1 no modifications are necessary, but in the proof the sequence $\{\bar{t}_n\}$ must be chosen a little more carefully. This is the setup in [14] where all the necessary details can be found, as well as the proof that (ii) implies (d').

Some more adjustments are necessary to deduce convergence of optimal controls. The conclusions of Theorem 6.3 are the same, except that the convergence in (6.12) is merely strong, and the sequences $\{t_n\}$ and $\{s_n\}$ must be modified. The basic convergence relation (7.3) is then true *in the weak topology*; however, since, by (ii), B^* (equivalently, B) is compact, we have

$$B^*S(\bar{t}_n,s; \bar{u}^n)^*y^n \to B^*S(\bar{t},s; \bar{u})^*y ,$$

which yields Theorem 7.2. Assumption (ii) is of course satisfied in the case (important in applications) where F is finite dimensional. For more details on this kind of approach see [14], where in fact the results, although restricted to the time optimal problem are more general; condition (ii) is not used there.

REMARK 7.4. In proving Assumption (d) (or its weak counterpart (d')) it is essential to assume that U is *convex*. However, the methods in this paper work as well in certain nonconvex situations. We illustrate this with an example.

Suppose that the control set U is closed and bounded but not convex, and let $V = \overline{\text{conv}}\,(U)$ be the closed convex hull of U. Then it is obvious that (d) will hold, but with $\bar{u} \in W(0,t_0, V)$. Theorem 3.1 and the two convergence principles remain true. Theorem 7.2, however, cannot be applied as it stands, since condition (7.5) will not hold. (Figure 2.)

Assume, however that (7.5) holds for all vectors z, $\|z\| = 1$ except for a (at most countable) number $z_1,z_2,\ldots$ Then the convergence conclusions of Theorem 7.2 will still hold in the set $d' = d\backslash(\cup_{n=1}^{\infty} p_n)$, where p_n is the set of all s in $0 \le s \le \bar{t}$ where $B^*S(\bar{t},s;u)^*y$ is parallel to z_n. In some cases, it is possible to ascertain *a priori* that d has full measure in $0 \le s \le \bar{t}$ and that each p_n is a null set, so that d' has full measure in $0 \le s \le \bar{t}$; in others, this is not possible and the sequence $\{u^n\}$ may not be convergent. However, the arguments in

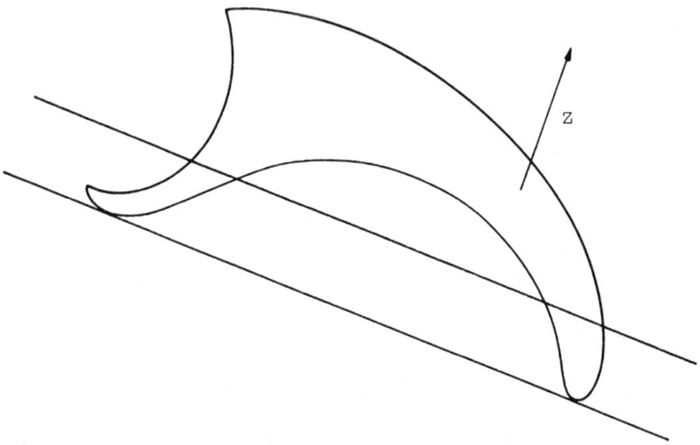

FIGURE 2

Theorem 7.2 can be used to show convergence to a relaxed control. We hope to treat these situations in a forthcoming paper.

8. CONVERGENCE RESULTS (CONTINUATION).

Example 8.1. We consider this time the cost functional

$$\int_0^t (1 + \eta\|u(\sigma)\|^2)\, d\sigma \tag{8.1}$$

with $\eta > 0$; we assume as well that U is the unit sphere of F. It is easily shown that

$$\xi(t,s,u,v,w) = \eta(\|v\|^2 - \|w\|^2)\ , \tag{8.2}$$

and that (1.3) holds at every Lebesgue point of $\|u(\hat{s})\|$. Hence, (3.9) is

$$\mu_n\eta(\|v\|^2 - \|\bar{u}^n(s)\|^2) + \langle B^*S(t_n,s,\bar{u}^n(s))^*y^n,\ v - \bar{u}^n(s)\rangle \geq -\ \delta_n \to 0$$

$$(v \in U,\ s \in e \cap (0,\bar{t}_n))\ , \tag{8.3}$$

where the sequence $\{(\mu_n,y^n)\}$ is weakly convergent to $(\mu,y) \neq (0,0)$ in $\mathbf{R} \times E$. Two possibilities arise:

(a) $\mu_n \to 0$. In this case, $y^n \to y \neq 0$ weakly in E, and the argument pertaining to the time optimal problem applies verbatim, producing the same conclusions.

(b) $\mu_n \to \mu \neq 0$. We have two subpossibilities.

(b1) $y^n \to 0$ weakly in E. Here, we take limits in (8.3) and obtain (with a different δ_n),

$$\|v\|^2 - \|\bar{u}^n(s)\|^2 \geq -\ \delta_n\ (v \in U, s \in e \cap (0,\bar{t}_n))\ . \tag{8.4}$$

This obviously implies that

$$\|\bar{u}^n(s)\| \to 1 \tag{8.5}$$

uniformly in e. Assume we know that

$$\|\bar{u}(s)\| = 1 \tag{8.6}$$

a.e. Then

$$\int_0^T \|u^n(s)\|^p \, ds \to \bar{t} = \int_0^T \|\bar{u}(s)\|^p \, ds \, . \tag{8.7}$$

Since, on the other hand, $u^n(\hat{t}) \to \bar{u}(\hat{t})$ weakly in $L^p(0,\bar{t}; F)$, (8.7) implies that $u^n \to \bar{u}$ strongly in $L^p(0,\bar{t}; F)$, $1 \leq p \leq \infty$.

The second subpossibility is

(b2) $y^n \to y \neq 0$ weakly in E. In this case, we subject (8.3) to the same treatment meted out previously to (7.1). The result is

$$\mu\eta(\|v\|^2 - \|\bar{u}^n(s)\|^2) + \langle B^*S(\bar{t},s; \bar{u})^*y, v - \bar{u}^n(s)\rangle \geq -\delta_n \to 0$$

$$(v \in U, \, s \in e \cap (0,\bar{t}_n)) \, . \tag{8.8}$$

Using (8.8) with $v - \bar{u}^n(s)$ orthogonal to $B^*S(\bar{t},s; \bar{u})^*y$ we obtain (8.5) anew, thus L^p convergence results again if we can prove (8.6). To this end, we use the maximum principle ([12], [13]) for the present optimal problem, which implies the existence of $(\mu,y) \in \mathbf{R} \times E$, $(\mu,y) \neq 0$ (not necessarily the weak limit of the sequence $\{(\mu_n,y^n)\}$ constructed above) such that

$$\mu\eta(\|v\|^2 - \|\bar{u}(s)\|^2) + \langle B^*S(\bar{t},s; \bar{u})^*y, \, v - \bar{u}(s)\rangle \geq 0$$

$$(v \in U, \, s \in e \cap (0,\bar{t})) \, , \tag{8.9}$$

where e is a set of full measure in $0 \leq s \leq \bar{t}$. We consider in relation to (8.9) the cases (a) $\mu = 0$, (b1) $\mu \neq 0$, $y = 0$, (b2) $\mu \neq 0$, $y \neq 0$ much in the same way similar possibilities for (8.3) were accounted for. Since we have no way to ascertain that $\mu \neq 0$, the only conclusion is that (8.6) will be true in all cases if

$$B^*S(\bar{t},s; \bar{u})^*y \neq 0 \text{ a.e. in } 0 \leq s \leq \bar{t} \, . \tag{8.10}$$

We have thus proved:

THEOREM 8.1 *Consider the optimal control problem for the cost functional (8.1) and* U *the unit sphere of* F. *Then, if the optimal control* $\bar{u}$ *is unique and (8.10) holds for arbitrary* $\bar{u}$ *and* $y \neq 0$, *any sequence* $\{u^n\}$ *of* (t_n,ϵ_n)*-suboptimal controls with* $\{t_n\}$ *bounded and* $\epsilon_n \to 0$ *converges in* $L^p(0,\bar{t}; F)$ $(1 \leq p < \infty)$ *to the optimal control* $\bar{u}$.

There is at least one case where (8.10) can be guaranteed *a priori*, namely, when $F = E$, $B = I$ (see (7.10) and following comments).

With convergence results in hand, the question remains of how suboptimal controls can be actually computed. One way is by using *penalty methods* of the type introduced by Hestenes for general optimization problems. Going back for a moment to the general system formulation, assume that $y_0(t,u)$ is bounded below. Let $\epsilon > 0$ and consider the function

$$F_\epsilon(t,u) = y_0(t,u) + \epsilon^{-2}\operatorname{dist}(y(t,u), Y)^2 \tag{8.11}$$

in the space of pairs $\{(t,u); t > 0, u \in W(-k,t; U)\}$ (the square in the second term is only used for Hilbert space convenience). Assume we can compute an ϵ-approximate minimum of F_ϵ, that is, an element $u^\epsilon \in W(-k,t_\epsilon; U)$ such that

$$F_\epsilon(t_\epsilon,u^\epsilon) \leq \inf F_\epsilon(t,u) + \epsilon\ , \tag{8.12}$$

the infimum in (8.12) taken over all $t > 0$ and $u \in W(-k,t; U)$. Now, assuming that an optimal control $\bar{u} \in W(-k,\bar{t}; U)$ exists, we must have

$$\inf F_\epsilon(t,u) \leq F_\epsilon(\bar{t},\bar{u}) = y_0(\bar{t},\bar{u}) = m\ , \tag{8.13}$$

thus it follows from (7.19) that

$$y_0(t_\epsilon,u^\epsilon) \leq F_\epsilon(t_\epsilon,u^\epsilon) \leq m + \epsilon\ , \tag{8.14}$$

$$\operatorname{dist}(y(t_\epsilon,u^\epsilon),Y) \leq \epsilon(m + \epsilon)^{1/2}\ . \tag{8.15}$$

Accordingly, u^ϵ is a (t,ϵ^*)-suboptimal control, where we have set $t = t_\epsilon$, $\epsilon^* = \max(\epsilon,\epsilon(m + \epsilon)^{1/2})$. Thus, even though m enters in the definition of suboptimal controls, no previous knowledge of the infimum is necessary for their computation.

Although the minimization problem above is "just" an infinite dimensional nonlinear programming problem, in order to calculate an ϵ-approximate minimum, $y(t,u)$ will have to be computed repeatedly. Since this may involve solving at each step a differential (or a more general) equation, methods that avoid this are of interest. One such is the "epsilon method" in [2], where, instead of computing $y(t,u)$, the deviation of $y(t)$ from $y(t,u)$ is penalized. To fix ideas, we consider only the case where the system is described by an initial value problem

$$y'(t) = \Phi(y(t),u(t)), \quad y(0) = y_0 , \tag{8.16}$$

(so that the formulation includes the systems in Sections 5 and 6). We minimize now the function

$$F_\epsilon(t,y,u) = y_0(t,u) + \epsilon^{-2}\,\mathrm{dist}(y(t),Y)^2 +$$

$$\epsilon^{-2}\int_0^t \|y'(\sigma) - \Phi(y(\sigma),u(\sigma))\|^2\,d\sigma \tag{8.17}$$

in the set of 3-ples $(t,y(\hat{t}),u(\hat{t}))$, where $t > 0$, $u \in W(0,t;U)$ and $y(\hat{t})$ belongs to the subset of the space $H^1(0,t;E)$ (of E-valued functions with distributional derivative in $L^2(0,t;E)$) consisting of all functions satisfying $y(0) = y_0$. The treatment of (8.17) is basically the same as that of (8.11). Assume we can compute an ϵ-approximate minimum (y^ϵ,u^ϵ), $y^\epsilon \in H^1(0,t_\epsilon;E)$, $u^\epsilon \in W(0,t_\epsilon;U)$ of the function $F_\epsilon(t,y,u)$. If $\bar{u} \in W(0,\bar{t};U)$ is an optimal control and we take $\bar{y}(t) = y(t,\bar{u})$ then we have

$$\inf F_\epsilon(t,y,u) \le F_\epsilon(\bar{t},\bar{y},\bar{u}) = y_0(\bar{t},\bar{u}) = m , \tag{8.18}$$

so that (8.14) and (8.15) follow and we have in our hands again a (t_ϵ,ϵ^*)- suboptimal control. Of course this argument assumes that $y(t,\bar{u}) \in H^1(0,\bar{t},E)$, which may not be the case for the systems studied in §5 and §6; if this happens, we may replace the third term in (8.18) by the following integral penalty term involving not (5.1) but its integrated version (5.3):

$$\epsilon^{-2}\,\|y(t) - S(t)y_0 - \int_0^t S(t-\sigma)f(\sigma,y(\sigma),u(\sigma))\,d\sigma\|^2 .$$

Or, we can use an approximate equation replacing the integrand in (8.17) by the expression

$$\epsilon^{-2}\,\|y'(t) - Ay(t) - P_k f(t,y(t),u(t))\|^2 ,$$

(P_k the operator in §5) with k large enough in function of ϵ. However, this is perhaps of minor importance, since, in practice, neither (8.11) nor (8.17) will be minimized directly but through adequate discretizations, these obtained using either finite differences or finite elements. An abstract setup for this situation is the following. Let $X_n : W(0,T;U) \to C(0,T;E)$ be a sequence of systems that *approximate* the system $X : W(0,T;U) \to C(0,T;U)$ in the sense that

$$y^n(t,u) \to y(t,u) \tag{8.19}$$

uniformly in $0 \le t \le T$ and in $u \in W(0,T;U)$, where $y^n(t,u) = X_n u$; the cost functional y_0 is approximated in the same way by cost functionals y_0. Obviously, a sequence $\{u^n\}$ of controls, each (t_n,ϵ_n)-suboptimal for X_n with respect to $y_0(t,u)$ will provide a sequence of (t_n,ϵ_n)-suboptimal controls for X with respect to $y_0(t,u)$. In practice, however, it is more natural to consider systems X_n where the output space E_n is different from E; in applications, E_n often is a (finite dimensional) space of "discrete" functions, defined only at certain grid points, or as linear combinations of finite elements. In this connection, a convenient setup seems to be Trotter's theory of approximation of Banach spaces, which has been extensively used in the abstract treatment of approximations of solutions of abstract differential equations (see [9, Chapter 5]). We hope to study this type of approximation in a forthcoming paper.

REMARK 8.2 The observations in Remark 7.3 apply here in full; in fact, they apply to any cost functional.

9. QUASILINEAR PARABOLIC EQUATIONS. We apply the theory in the last two sections to the operator

$$Au(x) = \Sigma\Sigma\, a_{jk}(x)\partial^j\partial^k u(x) + \Sigma\, b_j(x)\partial^j u(x) + c(x)u(x)\,, \tag{9.1}$$

in a domain Ω in m-dimensional Euclidean space $\mathbf{R}^m$; here $x = (x_1,x_2,\ldots,x_m)$, $\partial^j = \partial/\partial x_j$ and we require, as usual, that $a_{jk} = a_{kj}$. We assume the domain Ω bounded and of class $C^{(2)}$: the operator A is of class $C^{(1)}$, which means that the $a_{jk}(x)$ are continuously differentiable in $\bar{\Omega}$ and that the $b_j(x)$ and $c(x)$ are continuous in $\bar{\Omega}$. Functions in the domain of A will be asked to satisfy a boundary condition β on the boundary Γ, either of *Dirichlet type*,

$$u(x) = 0 \quad (x \in \Gamma)\,, \tag{9.2}$$

or of *variational type*

$$\partial^\nu u(x) = \gamma(x)u(x) \quad (x \in \Gamma)\,, \tag{9.3}$$

where ∂^ν denotes the *conormal derivative*

$$\partial^\nu = \Sigma\Sigma\, a_{jk}(x)\eta_j\partial^k\,,$$

$\eta = (\eta_1,\eta_2,\ldots,\eta_m)$ the outer normal vector on Γ. In case the boundary condition is (9.3), the function $\gamma(x)$ is assumed to be continuously differentiable on Γ. We assume that A is *uniformly elliptic*:

$$\Sigma\Sigma a_{jk}(x)\xi_j\xi_k \geq \kappa|\xi|^2 \quad (\xi \in \mathbf{R}^m,\ x \in \bar{\Omega}) \tag{9.4}$$

for some $\kappa > 0$. Under the assumptions above, the operator (which we shall name $A_p(\beta)$ to identify the boundary condition used and the space where A lives) is the infinitesimal generator of a strongly continuous (in fact, analytic) semigroup $S_p(t)$, $t \geq 0$ in $L^p(\Omega)$, $1 < p < \infty$; the domain of $A_p(\beta)$ is $W^{2,p}(\Omega)_\beta$, which is the subspace of the Sobolev space $W^{2,p}(\Omega)$ consisting of all y that satisfy the boundary condition β on Γ; of course, when β is the Dirichlet boundary condition, $W^{2,p}(\Omega)_\beta$ is the familiar space $W^{2,p}(\Omega)_0$. For each p, $A_p(\beta)$ satisfies (5.5) for some $\omega = \omega(p)$. Nonlinearities in the equation (5.1) are handled the easier the larger p is, thus the more advantageous course seems to be "to set

$p = \infty$". However, the operator $A(\beta)$ cannot be defined in any reasonable dense domain in $L^\infty(\Omega)$, thus, in the case of the Dirichlet boundary condition, we take $E = C_0(\overline{\Omega})$, the space of all continuous functions in $\overline{\Omega}$ that vanish at the boundary Γ; for a boundary condition of the form (9.3), E is the entire space $C(\overline{\Omega})$ of all continuous functions in $\overline{\Omega}$. (Both spaces are endowed with their usual supremum norm). We have

$$D(A(\beta)) = \left\{ y \in \bigcap_{p\geq 1} W^{2,p}(\Omega)_\beta \; ; Ay \in C(\overline{\Omega}) \right\}$$

($C_0(\overline{\Omega})$ in the case of the Dirichlet boundary conditions). The operator $A(\beta)$ so defined generates a strongly continuous, compact semigroup $S(\hat{t})$ and satisfies (5.5) in the case of the Dirichlet boundary condition: for boundary conditions of type (9.3), (5.5) may not hold, but will always be satisfied after renorming of the space $C(\overline{\Omega})$: the new norm is of the form

$$\|y\| = \max_{x\in\Omega} |u(x)|\rho(x) ,$$

where $\rho(x)$ is a continuous positive function in $\overline{\Omega}$. When $E = C(\overline{\Omega})$, E^* is the space $\Sigma(\overline{\Omega})$ of all finite Borel measures defined in $\overline{\Omega}$: elements μ of $\Sigma(\overline{\Omega})$ act on functions $y \in C(\overline{\Omega})$ by the formula

$$\langle \mu, y \rangle = \int_\Omega y(x)\rho(x)\mu(dx) . \tag{9.5}$$

and the duality set $\Theta(y)$ of an element $y \in C(\overline{\Omega})$ is the set of all measures $\mu(dx)$ with support contained in the set

$$m(y) = \{x \in \Omega;\ |y(x)| = \|y\|\} \tag{9.6}$$

and such that $y(x)\mu(dx)$ is a nonnegative measure with

$$\int_\Omega |\mu|(dx) = \|y\| ; \tag{9.7}$$

For all the necessary details, see [9, Chapter 4] and [10]. We shall consider the following nonlinear term:

$$(f(t,y(\hat{x}),u))(x) = g(t,y(x)) + Bu \,, \tag{9.8}$$

where B is a linear operator from F into E. Direct treatment of this nonlinearity in $L^2(\Omega)$ will force us to impose very strong assumptions on $g(t,y)$, thus we shall take $E = C(\overline{\Omega})$ or $C_0(\overline{\Omega})$ according to the boundary condition. We check easily that the conditions imposed on the nonlinearity in Section 5 will be satisfied if $g(t,y)$ is continuously differentiable: the Fréchet derivative $\partial_y f(t,y)$ is given by

$$(\partial f_y(t,y)h(x))(x) = \partial_y g(t,y(x))h(x) \,. \tag{9.9}$$

In view of the characterization of $\Theta(y)$ above, we see easily that condition (5.6) will be satisfied if

$$f(t,y) \le C(1 + |y|) \quad (0 \le t \le T, \, -\infty < y < \infty) \tag{9.10}$$

(so that no growth conditions where $f(t,y)$ is negative are imposed). The treatment of the Dirichlet boundary condition (9.2) is entirely similar: here $E = C_0(\overline{\Omega})$ and E^* is the subspace $\Sigma_0(\overline{\Omega})$ of $\Sigma(\overline{\Omega})$ consisting of all measures μ that vanish on the boundary Γ. Condition (9.10) is the same: of course, to make sure that $f(t,y)$ maps E into E we must assume

$$f(t,0) = 0 \,, \tag{9.11}$$

which implies that $y = 0$ is an equilibrium point of the system.

The theory in Section 5 and 6 shows that the controlled initial value problem

$$y'(t) = A(\beta)y(t) + f(t,y(t)) + Bu(t) \,, \tag{9.12}$$

$$y(0) = y_0 \,, \tag{9.13}$$

(y_0 arbitrary) produces a system in $E = C(\overline{\Omega})$ that satisfies Assumption (d) in §3. However, the convergence principles in §3 and §4 pertain to the Hilbert space case. This dilemma is easily solved noting that E is continuously imbedded in $L^2(\Omega)$. Obviously, postulates (a), (b), (c) and (d) survive if E is replaced by $L^2(\Omega)$ and we only have to take the initial condition y_0 in E. However, a serious difficulty appears in relation to the operator B; since B must be bounded from F into E, such natural choices as $F = L^2(\Omega)$, $B = I$ seem to be inadmissible. We solve this as follows. Note first that (9.10) guarantees as well that (5.6) will hold in $L^2(\Omega)$. Accordingly, the solution of (9.12)-(9.13) in $L^2(\Omega)$ (strictly speaking, the solution of

$$y'(t) = A_2(\beta)y(t) + f(t,y(t)) + Bu(t) , \tag{9.14}$$

$$y(0) = y_0 \in C(\overline{\Omega})) \tag{9.15}$$

will remain bounded wherever it exists; we then conclude that (9.14)- (9.15) is uniquely solvable in the whole interval $0 \leq t \leq T$. However, an essential difficulty still remains; namely that the operator $f(t,y,u)$ defined by (9.8) *may not be differentiable with respect to* y *in* $L^2(\Omega)$; for instance, this is the case if we take

$$g(t,y) = y^3 . \tag{9.16}$$

Accordingly, the proof in [13] that (9.14)-(9.15) generates a system breaks down and a subtler analysis is needed. An examination of the proof in [13, Example 2.3] shows that we can again argue in the space $C(\overline{\Omega})$ *if we can show that* $S(t) : L^2(\Omega) \to C(\overline{\Omega})$ $(t > 0)$ *and that*

$$\|S(t)\| \in L^1(0,T) , \tag{9.17}$$

(here $\|S(t)\|$ indicates the norm of $S(t)$ as an operator from $L^2(\Omega)$ into $C(\overline{\Omega})$). This is achieved by means of the theory of fractional powers of analytic semigroups as follows. Modulo a translation of $A_2(\beta)$ we may assume that the fractional powers $(-A_2(\beta))^\alpha$ $(-\infty < \alpha < \infty)$ exist; if $0 \leq \alpha \leq 1$ we have

$$\|(-A_2(\beta))^\alpha S(t)\| \leq Ct^{-\alpha} \quad (0 < t \leq T) . \tag{9.18}$$

Now, using interpolation theory we show that $D((-A_2(\beta))^\alpha)$ is boundedly imbedded in the Sobolev space $W^{2,2\alpha}(\Omega) = H^{2\alpha}(\Omega)$. In view of Sobolev's imbedding theorem for fractional exponents (see [1]), $H^{2\alpha}(\Omega)$ is continuously imbedded in $C(\bar{\Omega})$ for

$$2\alpha > m/2 \tag{9.19}$$

(m the dimension of the space). Since we may take α arbitrarily close to 1, it is clear that (9.17) will be satisfied if $m \leq 3$ and we have thus completed the proof that, for $m \leq 3$, (9.14)-(9.15) defines a system in $L^2(\Omega)$ if B is a bounded operator from F into $L^2(\Omega)$.

It remains to show that this system satisfies (d). A look at the proof of Theorem 6.2 (especially to (6.9)) shows that the argument applies verbatim since each $S(t)$ $(t > 0)$ is a compact operator in $L^2(\Omega)$ and the sequence $\{y_{n,m}(t); m = 1,2,\ldots\}$ is absolutely and uniformly convergent in $C(\bar{\Omega})$, thus in $L^2(\Omega)$.

10. CONVERGENCE OF SUBOPTIMAL CONTROLS. We apply the theory in §7 and §8 to the quasilinear parabolic systems described in the previous section; in all the examples below, $E = L^2(\Omega)$.

Example 10.1. We take $F = L^2(\Omega)$, $B = I$, U the unit sphere in $L^2(\Omega)$, so that the bound on the controls is

$$\int_\Omega |u(x,t)|^2\, dx \leq 1 \quad (0 \leq t \leq \bar{t}) . \tag{10.1}$$

It follows from the remarks at the end of Section 7 that, if $y \neq 0$,

$$S(\bar{t},s;\bar{u})^* y \neq 0 \text{ a.e. in } 0 \leq s \leq \bar{t} . \tag{10.2}$$

Hence, the results in §7 on the time optimal problem or those in §8 on the optimal problem relative to the cost functional

$$\int_0^t (1 + \eta\|u(\sigma)\|^2)\, d\sigma \tag{10.3}$$

apply in full; in both cases, assuming that it is known that the optimal control is unique, we obtain that if $\{u^n(t,x)\}$ is any sequence of (t_n,ϵ_n)-suboptimal controls with $\{t_n\}$ bounded and $\epsilon_n \to 0$, the sequence $\{u^n(t,x)\}$ converges to the optimal control $\bar{u}(t,x)$ in the L^2 norm of $\Omega \times [0,\bar{t}]$:

$$\int_0^{\bar{t}} \int_\Omega |u^n(t,x) - \bar{u}(t,x)|^2\, dx\, dt \to 0 . \tag{10.4}$$

We note, however, that the linearized system is *exactly* controllable to $D(A)$, so that results for the *point target* problem may also be available.

Example 10.2. We take again $F = L^2(\Omega)$, $B = I$, with U the unit sphere of $L^\infty(\Omega)$, that is the set of all $u \in L^2(\Omega)$ such that

$$|u(x)| \le 1 \text{ a.e. in } x \in \Omega , \tag{10.5}$$

so that the bound on the controls is

$$|u(t,x)| \le 1 \text{ a.e. in } t \ge 0,\ x \in \Omega . \tag{10.6}$$

Although Ω does not satisfy condition (7.5), convergence results can also be obtained in this case. We consider only the time optimal problem. Rewrite (7.4), setting

$$z(s,x) = (S(\bar{t},s,\bar{u})^* y)(x) , \tag{10.7}$$

obtaining the relation

$$\int_\Omega z(s,x)(v(s,x) - \bar{u}^n(s,x))\, dx \ge -\delta_n \to 0$$

$$(|v(s,x)| < 1 \text{ a.e. in } [0,t_n] \times \Omega,\ s \in e \cap (0,\bar{t}_n)) . \tag{10.8}$$

This is easily seen to imply that, for every $s \in e$, $u^n(s,x)$ converges in measure (uniformly with respect to s) to the optimal control

$$\bar{u}(s,x) = -\operatorname{sign} z(s,x)$$

in the subset $\Lambda(s)$ of Ω where $z(s,x) \neq 0$. Integrating (10.8) with respect to s in $0 \leq s \leq \bar{t}$ we deduce convergence in measure in

$$\Lambda = \bigcup_{0 \leq s \leq \bar{t}} \Lambda(s) .$$

However, no convergence conclusions are obtained in the complement of Λ, that is, where $z(s,x) = 0$. Obviously, we have L^p convergence of $u^n(s,x)$ to $\bar{u}(s,x)$ in Λ.

Unfortunately, it does not seem easy to ascertain *a priori* that the set Λ will have full measure in $\Omega \times [0,\bar{t}]$. This can be done in the *linear* case $(g = 0)$ when the coefficients of the operator A are analytic; in fact, under these conditions, $z(t,x)$ is analytic in $\Omega \times [0,\bar{t}]$, thus if Λ has not full measure there, $z(s,x)$ will be identically zero, which is impossible. In this particular case, the optimal problem considered in §8 can be treated as well.

Example 10.3. We consider the linear case $(g = 0)$ with $F = \mathbf{R}$, $U = [0,1]$, $B : \mathbf{R} \to L^2(\Omega)$ the operator

$$Bu = ub(x) , \tag{10.9}$$

where the Fourier coefficients of $b(x)$ with respect to the eigenfunctions of $A_2(\beta)$ are all nonzero. The linearized system coincides with the original control system. Let d be the set where

$$B^*S(\bar{t},\sigma,u)^*y = \langle S(\bar{t},\sigma,u)^*y,b\rangle = \langle S(\bar{t} - \sigma)^*y,b\rangle \neq 0 . \tag{10.10}$$

Since $S(\hat{t})$ (thus $S(\hat{t})^*$) is an analytic semigroup, it follows that (10.10) must happen almost everywhere (precisely, the set d where (10.10) holds must be a decreasing sequence $\{t_n\}$, either finite or with $\lim t_n = 0$.) Thus, the convergence results in §7 for the time optimal problem and in §8 for the cost functional (10.3) apply in full and we obtain L^p convergence of suboptimal controls, $1 \leq p < \infty$.

Example 10.4. We consider the time optimal problem in the nonlinear case ($g \neq 0$), with $F = L^2(\Omega_1)$ (Ω_1 a subset of Ω) and B the injection operator from $L^2(\Omega_1)$ into $L^2(\Omega)$, so that B^* is the restriction operator from $L^2(\Omega)$ into $L^2(\Omega_1)$. We assume that the coefficients $a_{jk}(x)$ of the principal part of A are C^∞ and that Ω_1 is a domain of class C^∞. Let $z(s,x)$ be given by (10.12). Then, if $B^*z(s,x) = 0$ for $t_0 \leq t \leq t_1$ we have

$$z(s,x) = 0 \quad (t_0 \leq s \leq t_1, \, x \in \Omega_1) \, .$$

It then follows from uniqueness theorems for partial differential operators in [15, Chapter 28] (specifically, Theorem 28.4.3) that $z(s,x)$ must vanish identically in $[t_0,t_1] \times \Omega$. Accordingly, the set d where

$$B^*S(\bar{t},s;\, u)^*y \neq 0$$

is an open dense set in $0 \leq s \leq \bar{t}$ where Theorem 7.2 applies.

REFERENCES

[1] R. A. ADAMS, *Sobolev Spaces*, Academic Press, New York, 1975.

[2] A. V. BALAKRISHNAN, On a new computing technique in optimal control, SIAM J. Control 6 (1968) 149-173.

[3] V. BARBU, *Optimal Control of Variational Inequalities*, Research Notes in Mathematics 100, Pitman, London 1984.

[4] V. BARBU, The time optimal problem for a class of nonlinear distributed systems, to appear in Proceedings of IFIP Workshop on Control Problems for Systems Described by Partial Differential Equations, Gainesville, February 1986.

[5] F. H. CLARKE, *Optimization and Nonsmooth Analysis*, Wiley, New York, 1983.

[6] I. EKELAND, Noncovex minimization problems, Bull. Amer. Math. Soc. 1 (NS) (1979) 443-474.

[7] H. O. FATTORINI, The time optimal control problem in Banach spaces, Appl. Math. Optimization 1 (1974) 163-188.

[8] H. O. FATTORINI, The time optimal problem for boundary control of the heat equation, F2Calculus of Variations and Control Theory, Academic Press, New York, 1976.

[9] H. O. FATTORINI, *The Cauchy Problem*, Encylopedia of Mathematics and its Applications vol. 18, Cambridge University Press, Cambridge 1983.

[10] H. O. FATTORINI, On the angle of dissipativity of ordinary and partial differential equations, Functional Analysis, Holomorphy and Approximation Theory II, Elsevier - North-Holland, 1984.

[11] H. O. FATTORINI, The maximum principle for nonlinear nonconvex systems in infinite dimensional spaces, Proceedings of the 2nd. Conference on Control Theory of Distributed Parameter Systems, Vorau, 1984, Springer Lecture Notes in Control and Information Sciences 75 (1986) 162-178.

[12] H. O. FATTORINI, The maximum principle for nonlinear nonconvex systems with set targets, Proceedings of the 24th. IEEE Conference on Decision and Control, Fort Lauderdale, 1985.

[13] H. O. FATTORINI, A unified theory of necessary conditions for nonlinear nonconvex control systems, to appear in Applied Mathematics & Optimization.

[14] H. O. FATTORINI, Optimal control of nonlinear systems: convergence of suboptimal controls, II, to appear in Proceedings of IFIP Workshop on Control Systems Described by Partial Differential Equations, Gainesville, February 1986.

[15] H. O. FATTORINI, Convergence of suboptimal controls for point targets, to appear in Proceedings of the Conference on Optimal Control with Partial Differential Equations, Oberwolfach, May 1986.

[16] H. O. FATTORINI, Convergence of suboptimal control: the point target case, to appear.

[17] H. O. FATTORINI, Some remarks on convergence of suboptimal controls, to appear.

[18] L. HORMANDER, *The analysis of Linear Partial Differential Operators IV*, Springer, Berlin, 1984.

[19] A. PAZY, *Semigroups of Linear Operators and Applications to Partial Differential Equations*, Springer, Berlin 1983.

[20] V. I. PLOTNIKOV and M. I. SUMIN, The construction of minimizing sequences in problems of control of systems with distributed parameters, Zh. Vychisl. Mat. mat. fiz. 22 (1982) 49-56.

Exponential Stabilization of the Wave Equations with Dirichlet Boundary Conditions

I. Lasiecka*
Department of Mathematics
University of Florida
Gainesville, Florida

1. INTRODUCTION

The present paper announces some recent results on exponential stabilization of nonlinear second order scalar hyperbolic equation with a stabilizing feedback acting as the Dirichlet boundary control. Here we shall only outline some basic ideas while for the complete proofs of the stabilization results we refer to the forthcoming papers [L-T-1], [L-2].

2. FORMULATION OF THE PROBLEM

Let Ω be an open bounded domain in R^n with the smooth boundary Γ

Consider the following second order hyperbolic mixed problem in $y(t,x)$; $x \in \Omega$ with <u>nonhomogeneous</u> Dirichlet data. $u(t,\sigma)$; $\sigma \in \Gamma$:

$$(1.1)\quad \begin{cases} y_{tt} = \Delta y \quad \text{in } (0,\infty)\mathrm{x}\Omega \equiv Q \\ y(0,x) = y_0(x) \in L_2(\Omega);\ y_t(0,x) = y_1(x) \in H^{-1}(\Omega) \\ y(t,\sigma) = u(t,\sigma) + G(y(t,\cdot),\ y_t(t,\cdot)) \quad \text{in } (0,\infty)\mathrm{x}\Gamma \equiv \Sigma \end{cases}$$

where $u \in L_2(\Sigma)$ and G (possibly nonlinear) is continuous from $L_2(\Omega)\mathrm{x}H^{-1}(\Omega) \to L_2(\Gamma)$

*Current Affiliation: University of Virginia, Charlottesville, Virginia

Recently developed regularity results for (1.1) (see [L-L-1], [L-T-2], [L-T-3]) guarantee that in the linear case ($G \equiv 0$) the solution $(y(t), y_t(t))$ lies in $L_2(\Omega) x H^{-1}(\Omega) \equiv E$ for any $u \in L_2(\Sigma)$. This fact motivates our choice of the state space E as
$E \equiv L_2(\Omega) x H^{-1}(\Omega)$

We shall study the question of existence and construction of a boundary feedback operator F based on the velocity

$$y_t \in H^{-1}(\Omega) \to F(y_t) \in L_2(\Gamma)$$

$$F(y_t) \in L_2(\Sigma)$$

such that the boundary feedback $u = F(y_t)$, inserted in (1.1) produces a feedback semigroup which is globally exponentially stable (linear case) and locally exponentially stable (for some classes of nonlinear peturbation G)

Due to the Dirichlet nature of the boundary conditions in (1.1), the above problem (even the linear case!) was, to the author's knowledge, open. The corresponding problem with u acting in the Neumann boundary condition and $G \equiv 0$, was solved only recently [Ch-1] [L-1]. Our motivation for studying the described problem is two fold

(i) On the one hand, this problem is one of the central questions in boundary feedback stabilization theory .

(ii) on the other hand, an affirmative solution to the above problem is a necessary pre-requisite for studying Algebraic Riccati Equations.

3. SEMIGROUP FORMULATION OF (1.1)

We begin with some preliminary background material, needed to both motivate our choice of the operator F and to state our main results.

Let A: $L_2(\Omega) \to L_2(\Omega)$ be defined as

$$Ay = -\Delta y \text{ for } y \in D(A) \equiv H_0^1(\Omega) \cap H^2(\Omega).$$

It is well known that the operator

$$(3.1) \quad \mathcal{A}\begin{pmatrix} y_1 \\ y_2 \end{pmatrix} = \begin{vmatrix} 0 & I \\ -A & 0 \end{vmatrix} \begin{pmatrix} y_1 \\ y_2 \end{pmatrix} \text{ with domain}$$

$$\mathcal{D}(\mathcal{A}) = D(A^{1/2}) \times L_2(\Omega)$$

generates a strongly continuous unitary group $e^{\mathcal{A}t}$ on $E \equiv L_2(\Omega) x H^{-1}(\Omega)$

Next we shall introduce the Dirichlet map D: $L_2(\Gamma) \to L_2(\Omega)$ defined by

$$\begin{cases} \Delta(Dv) = 0 \\ Dv\big|_{\Gamma} = v \end{cases}$$

It is well known [L-M] that

$$(3.2) \quad D \in L\,(H^s(\Gamma) \to H^{s+1/2}(\Omega)).$$

Now we are in a position to give an abstract semigroup model for (1.1). In fact, we recall from [L-T-2] that problem (1.1) can be model as an abstract second order equation

$$(3.3) \begin{cases} y_{tt} = -Ay + ADu + ADG(y) \quad \text{on } D(A)' \\ y(0) = y_0;\ y_t(0) = y_1 \end{cases}$$

or else as a first order abstract equation

$$(3.4) \begin{cases} \tilde{y}_t = \mathcal{A}\tilde{y} + \mathcal{B}u + \mathcal{B}G(\tilde{y}) \text{ on } \mathcal{D}(\mathcal{A})' \\ \tilde{y}(0) = (y_0, y_1) \end{cases}$$

where $\tilde{y} = (y, y_t)$ and

$$\mathcal{B}u \equiv \begin{pmatrix} 0 \\ ADu \end{pmatrix} \quad \in \mathcal{D}(\mathcal{A})'$$

Since $\mathcal{A}$ is skew-adjoint, eq (3.4) plainly suggests to take as a natural candidate $u = F(y_t) = -B^*(\tilde{y})$ since this choice then makes the corresponding feedback operator $\mathcal{A} - \mathcal{B}\mathcal{B}^*$ dissipative on E.

On the other hand, it can be verified that

$$\text{(3.5)} \qquad \mathcal{B}^*\tilde{y} = D^* y_t \quad \text{for } \tilde{y} \in \mathcal{D}(\mathcal{A})$$

(we recall that D is considered as a bounded operator from $L_2(\Gamma)$ into $L_2(\Omega)$).

By using Green formula one can also verify that

$$D^*Ay = \frac{\partial}{\partial y} y \qquad y \in D(A^*)$$

and consequently (3.5) can be rewritten equivalently as

$$\text{(3.6)} \qquad F(y_t) \equiv \mathcal{B}^*\tilde{y} = D^*y_t = \frac{\partial}{\partial \eta} A^{-1} y_t$$

With the above choice of the feedback operator, the open loop system (1.1) becomes the closed loop system

$$\text{(3.7)} \qquad \begin{cases} y_{tt} = -Ay - ADD^*y_t - ADG(y, y_t) \\ y(0) = y_0;\ y_t(0) = y_1 \end{cases}$$

equivalently

$$\text{(3.8)} \qquad \begin{cases} y_{tt} = \Delta y \\ y|_\Gamma = -\frac{\partial}{\partial \eta} A^{-1} y_t + G(y, y_t) \\ y(0) = y_0;\ y_t(0) = y_1 \end{cases}$$

Our main results concern the decay of the solutions to (3.7) or (3.8), either in the strong topology or else with additional assumptions imposed on the geometry of Ω in the uniform topology.

4. STATEMENT OF THE RESULTS IN THE LINEAR CASE ($G \equiv 0$)

Theorem 1 (strong stability)

Assume that $G \equiv 0$ in (3.8). Then for any $y_0 \in L_2(\Omega)$, $y_1 \in H^{-1}(\Omega)$

(i) the solution (y, y_t) of (3.8) exists and it is unique in E.

(ii) $u = F(y_t) \in L_2(\Sigma)$

(iii) $|\tilde{y}(t)|_E \equiv |y(t)|_{L_2(\Omega)} + |y_t(t)|_{H^{-1}(\Omega)} \to 0$

when $t \to \infty$.

The proof of Theorem 1 is functional analytic and it is based on the application of the Nagy-Foias theory (see [L-T-1]).

In order to obtain the stronger-uniform exponential decay for the system (3.8) we need to impose further assumptions on Ω.

We assume that there exists a vector field $\vec{h} = [h_1(x), \ldots h_n(x)] \in C^2(\bar{\Omega})$ defined on $\bar{\Omega}$, such that

H
(i) $\vec{h}$ is normal to Γ, at each point of Γ. i.e:
(ii) for the Jacobian $H(x)$ of h we have

$$H(x) = \begin{bmatrix} \frac{\partial h_1}{\partial x_1} & \cdots & \frac{\partial h_1}{\partial x_n} \\ \frac{\partial h_n}{\partial x_1} & & \frac{\partial h_n}{\partial x_n} \end{bmatrix} > 0$$

Theorem 2 [L-T-1] (exponential stabilization)

Assume that $G = 0$ and that assumption H holds. Then for any $y_0, y_1 \in E$, $t > 0$ we have

$$|\tilde{y}(t)|_E = |y(t)|_{L_2(\Omega)} + |y_t(t)|_{H^{-1}(\Omega)} \leq Ce^{-\alpha_0 t}[|y_0|_{L_2(\Omega)} + |y_1|_{H^{-1}(\Omega)}]$$

where $\alpha_0 > 0$ and $y(t), y_t(t)$ satisfy (3.8).

Remark

Although the full characterization of the domains satisfying condition H appears to be an open problem, it was shown however by Prof. E. De Giorgi that this condition holds indeed for all strictly convex and smooth domains Ω.

The proof of Theorem 2 lengthy and rather technical (based on pde methods), is given in [L-T-1]. It uses in an essential way the hyperbolicity of the underlined problem. In particular, the recently developed results on "exceptional" regularity of the traces of hyperbolic solutions (see [L-L-T]) are of crucial importance

Remark

In analyzing the results of Theorem 2 one should make a note of the following fact. It is well known that in the case of wave equation one can not stabilize exponentially the system by using compact feedback operators see ([G-1], (R-1)) or even more. By using A-bounded and finite rank operators [L-T-4]. On the other hand, in the special case when dim $\Omega = 1$, any feedback operator acting on the boundary is the finite rank operator. This fact may appear to be in contradiction with the statement of exponentive stability as in our Theorem 2. To explain this phenomenon, we must notice that the proposed feedback $Fy_t = {}^-D^*y_t$ although it might be finite rank, it would lead however to not- A bounded operator $\mathcal{B}\mathcal{B}^*$. In fact, it can be shown that

$A^{-1}\ \mathcal{B}\mathcal{B}^*$ is not bounded from $E \to E$. To see this, let us write (formally)

$$A^{-1}\ \mathcal{B}\mathcal{B}^*\tilde{y} = \begin{pmatrix} 0 & -A^{-1} \\ I & 0 \end{pmatrix} \begin{pmatrix} 0 \\ ADD^*y_t \end{pmatrix} = \begin{pmatrix} DD^*y_t \\ 0 \end{pmatrix}$$

To reach the conclusion it is enough to notice that DD^* is <u>not</u> bounded from $H^{-1}(\Omega) \to L_2(\Omega)$ (it is bounded however from $(H^1(\Omega))^1 \to L_2(\Omega)$).

5. LOCAL EXPONENTIAL STABILITY IN THE NONLINEAR CASE

Here we shall consider the effect of the nonlinear perturbation G on the stability of the system (3.7) or (3.8) with feedback F defined as in (3.6). We shall assume that the perturbation G satisfies the following conditions

$$(5.1)\quad \begin{cases} G \text{ is continuous from } E \equiv L_2(\Omega) x H^{-1}(\Omega) \to L_2(\Gamma) \\ G(0) = 0 \\ |G'(\tilde{y})|_{E \to L_2(\Gamma)} \to 0 \text{ when } |\tilde{y}|_E \to 0 \end{cases}$$

Let us point out that the main mathematical difficulty stems from the fact that our operator $\mathcal{B}$ (hence $\mathcal{B}$ G) is unbounded) on E. Even in the simplest linear case, unbounded perturbation added to the C_0-generators may destroy entirely the generation of the continuous semigroup. (see [T.1] for details) (this is in contrast with the analytic case where relatively bounded perturbation do not change the character of the generated semigroup). Therefore, in order to treat equation (3.8) we need to develop some sort of "unbounded perturbation" theory designed to work with general C_0 semigroups. To this end let us consider

$$(5.2)\quad \begin{cases} y_t = A_F y + \mathcal{B}\, G(y) \\ y(0) = y_0 \end{cases}$$

where A_F generates an exponentially stable semigroup on a Hilbert space H, $y_0 \in H$ B: $U \to D(A_F^*)'$, G: $U \to H$ is continuous and G satisfies

$$G(0) = 0$$

$$|G'(y)|_{H\to U} \to 0 \text{ when } |y|_H \to 0$$

With the above assumptions we have

Theorem 3 [L-2]

Assume that

(i) $A_F^{-1}B \in L(U\to H)$

(ii) $\int_0^\infty |B^* e^{(A_F^*+\alpha)t} x|_U \, dt \leq C|x|_H$ for all $\alpha < \alpha_0$

Where α_0 is the margin of stability of the semigroup A_F.

Then there exists $R > 0$ such that if $|y_0|_H < R$ then $y(t)$ solution of (5.2) satisfies

$$|y(t)|_H \leq Ce^{-\alpha t}|y_0|_H \quad \alpha < \alpha_0$$

Proof of Theorem 3 based on the application of the fixed point Theorem is given in [L.2]. In the remaining part of this section we shall show how to apply this theorem to our problem (3.8).

Remark

Notice that condition (ii) is always satisfied when B is bounded B or A_F is analytic and B is A_F bounded. Thus our results generalize to the case of general Co semigroup the known results for bounded perturbation or for relatively bounded perturbations of analytic semigroups.

To apply Theorem 3 to our problem (3.8) we set

$H \equiv E$; $U = L_2(\Gamma)$ $B = \mathcal{B}$; $A_F = \mathcal{A} - \mathcal{B}\mathcal{B}^*$.

From Theorem 2 we already know that for convex domains Ω, A_F generates exponentially stable semigroups. Thus, in order to apply Theorem 3 we need to verify hypothesis (i) and (ii)

Verification of (i)

Since

$$A_F^{-1}\mathcal{B}u = \begin{pmatrix} -DD^* & -A^{-1} \\ I & 0 \end{pmatrix}\begin{pmatrix} 0 \\ ADu \end{pmatrix} = \begin{pmatrix} Du \\ 0 \end{pmatrix}$$

boundedness of A_F^{-1} Bu as an operator acting from $L_2(\Gamma) \to L_2(\Omega)xH^{-1}(\Omega)$ follows immediately from (3.2).

Verification of (ii)

Let $\tilde{w}(t) \equiv e^{A_F^*t}\tilde{x}$; $\tilde{w} = (w,w_t)$; $\tilde{x} = (x_0,x_1)$

Then

$$(5.3)\quad \begin{cases} w_{tt} = -Aw - ADD^*w_t \\ w(0) = x_0 \\ w_t(0) = x_1 \end{cases}$$

On the other hand by virtue of (3.6) we have $B^*w(t) = D^*w_t = \frac{\partial}{\partial\eta} A^{-1}w_t$

Hence condition (ii) is equivalent to

$$(5.4)\quad \int_0^\infty \left|\frac{\partial}{\partial\eta} A^{-1}w_t(t)\right|_{L_2(\Gamma)} e^{\alpha t}dt \leqslant C[|w_0|_{L_2(\Omega)} + |w_1|_{H^{-1}(\Omega)}]$$

where w(t) satisfies (5.3)

Remark

Notice that the trace regularity of w(t) required by (5.7) does not follow from the interior regularity of the solutions to wave equation. In fact, with $(x_0,x_1) \in L_2(\Omega)xH^{-1}(\Omega)$ we obtain that $w_t \in C[0T;H^{-1}(\Omega)]$, consequently $A^{-1}w_t \in C[0T;H^1(\Omega)]$ and

$\frac{\partial}{\partial\eta} A^{-1} w_t \in ([0T; H^{-1/2}(\Gamma 1)$ instead of $L^2(\Sigma)$. Thus condition (5.4) requires certain "special" behavior of the traces to the hyperbolic solutions which is not a consequence of the trace theory and interior regularity.

To verify (5.4) we multiply both sides of (5.3) by $A^{-1} w_t e^{\alpha t}$. This yields

$$\tfrac{1}{2}\frac{d}{dt}[|A^{-1/2} w_t|^2_{L_2(\Omega)} e^{2\alpha t}] - \alpha |A^{-1/2} w_t|^2_{L_2(\Omega)} e^{2\alpha t}$$

$$= -\frac{1}{2}\frac{d}{dt} |w|^2_{L_2(\Omega)} e^{2\alpha t} + \alpha |w(t)|^2_{L_2(\Omega)} e^{2\alpha t} - |D^* w_t|_{L_2(\Gamma)} e^{2\alpha t}$$

Integrating the above equality from 0 to T gives

$$|A^{-1/2} w_t(T)|^2_{L_2(\Omega)} e^{2\alpha T} + |w(T)|^2_{L_2(\Omega)} e^{2\alpha T} + 2\int_0^T |D^* w_t(t)|^2_{L_2(\Gamma)} e^{2\alpha T} dt$$

$$= 2\alpha \int_0^T [|A^{-1/2} w_t(t)|^2_{L_2(\Omega)} + |w(t)|^2_{L_2(\Omega)}] e^{2\alpha t} dt$$

$$+ |A^{-1/2} x_1|^2_{L_2(\Omega)} + |x_0|^2_{L_2(\Omega)}$$

Since $|A^{-1/2} x|_{L_2(\Omega)} \sim |x|_{H^{-1}(\Omega)}$

by virtue of Theorem 2 we obtain (after passing to the limit when $T \to \infty$ and taking $\alpha < \alpha_0$)

$$\int_0^\infty |D^* w_t(t)|^2_{L_2(\Gamma)} e^{2\alpha t} dt \leq C[|x_0|^2_{L_2(\Omega)} + |x_1|^2_{H^{-1}(\Omega)}]$$

Using again that $\alpha < \alpha_0$ we conclude

(5.5) $$\int_0^\infty |D^*w_t(t)|_{L_2(\Gamma)} e^{2\alpha t}dt \leqslant C[|x_0|^2_{L_2(\Omega)} + |x_1|^2_{H^{-1}(\Omega)}]$$

(5.4) Hence part (ii) of Theorem 3 follows now from (5.5) and (3.6)

We summarize the discussion of this section in the following Theorem.

Theorem 4

Assume that $\Omega \subset R^n$ is a strictly convex domain. Moreover assume that G satisifes (5.1). Then the solution (y, y_t) to (3.8) decays exponentially and locally. i.e. $\exists$ R > 0 such that if

$$|y_0|_{L_2(\Omega)} + |y_1|_{H^{-1}(\Omega)} \leqslant R$$

then there exits $\alpha > 0$ such that

$$|y(t)|_{L_2(\Omega)} + |y_t(t)|_{H^{-1}(\Omega)} \leqslant Ce^{-\alpha t}[|y_0|_{L_2(\Omega)} + |y_1|_{H^1(\Omega)}.$$

REFERENCES:

[Ch-1] G. Chen. A note on boundary stabilization of the wave equation. SIAM J. Control and Opt. 18 (1981) 106-113

[G-1] J. S. Gibson. A note on stabilization of infinite dimensional linear oscillators by compact linear feedback. SIAM J. Contr. Optimiz. 18 (1980), 311-316

[L-1] J. Lagnese. Decay of solution of wave equations in a bounded region with boundary dissipation. J. Differ. Eq. 50, 2 (1983) 163-182

[L-2] I. Lasiecka. Boundary stabilization of hyperbolic and parabolic equations with nonlinearly perturbed boundary conditions, to appear in Recent Advances in System Science, Springer Verlag

[L-L-T] I. Lasiecka, J. L. Lions, R. Triggiani. Nonhomogeneous boundary value problems for second order hyperbolic operators, To appear in Journal de Math. Pure et Applique

[L-M] J. Lions, E. Magenes. Non-homogeneous Boundary Value Problems and Applications Vols I, II. Springer Verlag 1972

[L-T-1] I. Lasiecka, R. Triggiani. Uniform exponential delay in a bounded region with $L_2(0\infty; L_2(\Gamma))$ - feedback control in the Dirichlet boundary conditions. To appear in Journal Diff. Eq.

[L-T-3] I. Lasiecka, R. Triggiani. Regularity of hyperbolic equations under $L_2(0,t; L_2(\Gamma|)$ Dirichlet boundary terms. Appl. Math. Opt. 10, (1983), 275-286

[L-T-2] I. Lasiecka, R. Triggiani. A cosine operator approach to modelling $L_2(0T; L_2(\Gamma|)$ boundary input hyperbolic equations, Appl. Math. Opt. Vol V, 35-83 (1981)

[L-T-4] I. Lasiecka, R. Triggiani. Dirichlet boundary stabilization of the wave equation with damping feedback of finite range. J. of Math. Anal. and Applications. Vol. 87, No. 1, 1983 (p. 112-130)

[R-1] D. L. Russell. Decay rates for weddy damped systems in Hilbert spaces obtained via control-theoretic methods J. Diff. Eq. 19 (1975) 344-370

[T-1] R. Triggiani. A^{ε} finite rank perturbations of a s.c. group generators; A counter-example to generation and another condition for well-posedness. Lecture Notes in Mathematics, 1076, Infinite Dimensional Systems Springer-Verlag (1986) (227-254)

Stochastic Optimal Control Problem with Nonstandard Boundary Conditions

Sung J. Lee and A.N.V. Rao

Department of Mathematics
University of South Florida
Tampa, Florida 33620

1. INTRODUCTION

In this paper, the optimal control of a linear stochastic system with a quadratic cost function and with a set of constraints on the state, is investigated. More specifically, the dynamics is assumed to be described by the system of linear stochastic differential equations

$$dx(t) = A(t)\, x(t)\, dt + D(t)\, dW(t) + B(t)\, u(t)\, dt + \sum_{i=1}^{p} C_i(t)\, x(t)\, dW_i(t), \quad 0 \leq t \leq T$$

with the cost function

$$J = E\left\{\int_0^T (|U(t)\, u(t)|^2 + |V(t)\, x(t)|^2) dt + |G(x)|^2\right\}.$$

The system is to be optimized under the constraint that for any admissible control u, the corresponding response x is required to minimize the functional $E|F(x) - \gamma|^2$, where F is a given linear operator (to be made precise later) and γ, a given random vector. In some special cases, the above mentioned constraint leads to certain well-known boundary conditions of type $F(x) = \gamma$ w.p.1 discussued in the literature [AW1]. The optimal control will be synthesized as an open loop control and the results presented here extend to a stochastic setting, the earlier results obtained by one of the authors [L_1],[L_2]. The method used here is based on the theory of least-squares solutions of single-valued operator equation.

2. PRELIMINARIES

Let X and Y be real Hilbert spaces and let $X \oplus Y$ denote the product Hilbert space of all ordered pairs (x,y) with $x \in X$ and $y \in Y$. Throughout the paper unless otherwise stated, the inner products and norms in Hilbert spaces will be denoted by $\langle \cdot , \cdot \rangle$ and $||\cdot||$, respectively. Let M be a linear relation (also called a multi-valued linear operator) in $X \oplus Y$. Then

$$\begin{aligned}
\text{Dom } M &\equiv \{x: (x,y) \in M \text{ for some } y \in Y\} \\
\text{Range } M &\equiv \{y: (x,y) \in M \text{ for some } x \in X\} \\
M^{\perp} &\equiv \text{The orthogonal complement of } M \\
&= \{a : a \in X \oplus Y \text{ and } \\
&\qquad \langle m,a \rangle = 0 \text{ for all } m \in M\} \\
M^{*} &\equiv \text{Adjoint of } M \text{ in } X \oplus Y \\
&= \{(x,-y) : (y,x) \in M^{\perp}\} \\
\text{Null } M &\equiv \{x : (x,0) \in M\}
\end{aligned}$$

When Null M is closed, the orthogonal generalized inverse of M is defined by

$$M^{\#} = \{(x, (I - \mathcal{P})(z)) : x \in Y, z \in X, (z, x - \mathcal{P}^{+}(x)) \in M\}$$

where I is the identity operator and $\mathcal{P}$ (resp. $\mathcal{P}^{+}$) is the orthogonal projector of X (resp. Y) onto Null M. (rep. (Range $M)^{\perp}$). If S is a linear operator with graph $S \subset X \oplus Y$, then the adjoint of S, denoted also by S^*, is the multivalued operator such that graph $S^* = (\text{graph } S)^*$. Thus S^* is single-valued if and only if Dom S is dense in X. If S is a matrix, then its transpose conjugate is also denoted by S^*

If N is a $n \times m$ constant matrix then its Moore-Penrose generalized inverse will also be denoted by $N^\#$. The norm of N, denoted by $|N|$, is defined to be $(\text{tr } N N^*)^{1/2}$. Any element of $\mathbb{R}^k$ (k-dimensional real Euclidian space) is identified as a column vector. Throughout the paper T will denote a fixed positive real number. Let $(\Omega, \mathcal{A}, P)$ be a complete probability space where Ω is an underlying set, $\mathcal{A}$ a σ-algebra on Ω and P a probability measure on $\mathcal{A}$. Let W(t) $(t \geq 0)$ be a standard p-dimensional Brownian motion. Let $\{\mathcal{A}_t : t \geq 0\}$ be a family of σ-algebras in Ω such that $\mathcal{A}_t \subset \mathcal{A}_s \subset \mathcal{A}$ for all $0 \leq t \leq s$ and each W(t) is $\mathcal{A}_t$-measurable and for all $t, s \geq 0$, $W(t+s) - W(t)$ is independent of $\mathcal{A}_t$. For any positive integer k,

$$X_k \equiv L_2 \, ([0,T] \times \Omega, \, \mathbb{R}^k) \tag{2.1}$$

will denote the space of all real $\mathbb{R}^k$ - valued random processes x(t) such that

$$\int_0^T |x(t)|^2 \, dt < \infty \quad \text{w.p.1 and } E \int_0^T |x(t)|^2 \, dt < \infty \quad .$$

The inner product and the norm in the space X_k are defined, repectively, by

$$\langle f,g \rangle = E \int_0^T g^* (t) \, f(t) \, dt \quad ||f|| = \langle f,f \rangle^{1/2}.$$

Let m be a positive integer and let

$$\mathcal{U} = \{u : u \in X_m,\ u(t) \text{ is } \mathcal{A}_t\text{-measurable for for each } t \in [0,T]\}. \qquad (2.2)$$

Then it is clear that $\mathcal{U}$ is a closed subspace of X_m and thus is itself a Hilbert space. We shall further denote by $L_2(\Omega, \mathbb{R}^k)$ the real Hilbert space of all $\mathbb{R}^k$-valued random vectors α with $E|\alpha|^2 < \infty$. The inner product and norm of this space will be defined by

$$\langle \alpha, \beta \rangle = E(\beta^* \alpha) \text{ and } ||\alpha|| = \{E|\alpha|^2\}^{1/2}, \text{ respectively.}$$

3. STATE EQUATION AND CONSTRAINTS

In this section we will be concerned with the stochastic integral equation

$$x(t) = x(0) + \int_0^t A(s)\, x(s)\, ds + \int_0^t D(s)\, dW(s) \qquad (3.1)$$

$$+ \int_0^t B(s)\, u(s)\, ds + \sum_{i=1}^{p} \int_0^t C_i(s) x(s) (dW)_i(s), \quad 0 \leq t \leq T.$$

Here the state x is a $\mathbb{R}^n$ process; the control u is in the space $\mathcal{U}$ defined earlier in (2.2) and $A(t)$ (n×n), $D(t)$ (n × p), $B(t)$ (n × m) and $C_i(t)$ (n × n) are all $\mathcal{A}_t$-measureable matrix-valued processes, continuous w.p.1 and uniformily continuous in $t \in [0,T]$.

The process $(W(t))_i$ is the i-th component of the standard P-dimensional Brownian motion process $W(t)$, $0 \leq t \leq T$.

All integrals involving the process $W(t)$ (or $W(t))_i$) are to be understood as stochastic integrals in the Ito sense [G3]. By a solution of the equation (3.1), we mean a $\mathbb{R}^n$-process $x \in X_n$, such that $x(t)$ is $\mathcal{A}_t$-measureable for each

$t \in [0,T]$ and satisfies equation (3.1) for all $t \in [0,T]$ whenever $x(0)$ is $\mathcal{A}_0$-measurable and is in $L_2(\Omega,\mathbb{R}^n)$ and $u \in \mathcal{U}$. It is known that under the assumptions that we have made, equation (3.1) has a unique solution on $[0,T]$ for each $u \in \mathcal{U}$, ([GS_1]).

It is customary to denote the integral equation (3.1), by the differential equation

$$dx(t) = A(t)\, x(t)\, dt + D(t)\, dW(t) + B(t)\, u(t)\, dt + \sum_{i=1}^{p} C_i(t)\, x(t)\, (dW(t))_i, \quad 0 \leq t \leq T. \tag{3.2}$$

Equation (3.1) or (3.2) will be referred to as the state equation.

DEFINITION An element $x \in X_n$ is said to have a stochastic differential if there exists a $p\times1$ process $a(t)$ and an $n\times p$ process $b(t)$ such that both are $\mathcal{A}_t$ measurable, $\int_0^T |a(t)|^2 dt$, $\int_0^T b^*(t)\, b(t)\, dt$ exist and satisfy the differential equation $dx(t) = a(t)\, dt + b(t)\, dW(t)$, $0 \leq t \leq T$.

Let f_1 and f_2 be $\mathcal{A}_t$-measureable, $n\times d$ processes that are uniformly bounded in $t \in [0,T]$, w.p.1. Let us define a linear operator F with

$$\text{Dom } F = \left\{ x \in X_n \ :\ x \text{ has a stochastic differential} \right\}$$

and range in $L_2(\Omega,\mathbb{R}^d)$ by

$$F(x) = \int_0^T f_1^*(t)dx(t) + \int_0^T [f_2^*(t) - f_1^*(t)A(t)]\, x(t)\, dt. \tag{3.3}$$

Let γ be a given element of $L_2(\Omega,\mathbb{R}^d)$ which is $\mathcal{A}_0$ measurable. We shall now define a dynamical system $\mathcal{D}$ in $X_m \oplus X_n$.

DEFINITION Let $u \in X_m$ and $x \in X_n$. Then $(u,x) \in \mathscr{D}$ if and only if

(i) $u \in \mathscr{U}$

(ii) x is a process such that

$$dx(t) = A(t)x(t)\,dt + D(t)\,dW(t) + B(t)\,u(t)\,dt + \sum_{i=1}^{p} C_i(t)x(t)(dW(t))_i\,,\quad 0 \le t \le T$$

(iii) $E|F(x) - \gamma|^2 = \min_y E|F(y) - \gamma|^2$,

where the minimum is taken over all $y \in X_n$ such that

$$dy(t) = A(t)\,y(t)\,dt + D(t)\,dW(t) + B(t)\,u(t)\,dt + \sum_{i=1}^{p} C_i(t)\,y(t)\,(dW(t))_i,\quad 0 \le t \le T.$$

Let ϕ be the n×n process such that

$$d\phi(t) = A(t)\phi(t)\,dt + \sum C_i(t)\,\phi(t)\,(dW(t))_i,\quad 0 \le t \le T \tag{3.4}$$
$$\phi(0) = I_{n\times n}.$$

(See [A] for the existence of ϕ). Let us define a n×1 process ζ by

$$\zeta(t) = \phi(t)\int_0^t \phi^{-1}(s)[D(s)\,dW(s) - \sum_{i=1}^{p} C_i(s)\,(D(s))_i ds] \tag{3.5}$$

We note that $\zeta(t)$ is $\mathscr{A}_t$ - measurable and belongs to X_n. Define an operator H on $\mathscr{U}$ by

$$H(u) = \phi(t)\int_0^t (\phi^{-1}Bu)\,ds,\ 0 \le t \le T. \tag{3.6}$$

Then for any $u \in \mathscr{U}$ the solution x(t) of (3.2) with $x(0) = \alpha \in L_2(\Omega,\mathbb{R}^d)$ is given by

$$x = \phi\alpha + H(u) + \zeta. \tag{3.7}$$

Let us now define the operator Q_f from $L_2(\Omega,\mathbb{R}^n)$ into $L_2(\Omega,\mathbb{R}^d)$ by

Dom $Q_f = \{\alpha: \ \alpha \in L_2(\Omega,\mathbb{R}^n), \ \alpha$ is $\mathcal{A}_0$ measurable$\}$

and

$$Q_f\alpha := \left[\int_0^T f_1^* \, d\phi(t) + \int_0^T [f_2^* (t) - f_1^* (t) \, A(t)]\phi(t)dt\right]\alpha \quad (3.8)$$

Denote by Q_f^* the adjoint subspace of Q_t in $L_2(\Omega,\mathbb{R}^n) \oplus L_2(\Omega,\mathbb{R}^d)$.

It is clear that Dom Q_f is closed in $L_2(\Omega,\mathbb{R}^n)$ and so Q_f is <u>nondensely</u> defined and so Q_f^* multi-valued. Moreover, Q_f is continuous and has a closed graph. The following lemma will be needed for later use.

LEMMA 3.1 Assume that Range Q_f is closed then

$$\text{Dom } (Q_f^*)^\# = \text{Dom } (Q_f^\#)^* = L_2(\Omega,\mathbb{R}^n), \ \text{Dom } Q_f^\# = L_2(\Omega,\mathbb{R}^d).$$

Proof From Theorem 2 of [LN$_2$], we have

$$\text{Dom } Q_f^\# = \text{Range } Q_f \dotplus (\text{Range } Q_f)^\perp$$

$$\text{Dom } (Q_f^*)^\# = \text{Range } Q_f^* \dotplus (\text{Range } Q_f)^\perp.$$

It is well known [AG] that

$$L_2(\Omega,\mathbb{R}^d) = (\text{Range } Q_f)^c + (\text{Range } Q_f)^\perp$$

where the subscript c denotes the closure. A similar result

holds if Q_f is replaced by Q_f^*.

The following gives a sufficient condition for Range Q_f to be closed.

LEMMA 3.2 Assume that the eigenvalues $\lambda_i(w)$ of $(F(\phi))^*F(\phi)$ satisfy
$\lambda_i(w) \geq \delta > 0$ for all $w \in \Omega$, $i = 1,2,\ldots,n$ for some $\delta > 0$. Then range of Q_f is closed.

Proof Let $\{Q_f(\alpha_n)\}$ ($\alpha_n \in$ Dom Q_f) be a sequence converging to β in $L_2\ (\Omega,\mathbb{R}^d)$. Then $\{\alpha_n\}$ is a Cauchy sequence in Dom Q_f. For,

$$E(\beta_n - \beta_m)^*(\beta_n - \beta_m) = E[(\alpha_n - \alpha_m)^*(F(\phi))^*F(\phi)(\alpha_n - \alpha_m)]$$

$$\geq \delta\ E\ (\alpha_n - \alpha_m)^*(\alpha_n - \alpha_m).$$

Since Dom Q_f is closed, $\{\alpha_n\}$ converges to a $\alpha \in$ Dom Q_f. Thus $Q_f(\alpha_n) \to Q_f(\alpha)$ as $n\to\infty$, and so $\beta \in$ Range Q_f. This shows that Range Q_f is closed.

We now give the following characterization of $\mathcal{D}$.

THEOREM 3.3 Let the range of Q_f be closed. Then

(i) $(u,x) \in \mathcal{D}$ if and only if $u \in \mathcal{U}$ and

$$x = H(u) + \xi + \phi\ [Q_f^{\#}\{\gamma - F(H(u)) - F(\xi)\} + \alpha_0]$$

for some $\alpha_0 \in$ Null Q_f.

(ii) $(u,x) \in \mathcal{D}$ if and only if (u,x) satisfies the state equation and $F(x)-\gamma \in (\text{Range } Q_f)^{\perp}$. If, in particular, Null $(Q_f)^* = \{0\}$, then the boundary condition becomes $F(x) = \gamma$, w.p.1.

Proof (i) Assume $(u,x) \in \mathcal{D}$. Then from (3.7), we have

$$x = \phi\alpha + H(u) + \xi, \text{ for some } \alpha \in L_2(\Omega,\mathbb{R}^n) \tag{3.9}$$

which is $\mathcal{A}_0$ - measurable. Furthermore from the defininitions of $\mathcal{D}$ and Q_f, it follows that

$$E|Q_f(\alpha) + F(H(u)) + F(\xi) - \gamma|^2 \tag{3.10}$$

$$\leq E|Q_f(\beta) + F(H(u)) + F(\xi) - \gamma|^2 \text{ for all } \beta \in \text{Dom } Q_f.$$

Hence α is the least-squares solution of $Q_f(\beta) = \gamma - F(H(u)) - F(\xi)$ and therefore, since Range Q_f is closed,

$$\alpha = Q_f^{\#}\{\gamma - F(H(u)) - F(\xi)\} + \alpha \quad \text{for}$$

some $\alpha_0 \in$ Null Q_f. Using the above expression for α in (3.10), the result (i) of the Theorem follows. The converse is proved similarly.

(ii) Let $F(x) - \gamma \in (\text{Range } Q_f)^{\perp}$. Then for some $q \in (\text{Range } Q_f)^{\perp}$, we have

$$F(x) - \gamma = q. \tag{3.11}$$

Using (3.9) in (3.11), we obtain

$$Q_f(\alpha) = q - F(H(u)) - F(\xi) + \gamma \tag{3.12}$$

Since Dom $= L_2\ (\Omega,\mathbb{R}^d)$, it follws that

$$a = Q_f^{\#}\ (q) + Q_f^{\#}\ (\gamma - F(H(u)) - F(\xi)) + \alpha_0$$

for some $\alpha_0 \in$ Null Q_f. Now if $Q_f^{\#}\ (q) = 0$, then

$$\alpha = Q_f^{\#}\ (\gamma - F(H(u)) - F(\xi)) + \alpha_0$$

implying that α is a least squares solution of

$$Q_f(\beta) = \gamma - F(H(u)) - F(\xi)$$

which, from part (i), implies that $(u,x) \in \mathcal{D}$. Conversely, if $\{u,x\}$ satisfies the state equation (3.2) and $F(x) - \gamma \equiv q \in (\text{Range } Q_f)^{\perp}$, we can see easily that $\{u,x\} \in \mathcal{D}$. Now, if Null $Q_f^* = \{0\}$, then $(\text{Range } Q_f)^{\perp} = \{0\}$. Hence $F(x) - \gamma = 0$ in $L_2(\Omega, \mathbb{R}^d)$ implying that $F(x) = \gamma$, w.p.1. This completes the proof.

Similar to the deterministic cases considered in $[L_1]$, $[L_2]$ we shall demonstrate that the boundary operator can be reduced to a generalized two point operator.

THEOREM 3.4 Let h be a n×d random process such that for all $t \in [0,T]$, $h(t)$ is $\mathcal{A}_t$-measurable and h is uniformly bounded in $t \in [0,T]$ w.p.1. Let M and N be d×n random matrices such that $E(M^*M)$, $E(N^*N)$ exist. Then w.p.1 $F(x) = \int_0^T h^*(t)x(t)dt + M\,x(0) + N\,x(T)$ for all x in Dom F if and only if there exists $\Omega_1 \in \mathcal{A}$ with $P(\Omega - \Omega_1) = 0$ such that for any $w \in \Omega_1$ fixed,

$$f_1(\cdot,w) \in AC\,[0,T],\ \int_0^T |\dot{f}_1(t,w)|^2\,dt < \infty\ ,\ (\cdot = \frac{d}{dt}),$$

$$-\dot{f}_1(t,w) - A^*(t,w)f_1(t,w) = h(t,w) - f_2(t,w)$$

for all $0 \leq t \leq T$, and

$$-f_1^*(0,w) = M(w),\ f_1^*(T,w) = N(w).$$

Proof. Define deterministic differential operators S_0, S_1 in $L^2([0,T]\mathbb{R}^n)$ by

Dom $S_1 = \{f:[0,T] \to \mathbb{R}^n:\ f \in A\,C[0,T],\ \dot{f} \in L^2([0,T]\mathbb{R}^n)\}$,

$$\text{Dom } S_0 = \{f \in \text{Dom } S_1 : \ f(0) = f(T) = 0\},$$

$$S_1(f) = \frac{df}{dt}, \ f \in \text{Dom } S_1,$$

$$S_0(f) = \frac{df}{dt}, \ f \in \text{Dom } S_0.$$

Then it is well-known (see, for example, [L2]) that S_0, S_1 are densely defined closed operators in $L_2([0,T],\mathbb{R}^n)$ such that $S_0 \subset S_1$ and

$$\text{Dom } S_0^* = \text{Dom } S_1, \tag{3.13}$$

$$S_0^*(f) = -\frac{df}{dt}, \ f \in \text{Dom } S_0^*.$$

We now prove the "only if" part. Let $\Omega_1 \in \mathcal{A}$ with $P(\Omega-\Omega_1) = 0$ such that $w \in \Omega_1$ if and only if

$$F(x)(w) = \int_0^T h^*(t,w)x(t,w)dt + M(w)x(0,w)+N(w)x(w,T) \tag{3.14}$$

for all $x \in \text{Dom } F$. Let $w \in \Omega_1$ be fixed. Then (3.14) holds, in particular, for all deterministic $x \in \text{Dom } S$. Thus

$$\int_0^T f_1^*(t,w)\dot{x}(t)dt=\int_0^T \{h^*(t,w)-f_2^*(t,w)+f_1^*(t,w)A(t,w)\}x(t)dt \tag{3.15}$$

for all $x \in \text{Dom } S_0$. Thus by (3.13) $f_1(\cdot,w) \in \text{Dom } S_0^*$ and

$$\dot{f}_1(t,w) = -h(t,w) + f_2(t,w) - A^*(t,w)\ f_1(t,w)$$

for almost all $t \in [0,T]$. Returning to (3.14) we see that

$$M(w)x(0) + N(w)\ x(T) = f_1^*\ (T,w)x(T) - f_1^*(0,w)x(0)$$

for all $x \in \text{Dom } S_1$. Thus

$$M(w) = -f_1^*(0,w), \ N(w) = f_1^*\ (T,w). \tag{3.16}$$

This proves the "only if" part. We now prove the "if" part.

Let Ω_1 be given as in the Theorem. Let $x \in X_n$ be any element having an Ito differential. Then since $f_1(t,w)$ does not contain a term involving the Brownian motion W_t, we see that by [Theorem 3, p.22 of [GS1]] that

$$f_1^*(t,w)d_t x(t,w) = d_t\{f_1^*(t,w)x(t,w)\} - \dot{f}_1^*(t,w)x(t,w)dt$$

Using this fact we can show easily that (3.14) holds for all $w \in \Omega_1$ and $x \in X_m$ and having an Ito differential. This completes the proof.

4. MAIN RESULTS

In this section we will consider the main optimal control problem of the paper: minimize

$$J(u,x) = E\{\int_0^T (|U(t)\ u(t)|^2 + |V(t)x(t)|^2)dt + |G(x)|^2\} \quad (*)$$

subject to the constraint $(u,x) \in \mathcal{D}$, where

(i) $U(t)$, $V(t)$ are $m\times m$, $n\times n$ real matrix valued $\mathcal{A}_t$ measurable random processes such that $|U(t)|$, $|V(t)|$ are uniformly bounded in $t \in [0,T]$ w.p.1.

(ii) G is a linear operator from X_n into $L_2(\Omega,\mathbb{R}^d)$ defined by

$$G(x) = \int_0^T g_1^*\ (t)\ dx\ (t) + \int_0^T (g_2^*(t) - g_1^*(t)\ A(t))\ dx\ (t),$$

for all x having stochastic differentials. The functions $g_1(t)$ and $g_2(t)$ are real $n\times d$, $\mathcal{A}_t$-measurable processes that are uniformly bounded in $t \in [0,T]$. A pair $(u^+,x^+) \in \mathcal{D}$ that minimizes J, will be called an optimal pair; u^+ will be called an optimal control and x^+, a corresponding optimal response.

Let $||\cdot||$ also denote the norm in the product Hilbert space $X_m \oplus X_n \oplus L_2(\Omega,\mathbb{R}^d)$. Thus for (u,x,α) in this space

$$||(u,x,\alpha)||^2 = E\{\int_0^T (|u|^2 + |x|^2)dt + |\alpha|^2\}.$$

Define an operator $\mathcal{M}$ from $X_m \oplus L_2(\Omega,\mathbb{R}^n)$ into $X_m \oplus X_n \oplus L_2(\Omega,\mathbb{R}^d)$ by

$$\text{Dom } \mathcal{M} = \{(u,\alpha): u \in \mathcal{U},\ \alpha \in \text{Null } Q_f,\ F(H(u)) \in \text{Dom } Q_f^{\#}\}$$

$$\mathcal{M}(u,\alpha) = \{Uu, V[H(u) - \phi Q_f^{\#}F(H(u)),\ G[H(u) - \phi Q_f^{\#}F(H(u))]\} + \{0,\ V\phi\alpha, G(\phi(\alpha))\}. \tag{4.1}$$

Note that $\mathcal{M}$ is a nondensely defined bounded linear operator. Moreover if Range Q_f is closed, then the condition $F(H(u)) \in$ Dom $Q_f^{\#}$ becomes redundant.

We now prove the following lemma.

LEMMA 4.1 Assume that the range of Q_f is closed. Let

$$a = \xi + \phi\ Q_f^{\#}(\gamma - F(\xi)).$$

Then (u^+,x^+) is optimal if and only if

$u^+ \in \mathcal{U}$, $x^+ = \phi\ \alpha_0^+ - \phi\ Q_f^{\#}\ F(H(u^+)) + H(u^+) + a$ for some $\alpha_0^+ \in$ Null Q_f such that

$$\mathcal{M}(u^+,\ \alpha_0^+) + (0,\ Va,\ G(a)) \in (\text{Range } \mathcal{M})^{\perp}$$

Proof Let $(u,x) \in \mathcal{D}$. Then by Theorem 3.3,

$$x = H(u) + \xi + \phi\ [Q_f^{\#}\{\gamma - F(H(u)) - F(\xi)\} + \alpha], \tag{4.2}$$

for some $\alpha \in$ Null Q_f. Therefore

$$||J(u,x)||^2 \equiv ||(Uu, Vx,\ G(x))||^2$$
$$= ||\mathcal{M}(u,\alpha) + (0, Va,\ G(a))||^2$$

Hence, (u^+,x^+) is optimal if and only if $u^+ \in \mathcal{U}$ and

$$||\mathcal{M}(u^+,\alpha) + (0,Va,G(a))|| \le ||\mathcal{M}(u^+,\beta) + (0,Va,G(a))|| \tag{4.3}$$

for some $\alpha \in$ Null Q_f and all $\beta \in$ Null Q_f.

By Proposition 3.2 of [LN2], the previous inequality is equivalent to $(u^+,\alpha) \in$ Dom $\mathcal{M}$ and $\mathcal{M}\ (u^+,\alpha) + (0,Va,\ G(a)) \in$

(Range $\mathcal{M})^{\perp}$. This completes the proof.

In the following Theorem we discuss the existence of an optimal control for the optimization problem (*).

THEOREM 4.2 Assume that the Range $Q_f^{\#}$ is closed. Suppose that

(i) dim Null $Q_f < \infty$

(ii) for all $\epsilon[0,T]$, U^{-1}exists w.p.1 and $E\int_0^T |U^{-1}(t)|^2 dt<\infty$.

Then there exists an optimal control for the problem (*).

Proof. By the assumption on U, the map $u \to U\bar{u}^1$ defines a bounded linear operator form $\mathcal{U}$ into X_m. It therefore follows that the set $\{\mathcal{M}(u,0):u \in U\}$ is closed. Since dim Null $Q_f < \infty$, the dimension of $\{\mathcal{M}(0,\alpha) : \alpha \in \text{Null of } Q_f\}$ is finite. Since

$$\text{Range } \mathcal{M} = \{\mathcal{M}(u,0):u \in U\} \dot{+} \{\mathcal{M}(0,\alpha):\alpha \in \text{Null } Q_f\},$$

we see that the range of $\mathcal{M}$ is closed. Moreover

$$\text{Range } \mathcal{M} \dot{+} (\text{Range } \mathcal{M})^{\perp} = X_m \oplus X_n \oplus L_2(\Omega,\mathbb{R}^d).$$

By Proposition 3.1 of [LN2], the equation $\mathcal{M}(u,\alpha) = -(0,Va,G(a))$ has always a least-sqaures solution, say, (u^+,α^+).
But then

$$\mathcal{M}(u^+,\alpha^+) + (0,Va,G(a)) \in (\text{Range } \mathcal{M})^{\perp}.$$

Hence by Lemma 4.1, u^+ is an optimal control. This completes the proof.

In Lemma 4.1, we have characterized the optimal control in terms of the least-squares solutions of the equation

$\mathcal{M}(u,\alpha) = -(0, Va, G(a))$. These solutions are rather difficult to obtain because of the nature of $\mathcal{M}$. We will therefore describe the optimal controls by a system of integro-differential equations. To do this, it is convenient to introduce the operators H^+, H_W by

$$H^+(x) = \phi^{*-1}(t) \int_t^T \phi^*(s)x(s)ds \tag{4.4}$$

$$H_W(x) = \phi^{*-1}(t) \sum_{i=1}^{p} \int_t^T \phi^*(s)C_i^*(s)x(s)(dW(s))_i \quad 0 \leq t \leq T,$$

for $x \in X_n$ and $x(s)$ is $\mathcal{A}_s$-is measurable. Note that $H^+(x)$ and $H_W(x)$ are both $\mathcal{A}_T$-measurable. Recall that $\langle\alpha,\beta\rangle_\Omega : = E(\beta^*\alpha)$ for $\alpha,\beta \in L_2(\Omega,\mathbb{R}^k)$ and $\langle x,y\rangle : = E[\int_0^T y^*x \, dt]$ for $x,y \in X_k$. The following lemma will be needed later.

LEMMA 4.3 Let $x \in X_n$ and let y be a n×n process with its columns in X_n. Let $u \in \mathcal{U}$ and $\alpha \in L_2(\Omega,\mathbb{R}^d)$. Then

(i) $\langle x, H(u)\rangle = \langle V^*H^+(x), u\rangle$

(ii) If Range Q_f is closed, then

$$\langle x, y\, Q_f^{\#}F(H(u))\rangle = \langle B^*[f_1 + H^+(f_2) + H_W^+(f_1)](r), u\rangle$$

where

$$r: = \int_0^T (Q_f^{\#})^*(y^*x)(t)dt$$

(iii) $\langle\alpha, G(H(u))\rangle_\Omega = \langle B^*[g_1 + H^+(g_2) + H_W^+ (g_1)](\alpha), u\rangle$

(iv) If Range Q_f is closed, then

$$\langle\alpha, G(\phi)Q_f^{\#}F(H(u))\rangle_\Omega = \langle B^*[f_1 + H^+(f_2) + H_W^+(f_1)](Q_f^{\#})^*[(G(\phi))^*\alpha, u\rangle$$

Proof. We will prove (ii). Let $\ell: = \langle x, yQ_f^{\#}F(H(u))\rangle$. Since Range Q_f is closed, by Lemma 3.1, for all $t \in [0,T]$, $y^*(t)x(t) \in$ Dom $(Q_f^{\#})^*$. Thus

$$\ell = \int_0^T \langle x(t), [yQ_f^{\#}F(H(u))](t)\rangle_\Omega dt$$
$$= \int_0^T \langle (Q_f^{\#})^* y^*(t)x(t), F(H(u))\rangle_\Omega dt$$
$$= E \int_0^T (F(H(u))^* (Q_f^{\#})^* y^*(t)x(t)dt$$
$$= E[(F(H(w))^* r].$$

On the other hand,

$$F(H(u)) = \int_0^T f_1^*(t)d(H(u))(t) + \int_0^T (f_2^*(t) - f_1^*(t)A(t))(H(u))(t)dt$$
$$= \int_0^T [f_1^*(t)(d\phi(t))\int_0^t \phi^{-1}(s)B(s)u(s)ds] + \int_0^T f_1^*(t)B(t)u(t)dt$$
$$+ \int_0^T (f_2^*(t) - f_1^*(t)A(t))\phi(t) \int_0^t \phi^{-1}(s)B(s)u(s)ds\ dt$$
$$= I_1 + I_2 + I_3;$$

where

$$I_1 = \int_0^T f_1^*(t)[A(t)\phi(t)dt + \sum_{i=1}^p C_i(t)\phi(t)(dW(t))_i]\int_0^t \phi^{-1}(s)B(s)u(s)ds$$
$$= \int_0^T f_1^*(t)A(t)(\int_0^t \phi^{-1}(s)B(s)u(s)ds)dt$$
$$+ \int_0^T f_1^*(t) \sum_{i=1}^p [C_i(t)\phi(t)\int_0^t \phi^{-1}(s)B(s)u(s)ds](dW(t))_i$$
$$= \int_0^T [\int_{t=s}^{t=T} f_1^*(t)A(t)\phi(t)\phi^{-1}(s)dt]B(s)u(s)ds$$
$$+ \int_0^T [\int_{t=s}^{t=T} f_1^*(t) \sum_{i=1}^p C_i(t)\phi(t)\phi^{-1}(s)(dW(t))_i]B(s)u(s)ds$$
$$= \int_0^T (H^+(A^*f_1))^*(s)B(s)u(s)ds + \int_0^T (H_W^+(f_1))^* B(s)u(s)ds,$$

$$I_2 = \int_0^T f_1^*(t)B(t)u(t)dt,$$

$$I_3 = \int_0^T [\int_s^T \{f_2^*(t) - f_1^*(t)A(t)\}\phi(t)dt\phi^{-1}(s)]B(s)u(s)ds$$
$$= \int_0^T [\phi^{-1*}(s)\int_s^T \phi^*(t)(f_2(t) - A^*(t)f_1(t))dt]^* B(s)u(s)ds$$

$$= \int_0^T [H^+(f_2 - A^* f_1)]^* B(s)u(s)ds.$$

Thus

$$F(H(u)) = \int_0^T (H^+(A^* f_1))^* Bu\ ds + \int_0^T (H_W^+(f_1))^* Bu\ ds$$
$$+ \int_0^T f_1^* Bu\ ds + \int_0^T [H^+(f_2 - A^* f_1)]^* Bu\ ds$$
$$= \int_0^T (H_W^+(f_1))^* Bu\ ds + \int_0^T f_1^* Bu\ ds + \int_0^T (H^+(f_2))^*\ Bu\ ds$$
$$= \int_0^T [f_1 + H^+(f_2) + H_W^+(f_1)]^* Bu\ ds.$$

Thus

$$E[(F(H(u)))^* r] = E\{\int_0^T u^* B^* [f_1 + H^+(f_2) + H_W^+(f_1)] ds\ r\}$$
$$= \langle B^* \{f_1 + H^+(f_2) + H_W^+(f_1)]\ r,\ u\rangle.$$

This proves (ii). We now prove (iv). Let

$\ell = \langle \alpha, G(\phi) Q_f^{\#} F(H(u))\rangle_\Omega$. Since Range Q_f is closed, by Lemma 3.1, $(G(\phi))^* \alpha \in$ Dom $Q_f^{\#}$.

Thus

$$\ell = \langle (G(\phi))^* \alpha,\ Q_f^{\#} F(H(u))\rangle_\Omega = \langle (Q_f^{\#})^* (G(\phi))^* \alpha, F(H(u))\rangle_\Omega$$
$$= E\{F(H(u))\}^* (Q_f^{\#})^* (F(\phi))^* \alpha$$
$$= E\int_0^T \{u^* b^* [f_1 + H^+(f_2) + H_W^+(f_1)] ds\}\ (Q_f^{\#})^* (G(\phi))^* \alpha$$
$$= \langle B^* [f_1 + H^+(f_2) + H_W^+(f_1)](Q_f^{\#})^* [(G(\phi))^* \alpha], u\rangle$$

This proves (iv). As for (iii),

$$\langle\ \alpha\ ,\ G(H(u))\rangle_\Omega = E(G(H(u)))^* \alpha$$
$$= E \int_0^T u^* B^* [g_1 + H^+(g_2) + H_W^+(g_1)]\ ds\ \alpha$$
$$= \langle B^* [g_1 + H^+(g_2) + H_W^+(g_1)]\ \alpha,\ u\rangle.$$

This proves (iii). Part (i) is easy to check.

REMARK 4.1 In the proofs of (ii) and (iv) we have used the fact that Dom Q_f is nondensely defined. Its consequence is that Dom $(Q_f^{\#})^* = L_2(\Omega,\mathbb{R}^n)$. Now since $X \equiv \text{Dom } Q_f = \{\alpha \in L_2(\Omega,\mathbb{R}^n): \alpha \text{ is } \mathscr{A}_0\text{-measurable}\}$, is closed, we could have taken this as the whole space. Let us denote by $(Q_f^{\#})_*$ the adjoint of $Q_f^{\#}$ when the graph of Q_f is a linear relation in the space $X \oplus L_2(\Omega,\mathbb{R}^d)$. Then $\text{Dom}(Q_f^{\#})_* = X$. Consequently $(Q_f^{\#})_*(\alpha)$ for $\alpha \in L_2(\Omega,\mathbb{R}^n)$ will <u>not</u> be well-defined. This is the reason why we have treated Q_f as a non-densely defined operator.
We now prove the following lemma.

LEMMA 4.4 Let $f \in X_m$. Then $f \in X_m \ominus \mathscr{U} \equiv \{f \in X_m : f \perp \mathscr{U}\}$, if and only if $f \in X_m$ and $E[F(t)|\mathscr{A}_t] = 0$, for almost all $t \in [0,T]$.

Proof Let $f \in X_m$ and $u \in \mathscr{U}$. Then $u(t)$ is $\mathscr{A}_t$-measurable and $E(u^*(t)\, f(t))$ exists. Also

$$E\int_0^T (u^*(t)\ f(t)) = E\int_0^T [u^*(t)\ E(f(t)|\mathscr{A}_t)]\ dt.$$

This vanishes for all $u \in \mathscr{U}$ if and only if $E[f(t)|\mathscr{A}_t] = 0$ for almost all $t \in [0,T]$ as $E[f(t)|\mathscr{A}_t] \in \mathscr{U}$.

In the following Theorem we will describe an optimal pair by a set of integro-differential equations. To do this it is convenient to introduce the functions $\psi_i(x) \in X_n$ $(i = 1,2)$ by

$$(\psi_i(x))(t)=g_i(t)G(x)-f_i(t)\{\int_0^T (Q_f^{\#})^*[\phi^*(t)V^*(t)V(t)x(t)]dt + (Q_f^{\#})^*[(G(\phi))^*G(x)]\}. \tag{4.5}$$

THEOREM 4.5 Assume that the range of Q_f is closed. Then

(u^+, x^+) is optimal if and only if

(i) $(u^+, x^+) \in \mathcal{D}$

(ii) $\int_0^T \phi^*(t)V^*(t)V(t)x^+(t)dt + (G(\phi))^* G(x^+) \in (\text{Null } Q_f)^\perp$

(iii) $(U^* U u^+)(t) = -B^* \; E\,[z|\mathcal{A}_t]$, for almost all $(t,w) \in [0,T] \times \Omega$, where

$$z = H^+[V^* V x^+ + \psi_2(x^+)] + H_W^+[\psi_1(x^+)] + \psi_1(x^+).$$

Proof By Lemma 4.1, $\{u^+, x^+\}$ is optimal if and only if $u^+ \in \mathcal{U}$ and

$$x^+ = \phi\, \alpha_0^+ - \phi\, Q_f^{\#} F(H(u^+)) + H(u^+) + \mathbf{a} \tag{4.6}$$

for some $\alpha_0^+ \in$ Null Q_f such that

$$\{0, V\,\mathbf{a}\,,\, G(\mathbf{a})\} \in (\text{Range } \mathcal{M})^\perp \tag{4.7}$$

where

$$\mathbf{a} = \xi + \phi Q_f^{\#}(\gamma - F(\xi)). \tag{4.8}$$

We will characterize (4.6). Let u^+, x^+, α_0^+ be as above. Since

$$\mathcal{M}(u^+, \alpha_0^+) + \{0, V\mathbf{a}, G(\mathbf{a})\} = \{Uu^+, Vx^+, G(x^+)\}$$

the equation (4.6) can be rewritten as

$$0 = \langle Uu^+, Uu\rangle + \langle Vx^+, V[H(u^+) - \phi\, Q_f^{\#} F(H(u)) + \phi\, \alpha]\rangle \tag{4.9}$$

$$+ \langle G(x^+),\, G[H(u^+) - \phi\, Q_f^{\#} F(H(u)) + \phi\, \alpha]\rangle_\Omega$$

for all $u \in \mathcal{U}$ and $\alpha \in$ Null Q_f.

This is equivalent to

$$\langle Vx^+,\, V\phi\alpha\rangle + \langle G(x^+),\, G(\phi\alpha)\rangle_\Omega = 0 \tag{4.10}$$

for all $\alpha \in$ Null Q_f and

$$\langle Uu^+, Uu\rangle + \langle Vx^+, V[H(u) - \phi\, Q_f^{\#}\, F(H(u))]\rangle \tag{4.11}$$

$$+ \langle G(x^+), G[H(u) - \phi\, Q_f^{\#} F(H(u))]\rangle_\Omega = 0$$

for all $u \in \mathcal{U}$.

Using Lemma 4.3, equation (4.9) can be shown equivalent to

$$\int_0^T \phi^* V^* V x^+ \, dt + (G(\phi))^* G(x^+) \in (\text{Null } Q_f)^\perp \tag{4.12}$$

while (4.10) is equivalent to

$$U^* U\, u^+ + B^* z \in \mathcal{U}^\perp. \tag{4.13}$$

Noting that $(U^* U\, u^+)(t)$ is $\mathcal{A}_t$-measurable and using Lemma 4.4, it is easily verified that (4.12) is equivalent to (iii) of the Theorem and this completes the proof.

In the above theorem, the descriptions for the optimal pair involve the Ito integral operators H^+ and H_W^+ defined in (4.4). In the following we will replace these operators by Ito differentials. First we need the following.

LEMMA 4.6 The Ito differential of ϕ^{-1} exists and w.p.1

$$d\phi^{-1}(t) = -\phi^{-1}(t)[A(t)dt + \sum_{k=1}^{p} C_k(t)d(W(t))_k - \sum_{k=1}^{p} C_k^2(t)dt]$$

Proof. The proof can be carried out by the same idea which is used in §8.5 in [A].

THEOREM 4.7 Assume Range Q_f is closed. Then $\{u^+, x^+\}$ is optimal if and only if $\{u^+, x^+\} \in \mathcal{D}$,

$$\int_0^T \phi^*(t)V^*(t)V(t)x^+(t)dt + (G(\phi))^* G(x^+) \in (\text{Null } Q_f)^\perp$$

and there exists $z \in X_n$ which is $\mathcal{A}_T$ - measurable such that

(i) $(U^{*}Uu^{+})(t) + B^{*}(t)\epsilon[z|\mathcal{A}_t] = 0$

for a.a. $(t,w) \in [0,T] \times \Omega$

(ii) $z - \psi_1(x^{+})$ is continuous in $t \in [0,T]$,

$(z - \psi_1(x^{+}))(T) = 0$, and

$$d(z - \psi_1(x^{+}))(t) = -[A^{*}(t) - \sum_{k=1}^{p}(C_k^{*}(t))^2](z - \psi_1(x^{+}))(t)dt$$

$$+\{\sum_{k=1}^{p}(C_k^{*})^2\psi_1(x^{+}) - V^{*}Vx^{+} - \psi_2(x^{+})\}(t)dt$$

$$-\sum_{k=1}^{p} C_k^{*}(t)z(t)(dW(t))_k$$

Proof. In the previous theorem

$$z - \psi_1(x^{+}) = H^{+}(V^{*}Vx^{+} + \psi_2(x^{+})) + H_W^{+}(\psi_1(x^{+})).$$

Now use the properties of H^{+}, H_W^{+}, $\psi_1(x^{+})$, $\psi_2(x^{+})$ defined in (4.4), (4.5) and Lemma 4.6.

REMARK 4.2 In the above theorem, z may not be continuous in $t \in [0,T]$, so that $z(T)$ becomes meaningless. However, if $dg_1(t)$ and $df_1(t)$ exists on $[0,T]$, then z is continuous in t and the boundary condition and the differential equation in (ii) of the above theorem becomes:

$z(T) = (\psi_1(x^{+}))(T)$ and

$$dz(t) = (dg_1(t))G(x^{+}) + (A^{*}g_1 - g_2)(t)\ G(x^{+})dt - (df_1)(t)q$$
$$+ (f_2 - A^{*}f_1)(t)q\ dt - V^{*}(t)V(t)x^{+}(t)dt$$

where

$$q = \int_0^T (Q_f^{\#})^{*}[\phi^{*}(t)V^{*}(t)V(t)x^{+}(t)]dt + (Q_f^{\#})^{*}[(G(\phi))^{*}G(x^{+})].$$

REMARK 4.3 Suppose further that $F(x) = x(0)$ and dg_1 exists. Thus $d = n$ and $Q_f(\alpha) = \alpha$ for all $\alpha \in L_2(\Omega,\mathbb{R}^n)\cap\mathcal{A}_0$. In particular, Range $Q_f = L_2(\Omega,\mathbb{R}^n)\cap\mathcal{A}_0$ is closed. Thus $\{u,x\} \in \mathcal{D}$ if and only if $u \in U$ and $\{u,x\}$ satisfies the state equation (3.1) and $x(0) = \gamma$. Now $Q_f^{\#}$ is the orthogonal projector of $L_2(\Omega,\mathbb{R}^n)$ onto $l_2(\Omega,\mathbb{R}^n)\cap\mathcal{A}_0$, and by Theorem 3.4,

$$\dot{f}_1 + A^* f_1 = f_2 \qquad \text{w.p.1},$$
$$f_1(0) = -I_n,\ f_1(T) = 0.$$

Therefore using Remark 4.2 it follows from Theorem 4.7 that $\{u^+,x^+\}$ is optimal if and only if $\{u^+,x^+\} \in \mathcal{D}$ and there exists $z \in X_n$ which is $\mathcal{A}_T$-measurable such that

$$(U^*Uu^+)(t) + B^*(t)\, E[z|\mathcal{A}_t] = 0, \text{ a.a. } (t,w) \in [0,T] \times \Omega,$$
$$dz(t) = -V^*(t)V(t)x^+(t)dt + (dg_1(t))G(x^+) + (A^*g_1 - g_2)(t)G(x^+)dt,$$
$$z(T) = g_1(T)G(x^+) + \int_0^T (Q_f^{\#})[\phi^* V^* Vx](t)dt + (Q_f^{\#})(G(\phi))^* G(x^+).$$

REFERENCES

[A] Arnold, L., Stochastic Differential Equations, Theory and Applications, John Wiley & Sons, 1974.

[AG] Akhiezer, N. I. and Glazeman, I. M., Theory of Linear Operators, Vol. I, Unger Publishing Co., 1966.

[GS] Gikhman, I. and Skorohod, A., Introduction to the Theory of Random Processes, W. B. Saunders Co., 1965.

[AW1] Adams, M. B., and Willsky, A. S. and Levy, B. G., Linear Estimation of Boundary Value Stochastic Process - Part 2:1 - d Smoothing Problems, I.E.E.E. Trans. on Auto. Control, Vol. AC-29, No. 9 (1984), 811-821.

[C] Casti, J., The Linear Quadratic Control Problems: Some recent results and outstanding problems, SIAM Rev., Vol. 22, No. 4 (1980), 459-485.

[L1] Lee, S.J., A Multi-Response Quadratic Control Problem, SIAM J. Control Opt., Vol. 24, No. 4 (1986), 771-788.

[L2] Lee, S. J., A Class of Singular Quadratic Control Problems with Non-Standard Boundary Conditions, Honam Math. J., (In press).

[LN1] Lee, S. J., and Nashed, M. Z., Generalized Inverses for Linear Manifolds and Applications to Boundary Value Problems in Banach Spaces, Math. Rep. Acad. Sci. Canada, Vol. 4, No. 6 (1982), 347-352.

[LN2] Lee, S. J., and Nashed M. Z., Least Squares Solutions of Multi-Valued Linear Operator Equations in Hilbert Spaces, J. Approx. Theory, Vol. 38, No. 4 (1983), 380-391.

Nonsmooth Analysis and Differential Games

Jack W. Macki

Department of Mathematics
University of Alberta
Edmonton, Alberta
Canada T6G 2G1

Pietro Zecca

Dipartimento di Sistemi e Informatica
Università di Firenze
Via S. Marta 3
50139 Firenze (Florence), Italy

1. In this paper we demonstrate the application of nonsmooth analysis to certain problems in the theory of differential games. Since we are primarily interested in demonstrating necessary conditions, we will not discuss in depth the existence of value. The key idea is to recognize that a saddle point for a two-player differential game can be thought of as an optimal operating point for two different individual control problems. Thus we can directly transcribe results from the nonsmooth analysis of control problems. This idea goes back (at least) to a paper of Berkowitz [1], who related certain differential games to problems in the calculus of variations.

2. THE GAME

Two players, J_1 and J_2, have states $x_i(t) \in R^{n_i}$ $(i = 1,2,)$, determined on a fixed time interval $[0,T]$ as absolutely continuous solutions of the following constrained differential inclusions:

$$\dot{x}_i(t) \in F_i(t,x_i(t)) \quad \text{a.e.}, \tag{1}$$

$$x_i(0) \in C_i^0 \subset R^{n_i}, \quad C_i^0 \quad \text{compact}; \tag{2}$$

$$g_1(t,x_1(t)) \leq 0, \quad g_2(t,x_2(t)) \leq 0. \tag{3}$$

We assume that $F_i: \Omega_i \to R^{n_i}$ is measurably Lipschitz in x_i, integrably bounded, closed and convex, where Ω_i will be a tube about a given saddle point $\{(x_1^*(t),x_2^*(t)) \mid t \in [0,T]\}$ (Clarke [3] (pp. 113-114) gives precise definitions of the above-mentioned requirements). The functions g_i will be assumed lower semicontinuous. An admissible motion (henceforth, motion) is a pair $(x_1(\cdot),x_2(\cdot))$ solving (1)(2)(3). The components $x_1(\cdot)$ and $x_2(\cdot)$ of a motion will be called (admissible) trajectories for J_1, J_2 respectively. The set of all possible trajectories for J_i will be denoted X_i.

At time T a payoff can be calculated for any motion:

$$P = h(x_1(T),x_2(T)) + \int_0^T f^0(t,x_1(t),x_2(t))dt, \tag{4}$$

where f^0 is continuous, while f^0 and h are locally Lipschitz in (x_1,x_2). J_1 wishes to maximize the payoff, J_2 to minimize it. In an attempt to accomplish his goal, each player adopts a strategy, which at each time t examines the motion to date $\{(x_1(s),x_2(s)) \mid 0 \leq s \leq t\}$, and then makes a selection in (1). More precisely, we define a strategy for J_i as a map:

$$\text{for } J_1; \quad \alpha_1 : X_2 \to X_1,$$

$$\text{for } J_2; \quad \alpha_2 : X_1 \to X_2,$$

which is nonanticipating:

$$\text{if } x(t) = y(t) \text{ on } [0,t_*] \subset [0,T], \text{ then } \alpha_i x(t) = \alpha_i y(t) \text{ on } [0,t_*].$$

An outcome for a strategy pair (α_1,α_2) is a pair of functions $(z_1(\cdot),z_2(\cdot))$ for which there exists a sequence of motions $(x_1^k(\cdot),x_2^k(\cdot))$ satisfying

$$\lim_{k\to\infty} \alpha_1 x_2^k = \lim_{k\to\infty} x_1^k = z_1 \ , \ \lim_{k\to\infty} \alpha_2 x_1^k = \lim_{k\to\infty} x_2^k = z_2$$

uniformly on $[0,T]$. Under our assumptions an outcome is a motion for (1)(2)(3). However, one or both of the equalities $\alpha_1 z_2 = z_1$, $\alpha_2 z_1 = z_2$ may fail, since we make no assumptions about the continuity of the α_i. If a strategy pair has an outcome (or several outcomes) we can compute the associated payoff(s) $P(\alpha_1, \alpha_2)$ from (4). A given strategy pair may not have an outcome, as the following example indicates.

EXAMPLE 1 (Based on an example of Zaremba [8])

Let the dynamics be defined by

$$\dot{x}_i(t) \in [-2,2] \ , \ x_i(0) = 0 \ , \ g_i(t, x_i) = t - |x_i| \ , \ i = 1,2, \quad (n_i = 1).$$

The trajectories for J_i divide into two disjoint closed classes (see sketch), X_i^+ and X_i^-, isomorphic via the reflection map $x_i(t) \to -x_i(t)$. We define

$$\alpha_1 : \begin{cases} X_2^+ \to X_1^+ \\ X_2^- \to X_1^- \end{cases} ; \quad \alpha_2 : \begin{cases} X_2^+ \to X_1^+ \\ X_2^- \to X_1^- \end{cases} ;$$

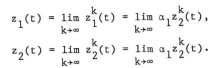

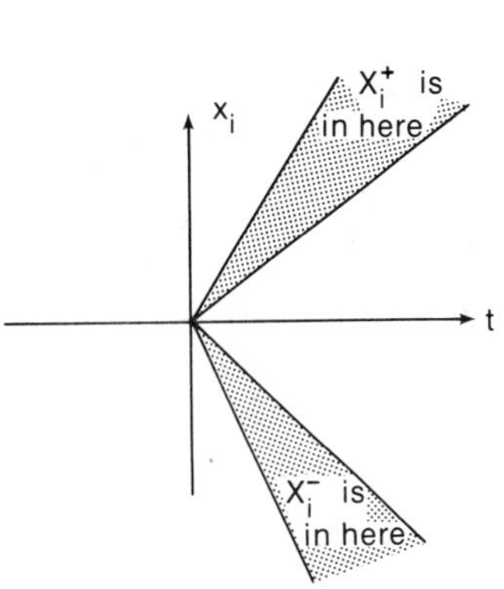

by assigning to $x_i(\cdot)$ its isomorphic twin in the image set. If there were an outcome $(z_1(\cdot), z_2(\cdot))$ then we would have

$$z_1(t) = \lim_{k\to\infty} z_1^k(t) = \lim_{k\to\infty} \alpha_1 z_2^k(t),$$

$$z_2(t) = \lim_{k\to\infty} z_2^k(t) = \lim_{k\to\infty} \alpha_1 z_2^k(t).$$

Suppose for example that $\{z_1^k(\cdot)\} \subset X_1^+$ (if necessary, using a subsequence). Then $\alpha_2 z_1^k(\cdot) \in X_2^-$, so $z_2(t) = \lim_{k\to\infty} \alpha_2 z_1^k(t) \in X_2^-$. Since X_2^- and X_2^+ are closed, we must have had $z_2^k(\cdot) \in X_2^-$ (if necessary, we again pass to a subsequence). But then $\alpha_1 z_2^k(\cdot) \in X_1^-$, so $\lim_{k\to\infty} \alpha_1 z_2^k(\cdot) = z_1(\cdot) \in X_1^-$, a contradiction. Other cases are argued similarly. A reader who wishes a more direct verification of the "impossibility" of these strategies need only try to play it and observe that one is forced to "jump" from X_i^- to X_i^+ violating the absolute continuity of trajectories.

The above example illustrates that we need to place additional requirements on our strategy pairs. The most natural, due in essence to Varaiya-Lin [7] and recently discussed by Kalton [5], is to require that at least one of the players have a reaction time $\delta > 0$:

$$\begin{gathered}\text{If } x(t) = y(t) \text{ on } [0,t_*] \subset [0,T] \text{ then} \\ \alpha x(t) = \alpha y(t) \text{ on } [0,\min\{T,t_*+\delta\}].\end{gathered} \tag{5}$$

Here α is any allowable strategy for the indicated player and δ may depend on α.

Under this assumption the following theorem holds:

THEOREM 1. If at least one player has a reaction time, then each strategy pair has an outcome. If the initial sets C_i^0 are singletons, then the outcome is unique.

Proof: Suppose J_1 has a reaction time. Given strategies α_1 and α_2, with delay δ in α_1, we can construct the unique outcome over successive time intervals of length δ. Consider a fixed initial situation, $x_i(0) = x_{i0}$, $i = 1,2$. Then by (5), with $t_* = 0$, it follows that $(\alpha_1 x_2)(t)$ is the same function on $[0,\delta]$, for <u>any</u> motion $x_2(t)$ satisfying $x_2(0) = x_{20}$. Call this function $x_1(t)$, $0 \le t \le \delta$. Because α_2 is nonanticipating, $(\alpha_2 x_1)(t)$ is now well defined on $[0,\delta]$. Call it $x_2(t)$, $0 \le t \le \delta$. Continuing this argument from $t_* = \delta$, we generate a unique motion $(x_1(t),x_2(t))$ on $[0,T]$.

REMARK: The above simple theorem can be extended to:

> "Suppose that X_1 and X_2 are compact in $C[0,T]$, and that for each $\delta > 0$ sufficiently small and each strategy α_1 there is a strategy α^δ_1 with delay δ, such that

$$\sup\{|(\alpha_1 x_2)(t)-(\alpha_1^\delta x_2)(t)| : x_2(\cdot) \in X_2, t \in [0,T]\} = o(1)$$

as $\delta \to 0$. Then there is at least one outcome for each pair (α_1,α_2) of strategies."

The proof (due to Zaremba [8]) follows easily from Theorem 1 (for a sequence $\delta_k \to 0$) noting the compactness of C_i^0 and X_i.

Example 1 is now excluded because all motions start at $x_i(0) = 0$ and hence if a player had a reaction time he would have a fixed single trajectory on the interval $[0,\delta(\alpha)]$, regardless of what the other player does. Clearly the given strategies do not allow this.

Having developed a clear notion of strategy and outcome, we can define the payoff $P(\alpha_1,\alpha_2)$ and then define upper and lower values

$$U = \sup_{\alpha_1} \inf_{\alpha_2} P(\alpha_1,\alpha_2) , \quad L = \inf_{\alpha_2} \sup_{\alpha_1} P(\alpha_1,\alpha_2)$$

where (α_1,α_2) ranges over all strategies with outcome(s).

We emphasize that one need not adopt our particular approach to strategy and outcome, there are several different approaches and any of them will do. For example, if in the definition of L we were to replace α_2 by $x_2(\cdot) \in X_2$ (and dually, to replace α_1 by $x_1(\cdot) \in X_1$ in the definition of U), then we would have the Elliott-Kalton definition of values. This corresponds, in the case of U, to (possibly) increasing J_2's strategies (given α_1) so as to allow any trajectory $x_2(\cdot)$ emanating from the given state $x_2(t)$ at time t. Thus the first infemum in U is smaller, and so the Elliott-Kalton upper value could be smaller than ours. However, in the case when any pair of strategies has an outcome (e.g., if we have a reaction time) the sets X_2 and $\{\alpha_2 x_1(\cdot) | \alpha_2 \text{ a strategy}\}$ must coincide. This is because J_2 can always select a fixed $x_2(\cdot)$ and define $\alpha_2 x_1(\cdot) = x_2(\cdot)$ for all $x_1(\cdot)$. In this case our upper and lower values coincide with the Elliott -Kalton values.

When $L = U$ we shall say that our game has a value $W = L = U$.

3. NECESSARY CONDITIONS FOR A SADDLE POINT

We say that our game has a saddle point if there are strategies (α_1^*,α_2^*) with outcome $(x_1^*(\cdot),x_2^*(\cdot))$ satisfying

$$P(\alpha_1,\alpha_2^*) \leq P(\alpha_1^*,\alpha_2^*) \leq P(\alpha_1^*,\alpha_2)$$

for all strategies α_1 (α_2) for which the relevant payoff can be computed.

THEOREM 2. Suppose that (α_1^*,α_2^*) is a saddle and that the set of outcomes includes the sets $\{(x_1^*(\cdot),x_2(\cdot))|x_2(\cdot) \in X_2\}$ and $\{(x_1(\cdot),x_2^*(\cdot))|x_1(\cdot) \in X_1\}$. Then there exists a <u>multiplier</u> in the sense of Clarke ([3], p. 121), i.e., $[p_i(t),p_i^0(t),\gamma_i(t),\mu_i,\zeta_i,\lambda_i]$ $(i = 1,2)$, such that $p_i(t)\colon R \to R^{n_i}$ is absolutely continuous P_i^0 is zero or one, $\gamma_i(t)\colon R \to R^{n_i}$ is measurable, μ_i is a nonnegative Radon measure on $[0,T]$, $\zeta_i \subset R^{n_i}$ is a constant vector, and λ_i is a constant, satisfying

$$\zeta_i \in \partial_{x_i} h((x_1^*(T),x_2^*(T)), \quad \gamma_i(t) \in \partial_{x_i}^{>} g_i(t,x_i^*(t)) \quad \mu_i\text{-a.e.},$$

$$\operatorname{supp}(\mu_i) \subseteq \{t \in [0,T]|\ \partial_{x_i}^{>} g_i(t,x_i^*(t)) \neq \emptyset\},$$

$$\begin{bmatrix} -\dot{p}_1 \\ \dot{x}_1 \end{bmatrix} \in \partial_{(x_1,p_1)} H_1(t,x_1^*(t),p_1^0,p_1(t) + \textstyle\int_{[0,t)} \gamma_1(s)\mu_1(s)ds) + \tag{6a}$$

$$+ \begin{bmatrix} \partial_{x_1}[-\lambda_1 x_1^{0*}(T) f^0(t,x_1,x_2^*(\cdot))] \\ 0 \end{bmatrix}$$

$$\begin{bmatrix} -\dot{p}_2 \\ \dot{x}_2 \end{bmatrix} \in \partial_{(x_2,p_2)} H_2\left(t, x_2^*(t), p_2^0, p_2 + \int_{[0,t)} \gamma_2(s)\mu_2(s)ds\right) + \begin{bmatrix} 0 \\ \partial_{x_2}\left[-\lambda_2 x_2^{0*}(T) f^0(t, x_1^*(t), x_2)\right] \end{bmatrix} \tag{6b}$$

$$p_i(a) \in \partial d_{C_i^0}(x_i^*(0)), \quad p_i(a) = 0 \text{ if } x_i^*(0) \in \text{Int } C_i^0 ,$$

$$\lambda_i \rho_i - p_0 + p_i(T) + \int_{[0,T]} \gamma_i(s)\mu_i(ds) = 0 \quad , \lambda_i + \|\mu_i\| = 1.$$

Here, $H_1(t,x_1,p_1^0,p_1) = p_1^0 f^0(t,x_1,x_2^*(t)) + \max\{p_1 \cdot v_1 \mid v_1 \in F_1(t,x)\}$, and H_2 is defined analogously.

Proof: We imbed f^0 into the dynamics of our problem in two ways by defining

$$\dot{x}_1^0 = f^0(t,x_1,x_2^*(t)) \; , \; \dot{x}_2^0 = f^0(t,x_1^*(t),x_2)$$

We observe that $(x_1^*(\cdot),x_1^{0*}(\cdot))$ maximizes the cost $h(x_1(T),x_2^*(T)) + x_1^0(T)$ for the "classical" (n+1)-dimensional control problem $\dot{x}_1 \in F_1(t,x_1)$, $\dot{x}_1^0 = f^0(t,x_1(t),x_2^*(t))$, $x_1(0) \in C_1^0$, $x_1^0(0) = 0$, $g_1(t,x_1(t)) \le 0$, on $[0,T]$. Clearly $(x_1^*(\cdot),x_1^{0*}(\cdot))$ minimizes the cost $h(x_1^*(T),x_2(T)) + x_2^0(T)$ for the obvious dual problem. We may apply Theorem 3.2.6 from Clarke to each problem in turn to conclude the existence of multipliers as claimed. One must carry out some generalized gradient calculus to eliminate x_i^0. For example, Theorem 3.2.6 asserts the existence of a vector $\tilde{\zeta}_1 \in R^{n_i+1}$ such that $\tilde{\zeta}_1 \in \partial_{\tilde{x}_1}[h(x_1(T),x_2^*(T)) + x_1^0(T)]$ where $\tilde{x}_1 = (x_1,x_1^0)$. Thus $\tilde{\rho}_1 \in \{(\rho_1,1) \mid \rho_1 \in \partial_{x_1} h(x_1(T),x_2^*(T))\}$. Then Theorem 3.2.6 asserts that

$$(*) \qquad \lambda_1\tilde{\zeta}_1 + \tilde{p}_1(T) + \int_{[0,T]} \tilde{\gamma}_1(s)\mu_1(ds) = 0.$$

But $\tilde{\gamma}_1(t) \in \partial^{>}_{\tilde{x}_1} g_1(t,x_1)$, so $\gamma_1^0(t) = 0$. Therefore, (*) implies that $(\tilde{p}_1 = (p_1, p_0))$

$$\lambda_1\zeta_1^0 + p_1^0(T) = 0 \quad\text{and}\quad \lambda_1\zeta_1 + p_1(T) + \int_{[0,T]} \gamma_1(s)\mu_1(ds) = 0.$$

The first equation says that $p_1^0(T) = -\lambda_1$. Therefore, Theorem 3.2.6 asserts that

$$\frac{d}{dt}\begin{bmatrix} -\tilde{p}_1 \\ \tilde{x}_1 \end{bmatrix} \in \partial_{(\tilde{x}_1,\tilde{p}_1)} H_1(t,\tilde{x}_1(t),\tilde{p}_1(t) + \int_{[0,t)} \tilde{\gamma}_1(s)\mu_1(ds)) ,$$

thus

$$\frac{d}{dt}\begin{bmatrix} -\tilde{p}_1 \\ \tilde{x}_1 \end{bmatrix} \in \partial_{(\tilde{x}_1,\tilde{p}_1)} [p_1^0 f^0(t,(x_1,x_2^*(t))]$$

$$+ \partial_{(\tilde{x}_1,\tilde{p}_1)} \max\{\langle p_1 + \int_{[0,t)} \gamma_1(s)\mu_1(ds), v_1\rangle \mid v_1 \in F_1(t,x_1)) \tag{7}$$

The last expression in brackets on the right does not involve p_1^0, and no term on the right involves x_0, so we conclude $-\dot{p}_1^0 = 0$, $\dot{x}_1^0 = f^0$. Thus we have $p_1^0(t) = -\lambda_1$. Reducing the equality (7) using the above observations yields (6_1). Note that one cannot always assert that $\partial_{(x,y)}f(x,y) = \{(a,b) \mid a \in \partial_x f(x,y),\ b \in \partial_y f(x,y)\}$. However the left side is always contained in the right side, and under certain conditions one has equality. See Theorems 2.3.15, 2.3.16 in Clarke [3].

The multiplier rule above is complicated, due in large part to its generality. To use this rule effectively, one must be prepared to use it in conjunction with the (nonsmooth) Pontryagin Maximum Principle and the nonsmooth version of the Bolza problem in the calculus of variations. These two approaches are easily extended from their control theory and calculus of variations contexts to the differential games context, following the model of Theorem 2. The combined use of these approaches is well illustrated by the Example of section 5.2 of Clarke [3]. We present instead a simple set of examples to illustrate the determination of multipliers in a nonsmooth setting.

EXAMPLE 2. Let x_1 , x_2 be scalars, with

$$\dot{x}_1(t) \in \{x_1 + r \mid r \in [0,1]\},\ x_1(0) = 0;$$

$$\dot{x}_2(t) \in \{|x_2| + s \mid s \in [-1,1]\},\ x_2(0) = 0;$$

and $P(\alpha,\beta) = |x_1(T) + x_2(T)|$. A straightforward computation shows that $x_1^*(t) = e^t-1$, $x_2^*(t) = e^{-t}-1$ is a saddle. Our theorem states that if a saddle exists, then there are multipliers such that

$\zeta_i \in \partial_{x_i} |x_1(T) + x_2(T)|$. Thus

$$\zeta_1 = \begin{cases} 1 + x_2(T) & \text{when } x_1(T) + x_2(T) > 0; \\ -1 - x_2(T) & \text{when } x_1(T) + x_2(T) < 0; \end{cases}$$

$$\zeta_1 \in \{s[1 + x_2(T)] \mid s \in [-1,1]\} \quad \text{when } x_1(T) + x_2(T) = 0;$$

and ζ_2 has a dual formulation. By definition

$$H_1(t,x_1,p_1) = \max\{p_1(x_1+r) \mid r \in [0,1]\} = p_1x_1 + \begin{cases} 0 & , \ p_1 \le 0 ; \\ p_1 & , \ p_1 > 0 . \end{cases}$$

$$H_2(t,x_2,p_2) = \max\{p_2(s+|x_2|) \mid s \in [-1,1]\} = p_2|x_2| + |p_2|.$$

Thus $\partial_{x_1} H_1 = p_1$,

$$\partial_{p_1} H_1 = x_1 + \begin{cases} 0 & , \ p_1 < 0 ; \\ [0,1] & , \ p_1 = 0 ; \\ 1 & , \ p_1 > 0 . \end{cases}$$

and

$$\partial_{x_2} H_2 = \begin{cases} p_2 & , \ x_2 > 0 ; \\ [-|p_2|,|p_2|] & , \ x_2 = 0 ; \\ -p_2 & , \ x_2 < 0 ; \end{cases}$$

$$\partial_{p_2} H_2 = |x_2| + \begin{cases} 1 & , p_2 > 0 ; \\ -1 & , p_2 < 0 ; \\ [-1,1] & , p_2 = 0 . \end{cases}$$

γ_i and μ_i are zero, and p_i^0 does not exist because $f^0 \equiv 0$. Then $\lambda_i = 1$ ($\mu_i = 0$) and $\lambda_i \zeta_i + p_i(T) = 0$, so $p_i(T) = -\zeta_i$. Also

$$-\dot{p}_i = \partial_{x_1} H_1 = p_1 , \quad -\dot{p}_2 = \partial_{x_2} H_2 , \text{ so}$$

$$p_1(t) = p_1(T)e^{T-t} \quad \text{with} \quad p_1(T) = \begin{cases} 1 + x_2(T) & , x_1(T) + x_2(T) > 0; \\ (1+x_2(t))[-1,1] & , x_1(T) + x_2(T) = 0; \\ -1 - x_2(T) & , x_1(T) + x_2(T) < 0 . \end{cases}$$

Finally $p_i(0) \in r\partial d_{C_0}(\overset{*}{x}_i(0))$ for some $r > 0$. Since $\overset{*}{x}_i(0) = 0 = C_0$, we have $d_{C_0}(x) = |x|$, so $p_i(0) \in rB$, B the unit ball. Thus this observation provides no useful information.

It is useful to use the Pontryagin Maximum Principle on the above example to fill in more information. To this end, note that the solutions of our dynamical equations can be written (8) $x_i(t) = \int_0^t e^{t-u} r(u)du$, $x_2(t) = \int_0^t e^{\int_u^t \operatorname{sgn} x(v)dv} s(u)du$, where $r(u) \in [0,1]$, $s(u) \in [-1,1]$ are measurable selections for $0 \leq u,v \leq T$. Since $\overset{*}{x}_2(t)$ minimizes P at a saddle (with $x_1 = \overset{*}{x}_1(t)$ fixed) we know that the Pontryagin Hamiltonian H_2^P is maximized by the optimal direction $\dot{x}_2(t)$, i.e.,

$$\max_{v_2} H_2^P(t,v_2) \cdot\equiv\cdot \max_{v_2} \{p_2 \cdot v_2 - \lambda_2 f^0(t,\overset{*}{x}_1(t),\overset{*}{x}_2(t)) | \; v_2 \in F_2(t,\overset{*}{x}_2(t))\}$$

is attained for $v_2 = \dot{\overset{*}{x}}_2(t)$ a.e.. The situation with constraints of the form (3) is more complex - see Clarke [3], Theorem 5.1.2. On the other hand, since $\overset{*}{x}_1$ maximizes $P(\alpha,\beta)$, it minimizes $-P(\alpha,\beta)$ so

$H_1^P \cdot\equiv\cdot p_1\cdot v_1 + \lambda_1 f^0(t,\overset{*}{x}_1(t),\overset{*}{x}_2(t))$, $v_1 \in F_1(t,\overset{*}{x}_1(t))$, is maximized by $v_1 = \dot{\overset{*}{x}}_1(t)$ a.e.. As usual, $\lambda_i = 0$ or 1. Clearly H_2^P is maximized by using -1 if $q_2(t) < 0$ and $+1$ if $q_2(t) > 0$, thus $r(u)$ is 0 or 1 in (8); a similar argument shows $s(u)$ is -1 or $+1$ in (8). This is of course just the bang-bang principle.

EXAMPLE 3. (Constraints) Using the dynamics, initial conditions, and cost as Example 2, we now add the constraint

$$g_1(t,x_1) = x_1(t) - t \leq 0,\ g_2(t,x_2) = x_2(t) - t \leq 0 \quad \text{on} \quad [0,T].$$

Then $\gamma_1(t) \in \partial^{>}_{x_1} g_1(t,x_1) = \{1\}$,

$$\gamma_2(t) \in \partial^{>}_{x_2} g_2(t,x_2) = \{-1\},$$

and $\text{supp}(\mu_i) \subseteq [0,T]$. Notice that the saddle of Example 2 does not satisfy the constraint. An obvious replacemnt is $\overset{*}{x}_1(t) \equiv \overset{*}{x}_2(t) \equiv 0$. Notice also that $x_1(t) \equiv 0$ is the only admissible trajectory of player 1. The equations for the costates are now more complex. If we write $q_i = p_i + \int_{[0,t)} \gamma_i(s)\mu_i(ds)$, then

$$-\dot{p}_i \in \partial_{x_1} H_1(t,x_1,q_1) = \partial_{x_1}[x_1 q_1 + \begin{cases} 0 &, q_1 \leq 0; \\ q_1 &, q_1 > 0 \end{cases}]$$

$$= p_1 + \int_{[0,t)} \gamma_1(s)\mu_1(ds).$$

$$-\dot{p}_2 \in \partial_{x_2} H_2(t,x_2,q_2),\ -\dot{p}_2 \in \begin{cases} p_2 - \int \mu_2(ds) &, x_2 > 0; \\ [-|p_2 - \int \mu_2(ds)|,\ |p_2 + \int \mu_2 ds|], & x_2 = 0; \\ -p_2 + \int \mu_2(ds) &, x_2 < 0. \end{cases}$$

4. THE VALUE FUNCTION

Let the strategies (α_1^*, α_2^*) be a particular saddle point for our game, with outcome $(x_1^*(t), x_2^*(t))$. Define a <u>tube</u> about this outcome as a set of the form

$$\{(t, x_1, x_2) \mid t \in [0,T],\ |x_i - x_i^*(t)| < d_i,\ i = 1,2,\}$$

Given any initial state (t^0, x_1^0, x_2^0) in this tube, we can start our game at this point giving rise to upper and lower values. Under our assumptions these value functions are Lipschitz in (t^0, x_1^0, x_2^0), which implies that they have Clarke genralized gradients.

If the upper and lower values coincide for each initial point in some tube, then their common value, the value function $W(t^0, x_1^0, x_2^0)$, is Lipschitz.

Sufficient conditions for the existence of a value function invariably include the Issacs condition:

$$\max_{v_1 \in F_1(t,x_1)} \ \max_{v_2 \in F_2(t,x_2)} (s_1 \cdot v_1 + s_2 \cdot v_2) = \max_{v_2 \in F_2(t,x_2)} \ \max_{v_1 \in F_1(t,x_1)} (s_1 \cdot v_1 + s_2 \cdot v_2) \tag{8}$$

holds for all $s_i \in R^{n_i}$. This condition is obviously true in our case. However, in general one must be careful when quoting results from the literature, since often the cost function f^0 is imbedded into the dynamics as in the proof of Theorem 1. This requires the addition of a term $s_0 f_0$ to each side of (8). In our case this additional term is not a problem since our cost function $f(t, x_1(t), x_2(t))$ does not explicitly depend on v_1, v_2.

Under our assumptions, then, we can assert the existence of a value function $W(t, x_1, x_2)$ (dropping the superscript zero), Lipschitz in (t, x_1, x_2). This is proved in essence in Zaremba [9] (but without asserting that a saddle exists). We could just as well have defined our strategy in the style of Krassovkii-Subbotin [6] , Elliot-Kalton [4], or Berkowitz [2]; under our assumptions we would have had a Lipschitz value function.

THEOREM 3. If there exists a saddle $(x_1^*(\cdot),x_2^*(\cdot))$ with a value function $W(t,x_1,x_2)$ defined and Lipschitz in some tube about this saddle, then in this tube $W(t,x_1,x_2)$ satisfies

$$0 \in \langle \zeta, [1, F_1(t,x_1), F_2(t,x_2)\rangle$$

for all $\zeta \in \partial W(t,x_1,x_2)$.

Proof: There is a multitude of proofs that

$$W_t + \langle W_{x_1}, x_1^* \rangle + \langle W_{x_2}, x_2^* \rangle = 0$$

at all points where the gradient of W with respect to (t,x_1,x_2) exists (for example, see Elliott-Kalton [4] or Berkowitz [2]). Therefore, $0 \in \{W_t + \langle W_{x_1}, F_1\rangle + \langle W_{x_2}, F_2\rangle\}$ for all such points, where the set is defined in the obvious way. Thus $0 \in \langle \nabla W, [1,F_1,F_2]\rangle$ at these points. Since W is Lipschitz in (t,x_1,x_2), it has a gradient almost everywhere, and the generalized gradient at any point (t,x_1,x_2) can then be defined as the set of all possible limits

$$\lim_{(t^i,x_1^i,x_2^i)\to(t,x_1,x_2)} \nabla W(t_i,x_1^i,x_2^i)$$

taken using points (t^i,x_1^i,x_2^i) where ∇W exists. This completes the proof. Notice that we can conclude

$$0 \in \partial_t W + \langle \partial_{x_i} W, F_1\rangle + \langle \partial_{x_2} W, F_2\rangle.$$

ACKNOWLEDGEMENTS

The research for this paper was supported in part, by grants from the G.N.A.F.A. of the C.N.R. of Italy, and the Canadian NSERC under grant A-3053.

REFERENCES

1. L.D. Berkowitz, A variational approach to differential games, Annals of Mathematics, Study No. 52, Princeton Univ. Press, Princeton, 1964.

2. L.D. Berkowitz, The existence of value and saddle point in games of fixed duration, SIAM J. Control and Optimization, 23(1985), 172-196.

3. F.H. Clarke, Optimization and Nonsmooth Analysis, Wiley, Toronto, Canada, 1983.

4. R.J. Elliot and N.J. Kalton, The Existence of Value in Differential Games, Mem. Amer. Math. Soc. 126, Providence Rhode Island, 1972.

5. N.J. Kalton, Differential games of survival, in the Theory and Applications of Differential Games (pp. 45-61), ed. J.D. Grote, Reidel, Dordrecht, Holland, 1975

6. N. Krassovskii and A. Subbotin, Positional Differential Games (in Russian), Nauka, Moscow USSR, 1974; available in French translation as Jeux Differentiels, Mir, Moscow, USSR, 1977.

7. P. Varaiya and J. Lin, Existence of saddle points in differential games, SIAM J. Control, 7 (1969), 141-157.

8. L.S. Zaremba, On the existence of value in the Varaiya-Lin sense in differential games of pursuit and evasion, JOTA, 29 (1979), 135-145.

9. L.S. Zaremba, Existence of value in generalized pursuit - evasion games, SIAM J. Control and Optmization, 22 (1984), 894-901.

Operator Extremal Theory of Compensation and Representation of Systems and Control Problems

M. Z. Nashed

Department of Mathematical Sciences

University of Delaware

Newark, Delaware

1. INTRODUCTION

This paper addresses several problems in systems and control theory and related operator-theoretic extremal problems. The problems can be said to arise from compensation, representation, stabilization and approximation of singular and/or unstable systems. In this paper we represent linear systems by input-output mappings and confine our analysis to a few results in this setting, while addressing some related extremal, ordering, and equality problems in generalized inverse operator theory.

In many problems of systems and control theory, (and more generally for problems in operator theory), one often needs a notion of a "metric" for measuring the distance between two systems and/or a notion of "ordering" or "partial ordering" between two systems. As examples, we mention the following types of problems that require such notions:

(i) Extremal problems in control and systems, where the extremization is over a class of controls or subsystems, or over a set of parameters that occur in a description of admissible systems.

(ii) The approximation of one system by another.

(iii) Sensitivity analysis and perturbation problems.

(iv) The behavior of a closed loop system in terms of an open loop system. Mathematically, this problem is equivalent to studying the effect of unstable perturbations in a closed linear operator K on the (bounded) inverse, or the generalized inverse, of the operator $I + K$.

(v) The representation of a singular system and the covariance defect for such systems.

(vi) Compensation problems, where one is interested in connecting a given system to another subsystem in order to achieve a desired performance, or to come as close as possible to it. The connection may be in series or parallel and may involve a scaling of the original system.

For stable systems the operator norm of the input-output mapping is often used as a metric. The same is true for bounded linear operators on a normed space. However, this norm cannot be used for comparison or approximation of unstable systems or unbounded linear operators since the domains may be different and the norm is unbounded. For such problems several notions are useful: the gap between closed subspaces, partial ordering in terms of Hermitian forms (also called Loewner ordering), a class of Schatten norms (including the Hilbert-Schmidt and the trace norm), and variants of these concepts.

All the problems considered in this paper are related in one way or another to generalized inverse operators. Section 2 provides mathematical preliminaries and an overview of various extremal properties of generalized inverses of operators with not necessarily closed range. In Section 3 we define notions of system pseudoinverses and system partial inverses which extend well-known notions (e.g., Zadeh and Desoer [28]) and analyze several compensation problems in systems and control theory in terms of operator extremal characterizations of generalized inverses. In Section 4 we consider generalizations of an operator inequality (relative to Hermitian ordering) which states that the inverse function is "convex" on the convex set of positive definite symmetric matrices. The generalizations hold for not necessarily bounded operators and do not require equality of the ranges of the operators. In Section 5 we address the "covariance condition" for the generalized inverse, namely $(T^{-1}AT)^{\dagger} = T^{-1}A^{\dagger}T$, and show that this problem is related to the representation of a singular system relative to a basis. We provide a new outlook on the "covariance condition" and show that the condition always holds relative to a transformed pair of projectors. Finally, we pose a problem, arising from the "covariance condition" in Banach spaces, that is still open in the case of an arbitrary linear operator on an infinite dimensional space.

This paper is an expansion of a talk at a Special Session on Operator Methods of Optimal Control Problems. Some of the results were also presented at a Conference on Information Sciences and Systems at Princeton University.

2. OPERATOR EXTREMAL CHARACTERIZATIONS OF GENERALIZED INVERSES WITH RESPECT TO HERMITIAN ORDER AND SCHATTEN NORMS

2.1. Three Extremal Characterizations of the Moore-Penrose Inverse

The Moore-Penrose inverse $A^{\dagger}$ of an $m \times n$ (real or complex) matrix A is the unique solution of the following equations:

$$AXA = A \tag{2.1}$$

$$XAX = X \tag{2.2}$$

$$(AX)^* = AX \tag{2.3}$$

$$(XA)^* = XA \tag{2.4}$$

where $*$ denotes conjugate transpose. From a function-theoretic viewpoint $A^\dagger$ is the linear transformation

$$A^\dagger = \begin{cases} (A\,|N(A)^\perp)^{-1} & \text{on } R(A) \\ 0 & \text{on } R(A)^\perp \end{cases}$$

where $R(A)$ is the range of A and $A\,|N(A)^\perp$ is the restriction of A to the orthogonal complement of the nullspace of A. Equivalently, $A^\dagger = (A\,|N(A)^\perp)^{-1}Q$, where Q is the orthogonal projector of R^m onto $R(A)$. Here and in what follows R^m denotes the (real or complex) Euclidean m-space.

There are three *extremal characterizations* of $A^\dagger$, all of which were established by Penrose [21]:

(E1) For any $m \times p$ matrix B, the matrix $A^\dagger B$ is the unique matrix X which minimizes $||AX - B||_F$ and for which $||X||_F$ is minimal among all matrices which minimize $||AX - B||_F$. Here $||C||_F$ denotes the *Frobenius norm* of the matrix $C = [c_{ij}]$, i.e., $||C||_F = (\text{trace } C^*C)^{1/2} = \left(\sum_{i,j} |c_{ij}|^2\right)^{1/2}$.

In particular, if we take $B = I$ we obtain a characterization of $A^\dagger$ as the unique best-approximate solution (with respect to the Frobenius norm) of the (in general unsolvable) matrix equation $AX = I$. A similar result holds for $XA = I$.

(E2) For any $b \in R^m$, $A^\dagger b$ is the unique vector which minimizes $\|Ax - b\|$ (where $\|\cdot\|$ is the Euclidean norm) and is of minimal Euclidean norm among such minimizers. Note that (E2) follows immediately from (E1) by taking X and B to be $n \times 1$ and $m \times 1$ matrices, denoted as vectors x and b respectively.

(E3) Let $J(A)$ be the set of all inner inverses of A, i.e. $J(A) = \{X : AXA = A\}$. Then $A^\dagger$ is the unique matrix which minimizes $\|X\|_F$ over $J(A)$. Thus $A^\dagger$ is the unique inner inverse of minimal Frobenius norm.

The *vector version* of the extremal characterization of the generalized inverse, i.e. (E2), generalizes immediately to infinite dimensional spaces. In fact, it was established by Tseng for closed densely defined linear operators between Hilbert spaces several years before Penrose's paper appeared. Extremal properties have also been studied for generalized inverse in Banach spaces [19], [20]. Both (E2) and its Hilbert-space analog have found many significant applications in control and system theory, pattern recognition, statistics, communication theory, etc. (See the annotated bibliography in [14]). Unlike the vector version (E2), a straightforward generalization of (E1) or (E3) is not possible since if A is a bounded linear operator between infinite dimensional Hilbert spaces, then $\|AA^\dagger - I\|_2$ is infinite, where $\|\cdot\|_2$ is the analog of the Frobenius norm in Hilbert space, namely the Hilbert-Schmidt norm. Also if the range of the operator is nonclosed, as is often the case for many integral operators that arise in system and control theory, then $A^\dagger$ is unbounded and hence cannot be characterized as the minimizer of $\|X\|$ over $J(A)$.

Recently, Engl and the author have obtained infinite dimensional versions of Penrose's full results (i.e., (E1) and (E3)). Some of these new extremal characterizations will be briefly examined below.

We recall that matrices which satisfy subsets of the equations (2.1)-(2.4) play an important role in solvability and least-squares solvability of the equation

$$Ax = b \ . \tag{2.5}$$

In particular: (i) if $b \in R(A)$, then Xb is a solution of (2.5) for every inner inverse X; (ii) if $b \in R(A)$ and X is a $\{1, 4\}$-inverse, i.e. it satisfies (2.1) and (2.4), then Xb is the minimal norm solution of equation (2.5); (iii) if X is a $\{1, 3\}$-inverse, i.e. it satisfies (2.1) and (2.3), then Xb is a least-squares solution. These properties carry over to bounded linear operators with closed range in Hilbert space. For operators with nonclosed range and for the scope and limitations of analogous properties in Banach spaces see [18], [20].

2.2 Operator Extremal Characterizations of Generalized Inverse Operators in Hilbert Spaces

A natural and well-known partial order on spaces of linear operators on a Hilbert space is induced via the quadratic form associated with an operator. Let $\mathbf{L}(H)$ denote the space of all bounded linear operators on a Hilbert space H. For $T, S \in \mathbf{L}(H)$, we write

$$T \leq S \text{ if } (Tx,x) \leq (Sx,x) \text{ for all } x \in H \ .$$

We call "$\leq$" the Hermitian (partial) order. T and S are said to be *comparable* if either $T \leq S$ or $S \leq T$.

Let $A \in \mathbf{L}(H_1,H_2)$ with *closed* range and $B \in \mathbf{L}(H_3,H_2)$, where H_1, H_2 and H_3 are Hilbert space. The adjoint of A is denoted by A^*.

Theorem 2.1. Let $F : \mathbf{L}(H_3,H_1) \to \mathbf{L}(H_3)$ be defined by

$$F(X) := (AX-B)^* (AX-B) . \tag{2.6}$$

Then the set $\{F(X) : X \in \mathbf{L}(H_3,H_1)\}$ has a smallest element F_0 with respect to the Hermitian order on $\mathbf{L}(H_3)$. Let

$$M := \{X : F(X) = F_0\}$$

Then $X \in M$ if and only if $AX = QB$, where Q is the orthogonal projector onto $R(A)$. Moreover, the set $\{X^*X : X \in M\}$ has a smallest element F_1 with respect to the Hermitian order. The unique $X \in M$ with $X^*X = F_1$ is $X = A^\dagger B$, where $A^\dagger$ is the generalized inverse of A.

Note that if we let $H_3 = H_2$ and $B = I$, then the minimizers of F are precisely the $\{1, 3\}$-inverses of A; among them $A^\dagger$ is characterized as the unique minimizer of X^*X. Taking $H_3 = R$ in Theorem 2.1 we recover the well-known result that the set S_b of all least-squares solutions of $Ax = b$ coincides with the set of solutions of $Ax = Qb$, and the best-approximate solution is given by $x = A^\dagger b$, so $S_b = A^\dagger b + N(A)$.

We can replace the operator $AX - B$ by $XA - B$ in order to obtain a characterization of $\{1, 4\}$-inverses. The criteria will involve operators of the type XX^* instead of X^*X. This difference is essential in connection with Hermitian order.

Theorem 2.2. Let $A \in \mathbf{L}(H_1,H_2)$ have closed range, $B \in \mathbf{L}(H_1,H_3)$, and define

$$G(X) := (XA-B)(XA-B)^* . \tag{2.7}$$

Then the set

$$\{G(X) : X \in \mathbf{L}(H_2,H_3)\}$$

has a smallest element G_0 with respect to the Hermitian order on $\mathbf{L}(H_3)$. Let $M := \{X : G(X) = G_0\}$. Then $X \in M$ if and only if $XA = B(I-P)$, where P is the orthogonal projector onto $N(A)$. Furthermore, the set $\{XX^* : X \in M\}$ has a smallest element G_1 with respect to the Hermitian order; the unique $X \in M$ with $XX^* = G_1$ is $X = BA^\dagger$.

Theorems 2.1 and 2.2 can be easily modified to include the case when A is a closed operator. For proofs and detailed results, see [6]. For example, in Theorem 2.1 simply take X to vary in the set

$$Z(A) := \{X : R(X) \subseteq D(A),\ AX \text{ bounded}\} .$$

The case of operators with nonclosed range is technically more complicated. Among other things, one has to alter the set over which the minimization takes place, since A is unbounded and $A^\dagger B \notin Z(A)$ for all $B \in \mathbf{L}(H_3,H_2)$. However, $AA^\dagger$ is still bounded. We assume that B is such that

$$D(A^\dagger B) := \{y \in H_3 : By \in D(A^\dagger) = R(A) \dotplus R(A)^\perp\}$$

is dense in H_2 and choose the set of operators over which the minimization takes place to be the following set of closed operators:

$$\tilde{Z}(A) := \{X : D(X) = D(A^\dagger B),\ R(X), \subseteq D(A)\ , \text{ and } \ AX \ \text{ bounded}\} .$$

With these modifications, theorems analogous to Theorems 2.1 and 2.2 can be obtained (see [6]).

For a certain class of operators, one can phrase these characterizations in terms of a whole class of norms (including the Hilbert-Schmidt and trace norms, but *not* including the uniform operator norm) thus providing new extremal characterizations even in the matrix case. These characterizations in the case of infinite dimensional spaces have significant relevance to system and control theory, as we shall show in Section 3, since the class of operators considered subsumes important problems in the system theory. The class of operators for which one can find a relation between the Hermitian order and the size of certain operator norms, includes the Schatten classes (see [6]).

Let H_1 and H_2 be Hilbert spaces and let $T : H_1 \to H_2$ be a (nonzero) compact linear operator. Since T^*T is a nonnegative symmetric, compact linear operator on H_1 we have the spectral expansion

$$T^*Tx = \sum_k \lambda_k^2(x,\phi_k)\ \phi_k ,$$

where $\{\phi_k\}$ is an orthonormal set of eigenvectors of T^*T with

$$T^*T\phi_k = \lambda_k^2\ \phi_k \ \text{ and } \ \lambda_1 \geq \lambda_2 \geq \cdots > 0 .$$

Set $\psi_k := \lambda_k^{-1}\ T\phi_k$. Then the ψ_k's form an orthonormal set in H_2, and it is easy to show that

$$Tx = \sum_k \lambda_k(x,\phi_k)\ \psi_k . \tag{2.8}$$

This series is called the *singular value-singular function expansion* of T, and the collection of the triplets $\{\lambda_n;\phi_n,\psi_n\}$ is called a *singular system* for T. Let F_n denote the nth partial sum of this series. For a compact operator with an infinite-dimensional range, we merely have $\lambda_n \to 0$ and $\|T - F_n\| \to 0$ as $n \to \infty$. For $1 \leq p < \infty$, the set

$$\mathbf{C}_p = \left\{ T : \|T\|_p := \left(\sum_n \lambda_n^p \right)^{1/p} < \infty \right\} \tag{2.9}$$

is a Banach space with respect to the *Schatten norm* $\|T\|_p$. For $1 \leq p < \infty$, $\mathbf{C}_p$ are called *Schatten classes.* $T \in \mathbf{C}_p$ is equivalent to a quantitative statement about the rapidity with which $\|T - F_n\|$ can converge to zero. Precisely:

$T \in \mathbf{C}_p$ if and only if there exists a sequence $\{F_n\}$ of operators such that

$$\text{rank } (F_n) \leq n \text{ and } \sum_{n=1}^{\infty} \|T - F_n\|^p < \infty . \tag{2.10}$$

The space $\mathbf{C}_p$ may also be viewed as the completion of the space of finite rank operators with respect to the norm $\|\cdot\|_p$. For $p = \infty$, we get the space of all compact operators. $\mathbf{C}_2$ and $\mathbf{C}_1$ are particularly important and are called the spaces of ***Hilbert-Schmidt*** operators and ***nuclear*** (or trace) operators, respectively. If $T \in \mathbf{C}$ and $\{\phi_i\}$ is an orthonormal basis in H_1, the ***trace*** of T is $tr\ T = \sum_{i,j} <T\phi_i, \phi_j>$: it depends only on T, not on $\{\phi_i\}$. If $S, T \in \mathbf{C}_2$, then ST and $TS \in \mathbf{C}_1$. $\mathbf{C}_2$ is a Hilbert space with the inner product $[S, T] = tr\ T^*S$. For proofs of some of these statements, see [22].

Theorem 2.3. Let T and S be compact, $\|T\|_2 < \infty$. If T^*T and S^*S are comparable with respect to the Hermitian order, then for every Schatten norm

$$T^*T \leq S^*S \Leftrightarrow \|T\|_p \leq \|S\|_p .$$

Theorem 2.3 is not true for $p = \infty$. By combining this theorem with Theorems 2.1 and 2.2 and their generalizations to operators with nonclosed range we obtain extremal characterizations in terms of the Schatten norms (see [6] for details). In particular, we have:

Corollary 2.4. Let $R(A)$ have finite codimension. Then the $\{1, 3\}$-inverses of A are precisely those X for which $\|AX - I\|_p$ is minimal. Among them, $A^\dagger$ is the unique minimizer of X^*X with respect to the Hermitian order. (Note that if the minimization of X^*X is replaced by the minimization of a Schatten norm of X, the result does not hold unless $\dim H_2 < \infty$).

Corollary 2.5. Let $A \in \mathbf{L}(H_1,H_2)$ be a semi-Fredholm operator (see, e.g., [24]). The $\{1, 4\}$-inverses of A are precisely those X for which $\|XA - I\|_p$ is minimal, $1 \leq p < \infty$. Among them, $A^\dagger$ is the unique minimizer of XX^* with respect to Hermitian order.

We now give a generalization of the extremal property (E3) of Section 2.1.

Theorem 2.6. Let $A \in \mathbf{L}(H_1,H_2)$ have closed range and let J denote the set of all inner inverses of A. If $J \cap \mathbf{C}_2(H_2,H_1)$ is nonempty, then $A^\dagger$ is the unique inner inverse of minimal Hilbert-Schmidt norm.

The scope of the preceding results that require Hilbert-Schmidt operators can be extended via the use of an inner product on the space $\mathbf{L}(H_1,H_2)$ defined by

$$[A, B]_V = \text{trace}\,(A^*VB) \tag{2.11}$$

where V is a given positive definite nuclear operator on H_2. Suitable approximations can then be obtained by using a sequence of such operators $\{V_n\}$ that converges pointwise to the identity. The fact that each Schatten class is a two-sided ideal in the space of bounded linear operators enables us to use appropriate re-norming processes to obtain various extremal characterizations (relative to a fixed or a sequence of operators in $\mathbf{C}_p$) for bounded linear operators which do not necessarily satisfy the assumptions of the preceding theorems and corollaries. This would be particularly useful in contexts where V and V_n can have system-theoretic significance, as well as for the approximate realization of $A^\dagger$ by iterative and projection schemes.

The extremal theory also extends to generalized inverses relative to projections under seminorms induced by a nonnegative definite bilinear form (Mx,y), where M is a nonnegative definite symmetric continuous linear operator.

3. APPLICATIONS TO COMPENSATION PROBLEMS IN SYSTEMS AND CONTROL THEORY. SYSTEM PSEUDO INVERSES

3.1. Tandem Compensation

Consider the linear system S (represented by the dotted block) with input u and output y. This system is a tandem combination of two linear subsystems S_1 and S_2.

Given S_1, the problem is to determine S_2 such that the input-output relation of the overall system S is as "close" as possible (in a sense to be described) to that of a prescribed system S_0 which admits the same class of inputs as S_2. We assume the tandem combination of S_1 and S_2 to be initially free. We denote the zero states of S_1 and S_2 by θ' and that of $S = S_1S_2$ by θ'', with $\theta'' = (\theta',\theta')$. We can then describe the impulse response of S in terms of the impulse responses of S_1 and S_2. A convenient way for this description is obtained in terms of the weighting function. Let $\mathbf{L}_2^n[0,T]$ denote the space of square-integrable n-vector functions on $[0,T]$. If the space of inputs is taken to be $\mathbf{L}_2^n[0,T]$ and the space of outputs to be $\mathbf{L}_2^m[0,T]$, then the input-output response is given by

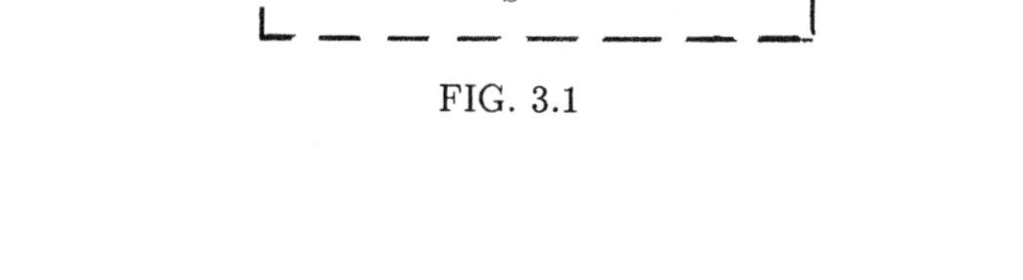

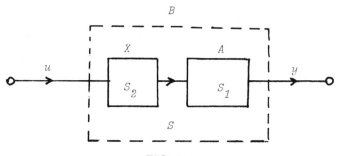

FIG. 3.1

$$y(t) = \int_{-\infty}^{t} W(t,\tau)\, x(\tau)\, d\tau, \tag{3.1}$$

where the weighting function (or kernel) is an $m \times n$ matrix whose entries are square-integrable functions, and $W(t,\tau) = 0$ for $t < \tau$ for physical realizability. Let $W_1(t,\tau)$ and $W_2(t,\tau)$ denote the kernel matrices for the systems S_1 and S_2 respectively. Then the *kernel matrix* or the *impulse response* of the system S is given by

$$W(t,\tau) = \int_{\tau}^{t} W_1(t,\theta)\, W_2(\theta,\tau)\, d\theta\ . \tag{3.2}$$

For zero-state time-invariant systems, this relation reduces to

$$w(t) = \int_{0}^{t} w_1(t-s)\, w_2(s)\, ds\ , \tag{3.3}$$

i.e., w is the convolution of the impulse responses of the systems S_1 and S_2.

The concepts of inverse and converse systems play a useful role in system theory (see [28]). The notion of an inverse system S^{-1} is tantamount to the inverse operator, if we restrict the initial states of the products SS^{-1} and $S^{-1}S$ to the states of the form (α,α^{-1}), where α is an arbitrary state of S and α^{-1} is its correspondent in S^{-1}. The system equivalent to a tandem combination of a system and its inverse has as its kernel the Dirac delta "function". Thus by virtue of (3.1) the kernel matrix of the inverse system, $W^{-1}(t,\tau)$, is the solution of the integral equation of the first kind

$$\delta(t-\tau) = \int_{\tau}^{t} W^{-1}(t,\theta)\, W(\theta,\tau)\, d\theta \tag{3.4}$$

where $W(t,s)$ is the kernel matrix of the system S, and $\delta(t-\tau)$ is the delta "function" at τ. The problem of finding a system inverse falls therefore within the theory of distributions. The solution may be in the form of an infinite series that may not be physically realizable, and may have to be approximated by truncating the series or by modifying the integral equation (3.4). If we restrict our consideration to proper systems (i.e. systems that do not have delta functions among their basis functions), e.g., consider only systems whose kernel matrix has square-integrable elements, then S^{-1} does not exist. This motivates introducing the notion of a *system pseudoinverse.*

Definition 3.1. Let S_1 be a linear system. A system $S_1^{\#}$ is said to be a *system pseudoinverse* of S_1 relative to a given criterion for comparison if for each tandem combination as in Figure 3.2, the input-output map of the system is as close as possible to that of S.

We shall assume that all tandem combinations are initially free and that the zero states of the system products are restricted, as usual, to the states of the components. We denote the input-output maps of the systems S_1, $S_1^{\#}$ and S by A, $A^{\#}$ and B, respectively. Then $A^{\#}$ renders $AX - B$ as small as possible relative to the given criterion for comparison. Criteria for comparison that can be used in Definition 3.1 include Hermitian order and ordering operators by the size of their norms, when possible. For Schatten norms and Hermitian order, the results of Section 2 would then relate system pseudoinverses to {1, 3}- and

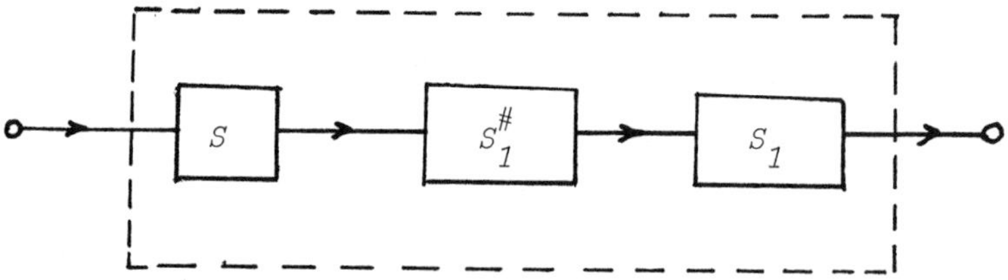

FIG. 3.2 System Pseudoinverse

{1, 4}-inverses and enable us to make a particular selection of these inverses that would minimize X^*X.

The setting and results of Section 2 can now be brought to bear on the tandem compensation problem. For example, if $A \in \mathbf{C}_p$, $X \in \mathbf{C}_q$ and $B \in \mathbf{C}_2$, where $p^{-1} + q^{-1} = 1/2$, then $AX - B$ is a Hilbert-Schmidt operator. In view of Theorem 2.3 and the analog of Theorem 2.1 for operators with nonclosed range, the minimizers of $\|AX - B\|_2$ exist and are precisely the {1, 3}-inverses of A. Among these $A^\dagger$ is the unique minimizer of X^*X with respect to the Hermitian order (we assume that X^*X has been extended to a bounded linear operator). In particular, if the systems A, B and X of Figure 3.1 are Volterra integral operators with square-integrable kernel matrices K_A, K_B and K_X, respectively, and if $0 \le t \le T$, then

$$\|AX-B\|_2^2 = \int_0^T\int_0^T tr\{[K_S(t,\tau)-K_B(t,\tau)]^*[K_S(t,\tau)-K_B(t,\tau)]\}\, dt\, d\tau$$

where

$$K_S(t,\tau) = \int_\tau^t K_A(t,\tau)\, K_X(\tau,s)\, d\tau$$

for $\tau < t$ and $K_S(t,\tau) = 0$ for $\tau \ge t$. Thus X minimizes the Hilbert-Schmidt norm of $AX - B$ if and only if the kernel of X is a ***least-squares*** solution of the integral equation of the first kind:

$$\int_\tau^t K_A(t,s)\, f(x,\tau)\, ds = K_B(t,\tau)\,. \tag{3.5}$$

Note that if $K_B(t,\tau)$ has a lower degree of smoothness than that of $K_A(t,\tau)$, then there will be no systems X with square-integrable integrals for which (3.5) has solutions, or equivalently for which $AX = B$ in Figure 3.1. In such cases, we seek a system pseudoinverse, if we do not wish to admit equations with distributions.

Let us compare this problem with the standard least-squares problem in linear control theory. Consider a system which is described by a linear system of differential equations

$$\dot{y} = B(t)y + C(t)u, \ \ y(0) = y_0 \tag{3.6}$$

where $y \in L_2^n\,[0,T] = H_2$, $u \in L_2^m\,[0,T] = H_1$, B is an $n \times n$ matrix and C is an $n \times m$ matrix. Under mild conditions on the coefficient matrices, it is well known that for each $u \in H_1$, there exists a unique solution y to (3.6) and that the relationship between y and u can be written in the form

$$y(t) = \int_0^t W(t,\tau)\ C(\tau)\ u(\tau)\ d\tau + W(t,0)\ y_0 \tag{3.7}$$

for all $0 \le t \le T$, where the system function $W(t,s) := \Phi(t)\Phi^{-1}(s)$ and $\Phi(t)$ is the fundamental matrix for the homogeneous equation $\dot{y} = B(t)y$, $\Phi(0) = I$. We assume without loss of generality that $y_0 = 0$. The problem of choosing a control u such that y is as close as possible to a desired response z can be easily described in terms of generalized inverses. Let A denote the linear operator on H_1 defined by

$$(Au)(t) = \int_0^t W(t,\tau)\ C(\tau)\ u(\tau)\ d\tau \tag{3.8}$$

Let $S_z = \{x \in H_1 : \inf \|Au - z\| = \|Ax - z\|\}$. Let L be a bounded or closed densely defined operator from H_1 to H_3. We consider the problem of finding $w \in D(L)$ such that

$$\|Lw\| = \inf \{\|Lx\| : x \in S_z\}\ . \tag{3.9}$$

Operator-theoretic and computational aspects of this problem have been considered by many authors. Generalizations to linear relations and multiresponse problem have been recently considered by Lee [7] and Lee and Nashed [9]. Finite element projectional methods coupled with regularization methods provide suitable approximations in the presence of data errors. Assume $R(L)$ is closed, $LN(A)$ is closed, and $N(L) \cap N(A) = \{0\}$. Then there exists a unique solution w that satisfies (3.9) and is denoted by $w = A_L^\dagger z$, where $A_L^\dagger$ is the induced generalized inverse that assigns to each $z \in \mathbf{R}(A) \dot{+} \mathbf{R}(A)^\perp$ the unique element in S_z of minimal $\|Lx\|$, see Nashed [14; pp. 234-236]. When $L = I$ this reduces to $A^\dagger$.

We can pose in the context of the least-squares control problem, compensation problems which are equivalent to the main problem of this section. This would lead to an integral equation for an unknown fundamental matrix. The details are similar to those described earlier. The extremal problems of Sections 2 and 3.1 can also be modified, under suitable assumptions on L, to find X for which $(LX)^* LX$ is minimal with respect to Hermitian order or for which $\|LX\|_p$ is minimal.

3.2. System Partial or Inner Inverses

Given a system S whose input-output map is $A : H_1 \to H_2$, and the set AH_1 of all realizable outputs, we seek a system *partial* (or inner) *inverse* X such that for each $y \in AH_1$, $u := Xy$ is an input that produces y and $\|X\|$ is minimal for some norm $\|\cdot\|$ on a subspace of the space $L(H_2,H_1)$.

The problem is identical with the extremal characterization considered in Theorem 2.6. However, it should be pointed out that the set of partial inverses for a system theory problem usually do not contain a Hilbert-Schmidt operator, unless the range of the system is finite dimensional. In the latter case, Theorem 2.6 provides a useful characterization.

3.3. System Compensation to Achieve a Desired Output

Given a system whose input-output map is A and a desired output z we wish to find an input $w \in H_1$ and a system $\tilde{X}$ from a specified class $\mathbf{S}$ of input-output maps such that

$$\inf_{\substack{X \in \mathbf{S} \\ u \in H_1}} \|AXu - z\| = \|A\tilde{X}w - z\| .$$

This can be treated as a hybrid of two optimization problems involving the vector version and the operator version of the extremal property of generalized inverse operators.

3.4. Tandem Compensation to Achieve a Desired Impulse Response at a Finite Number of Points

In some applications, the desired impulse response $K_B(t,\tau)$ of Section 3.1 is specified only at a finite number of points $0 \leq t_1 < t_2 < \cdots < t_r \leq T$ and $0 \leq \tau_1 < \tau_2 < \cdots < \tau_l \leq T$. In this case the integral equation (3.5) becomes:

$$\int_{\tau}^{t} K_A(t_i,s)\, f(s,\tau_j)\, ds = K_B(t_i,\tau_j)$$

$$1 \leq i \leq r, \; 1 \leq j \leq l .$$

This defines an integral operator with finite-dimensional range. In this case the minimization of X^*X can be replaced by the minimization of the Schatten norm of X; see [15] for generalized inverse aspects of moment-discretization of integral equations, which apply directly to this problem. See also [2] for an overview of linear inverse problems with discrete data.

Note that the system compensation and optimization problems considered here are to be distinguished from the problem (treated briefly for comparison purposes at the end of Section 3.1) of determining an input to a linear system such that the actual output is as close as possible to a desired output. Both of these problems are ill-posed in the L_2-setting. Thus the problem of computation or construction of the system pseudoinverse is generally unstable numerically; computational aspects have to be addressed separately. Note also that in this context we wish to compute a system pseudoinverse, or a kernel matrix in the problem considered, in contrast to the computation of $A^\dagger b$, the least-squares solution of minimal norm.

3.5. Impulse Response Identification and Control of Dimensionality

The identification of the impulse response $f(t)$ for a linear time-invariant system from observations of its output $g(t)$ to a known input $u(t)$, is governed by the integral equation

$$\int_0^t u(t-x)\, f(x)\, dx = g(t) \ .$$

This problem is ill-posed in $L_2\,[0,T]$. In alternate approaches of input-output representations (e.g., step response, ramp response) whose identification problem may be well posed, do not overcome the ill-posedness of the impulse response identification problem. For example, to obtain the impulse response from the step-function response, one has to perform numerical differentiation, which is an ill-posed problem. Similarly, using frequency response yields the Fourier transform of u. Numerical inversion of the Fourier transform is another ill-posed problem. The frequency-domain formulation of impulse-response identification, namely the problem of identifying the transfer function of a system by its poles and residues, is also numerically unstable when the input and output are contaminated. One approach to the resolution of ill-posed problems is to seek an approximation to the solution (the inverse or identification problem) by a linear combination of a "*small*" number of judiciously chosen functions. The problem then becomes approximation with a "few" parameters. Clever choices of functions for certain ill-posed problems abound in the literature. For example, in spectroscopy Voigt functions have been found effective for numerical deconvolution. A Voigt function is the convolution of the dispersion function $\frac{a}{x^2+a^2}$ and the Gauss (normal) distribution; hence it involves two parameters. The convolution of two Voigt Functions is a Voigt function. In this example if f and g are represented as a linear combination of several Voigt functions, then the coefficients of the Voigt function representation of u can be easily computed.

4. A HERMITIAN-ORDER INEQUALITY FOR GENERALIZED INVERSES

Let A and B be positive definite symmetric real matrices (or positive definite complex matrices) of order n. It is well known that the following inequality holds for all $0 \leq \alpha \leq 1$:

$$[\alpha A + (1-\alpha)B]^{-1} \leq \alpha A^{-1} + (1-\alpha)B^{-1} , \tag{4.1}$$

where the "ordering" is induced by Hermitian form (also called Loewner ordering), i.e.,

$$A \leq B \text{ if and only if } (Ax,x) \leq (Bx,x) \text{ for all } x .$$

There are over ten different proofs of (4.1). We sketch a proof based on simultaneous diagonalization of A and B, due to M. H. Moore [12]. We take A and B to be real matrices. Under the above assumptions, there exists a real nonsingular matrix S such that $A = SS^T$ and $B = SDS^T$, where D is a diagonal matrix whose diagonal elements λ_i are eigenvalues (necessarily real) of the generalized eigenvalue problem $Bx = \lambda Ax$. Since B is positive definite, $\lambda_i > 0$ for $1 \leq i \leq n$. Then

$$\alpha A + (1-\alpha)B = S\ [\alpha I + (1-\alpha)D]S^T = SE_\alpha S^T ,$$

where $E_\alpha := \alpha I + (1-\alpha)D$. Both D and E_α are nonsingular. Thus, letting $P := S^{-1}$, we have $A^{-1} = P^TP$, $B = P^TD^{-1}P$ and

$$\alpha A^{-1} + (1-\alpha)B^{-1} - [\alpha A + (1-\alpha)B]^{-1} = \alpha P^TP + (1-\alpha)P^TD^{-1}P - P^TE_\alpha^{-1}P$$

$$= P^T\ [\alpha I + (1-\alpha)D^{-1} - E_\alpha^{-1}]P = P^TG_\alpha P ,$$

where $G_\alpha := \alpha I + (1-\alpha)D^{-1} - E_\alpha^{-1}$ is a diagonal matrix. By the convexity of the function $f(x) = x^{-1}$ for $x > 0$, it follows that the diagonal entries of G_α are nonnegative, and hence $P^TG_\alpha P \geq 0$.

The inequality (4.1) may be interpreted as saying that with respect to the Loewner ordering the matrix inverse function is convex on the (convex) set of all positive definite symmetric real matrices. There is a wide literature on convexity for functions of matrices and for Hermitian operators. For a thorough perspective on this subject, one should refer to a paper by C. Davis [4].

It is natural to ask for an analog of (4.1) when A and B are nonnegative definite symmetric matrices, and more generally when A and B are nonnegative definite symmetric linear operators on a real Hilbert space (or nonnegative definite on a complex Hilbert space). It is also tempting in these cases to consider an analog of (4.1) in which the "inverse" is replaced by the "generalized inverse". What happens then? Kaffes (see [26]) proved that

$$[\alpha A + (1-\alpha)B]^\dagger \leq \alpha A^\dagger + (1-\alpha)B^\dagger \tag{4.2}$$

holds for the Moore-Penrose inverse, where A and B are nonnegative definite matrices of the same order provided that A and B have the same range. Siafarikas [26] proved that the above inequality holds in a Hilbert space H, where A and B are bounded self-adjoint operators defined on H such that $\mathbf{R}(A) = \mathbf{R}(B)$ and $\mathbf{R}(A)$ is closed. It is easy to construct 2×2 nonnegative definite symmetric matrices for which the *reverse* strict inequality in (4.2) holds if the two matrices do not have the same range. This might make the restriction $\mathbf{R}(A) = \mathbf{R}(B)$ plausible for a generalization of (4.1) to nonnegative definite matrices. Unfortunately, however, this is not the case. For with this restriction, $\mathbf{R}(\alpha A + (1-\alpha)B) = \mathbf{R}(A)$, $0 \leq \alpha \leq 1$, the inequality (4.2) is nothing more than an inequality for *inverses*. This follows from the function-theoretic definition of $A^\dagger$, namely $A^\dagger = \tilde{A}^{-1}Q$, where $\tilde{A}$ is the restriction of A to the orthogonal complement of the nullspace of A, and Q is the orthogonal projector onto the range of A.

What, then, is a *bona fide* generalization of (4.1) for nonnegative definite linear operators?

Nashed and Visick have recently proved the following theorem:

Theorem 4.1. Let A and B be nonnegative definite (symmetric) linear operators on a Hilbert space H. Let $P_\alpha := \alpha AA^\dagger + (1-\alpha)BB^\dagger$. Then for all $0 \leq \alpha \leq 1$, the following inequality holds:

$$P_\alpha \, [\alpha A + (1-\alpha)B]^\dagger \, P_\alpha \leq \alpha A^\dagger + (1-\alpha)B^\dagger \, . \tag{4.3}$$

The proof of this theorem along with related operator inequalities will appear elsewhere.

It should be noted that P_α is a convex combination of the orthogonal projector of H onto the range of A and the orthogonal projector of H onto the range of B. Thus if $\mathbf{R}(A) = \mathbf{R}(B)$, then P_α becomes the identity operator. The inequality (4.3) holds also for weighted or oblique generalized inverse in Hilbert space as defined in [20].

For systems described by input-output mappings, the inequality (4.3) is relevant to the compensation problem by parallel connections. For a given system A, $\alpha A + (1-\alpha)B$ represents a scaling of A by α and a compensation by a parallel connection to the system $(1-\alpha)B$. In the case when $\mathbf{R}(A) = \mathbf{R}(B)$, (4.2) is an ordering between the quadratic form of the overall compensated system and the combination of the quadratic forms of the original system and the compensating subsystem. The convexity inequality (4.1) has found several applications in statistics and matrix theory (see Marshall and Olkins [11] and references cited therein).

5. REPRESENTATION OF SINGULAR SYSTEMS AND THE COVARIANCE CONDITION OF THE GENERALIZED INVERSE

Let $A : V \to V$ be an invertible linear transformation on a finite dimensional vector space V and let σ is a basis for V. If $[A]_\sigma$ denotes the matrix representation of A with respect to the basis σ, then $[A^{-1}]_\sigma = [A]_\sigma^{-1}$, i.e., the matrix representation of A^{-1} with respect to the basis σ is the inverse of the matrix representation of A. Also, if the linear operator B is similar to A, say $B = T^{-1}AT$ for some invertible linear operator T and if A is invertible, then $(T^{-1}AT)^{-1} = T^{-1}A^{-1}T$. These relations do not necessarily hold for a noninvertible operator A if A^{-1} is replaced by a generalized inverse; for example, they do not necessarily hold for the Moore-Penrose inverse of a square matrix A. The following relation (when it holds)

$$(TAT^{-1})^\dagger = TA^\dagger T^{-1} \tag{5.1}$$

is called the *covariance condition*. Thus the problem arises to describe the class

$\mathbf{C}(A)$ of all those matrices T for which this condition is valid. Several authors have recently addressed this problem in the case of matrices (see Schwerdtfeger [25] and Robinson [23]).

It is clear that this problem is also of interest in connection with matrix representations of a singular linear system on a finite dimensional vector space. In this short section we will briefly indicate an alternate approach to this problem for linear transformations on a vector space.

Let V and W be (real or complex) vector spaces and A be a linear transformation from V into W. Let $\mathbf{M}$ be any algebraic complement to the nullspace of A, $\mathbf{N}(A)$, and let $\mathbf{S}$ be any algebraic complement to the range of A, $\mathbf{R}(A)$. We denote by P and Q the algebraic projectors onto $\mathbf{N}(A)$ and $\mathbf{R}(A)$, respectively, induced by these complements. The generalized inverse $A^{\dagger}_{P,Q}$ of A relative to these projectors is defined as the unique linear extension of the operator $(A|\mathbf{M})^{-1}$ to $\mathbf{R}(A) \dot{+} \mathbf{S}$ such that $\mathbf{N}(A^{\dagger}_{P,Q}) = \mathbf{S}$. (Here $A|\mathbf{M}$ denotes the restriction of A to $\mathbf{M}$).

The linear transformation $A^{\dagger}_{P,Q}$ fulfills the following equations, which could also be used as a definition of $A^{\dagger}_{P,Q}$:

$$\begin{cases} AA^{\dagger}_{P,Q}A = A \\ A^{\dagger}_{P,Q}AA^{\dagger}_{P,Q} = A^{\dagger}_{P,Q} \\ A^{\dagger}_{P,Q}A = I - P \\ AA^{\dagger}_{P,Q} = Q \end{cases} \tag{5.2}$$

Theorem 5.1. Let A be a linear transformation on a (real or complex) vector space V into V. Let $A^{\dagger}_{P,Q}$ be defined as above. Then for any invertible linear transformation T from V into V,

$$(T^{-1}AT)^{\dagger}_{P',Q'} = T^{-1}A^{\dagger}_{P,Q}T\ , \tag{5.3}$$

where

$$P' \ := T^{-1}PT \ \text{ and } \ Q' \ := T^{-1}QT\ . \tag{5.4}$$

Proof. The proof is by computation, i.e., we show that if $B := T^{-1}AT$ and $\mathbf{C} := T^{-1}A^{\dagger}_{P,Q}T$, then $\mathbf{C}$ is the generalized inverse of B relative to the projectors P' and Q' defined by (5.4). To this end, one can easily verify that $\mathbf{C}$ satisfies the four algebraic conditions in (5.2). We omit the simple algebraic verification.

Theorem 5.1 suggests that the natural approach to the "covariance problem" for generalized inverses is via the effect of the similarity transformation $A \to T^{-1}AT$. It is reasonable to expect that we should also subject the projectors P and Q to the same similarity transformation and consider the generalized inverse of $T^{-1}AT$ relative to the transformed projectors. This is confirmed by Theorem 5.1.

The covariance condition for the generalized inverse can then be restated in terms of the projectors as follows: Under what conditions do the projectors P and Q commute with T?

The problem can also be formulated in topological linear spaces and one then asks for continuous projectors that commute with T. To this end, we recall, for convenience, the definition of the generalized inverse in Banach spaces.

Let X and Y be Banach spaces and A be a densely defined closed linear operator from X into Y, or a bounded linear operator from X into Y. We assume that the nullspace of A, $\mathbf{N}(A)$, has a topological complement

M, and the closure of the range of A, $\overline{\mathbf{R}(A)}$, has a topological complement **S**. Let P and Q denote the projectors onto $\mathbf{N}(A)$ and $\overline{\mathbf{R}(A)}$, respectively, induced by these complements. (Note that P and Q are now topological projectors, i.e., they are *continuous* linear idempotents). We define the generalized inverse $A^{\dagger}_{P,Q}$ relative to these projectors as the unique linear extension of $(A\,|\mathbf{M})^{-1}$ to $\mathbf{R}(A) \dotplus S$ such that $\mathbf{N}(A^{\dagger}_{P,Q}) = \mathbf{S}$. The closed and densely defined operator $A^{\dagger}_{P,Q} =: A^{\dagger}$ satisfies the following equations, which could also be used as a definition of $A^{\dagger}_{P,Q}$, if we take as the domain the maximal domain for which these equations have a solution, namely $\mathbf{D}(A^{\dagger}_{P,Q}) := \mathbf{R}(A) \dotplus \mathbf{S}$:

$$\begin{cases} AA^{\dagger}A = A & \text{on } \mathbf{D}(A) \\ A^{\dagger}AA^{\dagger} = A^{\dagger} & \text{on } \mathbf{D}(A^{\dagger}) \\ A^{\dagger}A = I - P & \text{on } \mathbf{D}(A) \\ AA^{\dagger} = Q & \text{on } \mathbf{D}(A^{\dagger}) \end{cases} \tag{5.5}$$

Note that $A^{\dagger}$ is bounded if and only if $\mathbf{R}(A)$ is closed, in which case $\mathbf{D}(A^{\dagger}) = Y$, and clearly $\mathbf{D}(A) = X$ if and only if A is bounded.

We may now state a problem that arises from the covariance condition for the generalized inverse as follows (for simplicity we assume that A is bounded).

Problem. Let A be a bounded linear operator on a Banach space X into X and assume that both the nullspace of A and the closure of the range of A have topological complements, say **M** and **S**, respectively, in X. Let P be the projector of X onto **M** along $\mathbf{N}(A)$ and Q be the projector of X onto $\overline{\mathbf{R}(A)}$ along **S**. Characterize all invertible operators T which commute with P and Q:

$$TP = PT\ , \quad TQ = QT\ . \tag{5.6}$$

Equivalent formulations of this problem, along with some characterizations and other results related to Theorem 5.1 will be considered elsewhere. The covariance problem in the case of the Moore-Penrose inverse of a complex matrix A was solved by H. Schwerdtfeger [25] in case A is of rank 1 or 2 and by D. W. Robinson [23] for an arbitrary rank, $r < n$.

For a fixed A and for T belonging a class J of invertible operators, we define the *covariance defect* by

$$E(T) := (T^{-1}AT)^{\dagger} - T^{-1}A^{\dagger}T .$$

The problem of minimizing $E(T)$, over various classes J, with respect to different norms, or with respect to Hermitian order, is of related interest.

ACKNOWLEDGMENT

The research of the author was supported by the United States Army under contract No. DAAG-29-83-K-0109.

REFERENCES

[1] D. R. Audley and D. A. Lee, Ill-posed and well-posed problems in system identification, *IEEE Trans. Automatic Control,* AC-**19** (1974), 738-747.

[2] M. Bertero, C. De Mol and E. R. Pike, Linear inverse problems with discrete data. I: General formulation and singular system analysis, *Inverse Problems* **1** (1985), 301-330.

[3] F. J. Beutler and W. L. Root, The operator pseudoinverse in control and systems identification, in [14], pp. 397-494.

[4] C. Davis, Notions generalizing convexity for functions defined on spaces of matrices, in "Convexity," (V. Klee, ed.), pp. 187-201, Amer. Math. Soc., Providence, R.I., 1963.

[5] A. Deif, "Sensitivity Analysis in Linear Systems," Springer-Verlag, Berlin-Heidelberg - New York, 1986.

[6] H. W. Engl and M. Z. Nashed, New extremal characterizations of generalized inverses of linear operators, *J. Math. Anal. Appl* **82** (1981), 566-586.

[7] S. J. Lee, Multi-response quadratic control problems, *SIAM J. Control Optim.* **24** (1986), 771-788.

[8] S. J. Lee and M. Z. Nashed, Least-squares solutions of multi-valued linear operator equations in Hilbert spaces, *J. Approx. Theory* **38** (1983), 380-391.

[9] S. J. Lee and M. Z. Nashed, Constrained minimization problems for linear relations in Hilbert spaces, to appear.

[10] P. Z. Marmarelis and V. Z. Marmarelis, "Analysis of Physiological Systems: The White-Noise Approach," Plenum Press, New York, 1979.

[11] A. W. Marshall and I. Olkin, "Inequalities: Theory of Majorization and Its Applications," Academic Press, New York, 1979.

[12] H. H. Moore, A convex matrix inequality, *Amer. Math. Monthly* **80** (1973), 408-409.

[13] R. H. Moore and M. Z. Nashed, Approximations to generalized inverses of linear operators, *SIAM J. Appl. Math.* **27** (1974), 1-16.

[14] M. Z. Nashed (Ed.), "Generalized Inverses and Applications," Academic Press, New York, 1976.

[15] M. Z. Nashed, Moment-discretization and least-squares solutions of integral equations of the first kind, *J. Math. Anal. Appl.* **53** (1976), 359-366.

[16] M. Z. Nashed, On the perturbation theory for generalized inverse operators in Banach spaces, in "Functional Analysis Methods in Numerical Analysis," (M. Z. Nashed, ed.) Lecture Notes in Mathematics, Vol. 701, pp. 180-193, Springer-Verlag, New York, 1979.

[17] M. Z. Nashed, Operator-theoretic and computational approaches to ill-posed problems with applications to antenna theory, *IEEE Trans. Antennas and Propagation,* AP-**29** (1981), 220-231.

[18] M. Z. Nashed, Inner, outer, and generalized inverses in Banach and Hilbert spaces, *Numer. Funct. Anal. Optim.* **9** (1987), 261-326.

[19] M. Z. Nashed and G. F. Votruba, A unified approach to generalized inverses

of linear operators: II. Extremal and proximinal properties, *Bull. Amer. Math. Soc.* **80** (1974), 831-835.

[20] M. Z. Nashed and G. F. Votruba, A unified operator theory of generalized inverses, pp. 1-109, in [14].

[21] R. Penrose, On best approximate solutions of linear matrix equations, *Proc. Cambridge Philos. Soc.* **52** (1956), 17-19.

[22] J. Ringrose, "Compact Non-Self Adjoint Operators," van Nostrand Reinhold Company, London, 1971.

[23] D. W. Robinson, On the covariance of the Moore-Penrose inverse, *Linear Algebra and Appl.,* **61** (1984), 91-99.

[24] M. Schechter, "Principles of Functional Analysis," Academic Press, New York, 1971.

[25] H. Schwerdtfeger, On th covariance of the Moore-Penrose inverse, *Linear Algebra and Appl.* **52** (1983), 629-643.

[26] P. D. Siafarikas, An operator inequality, *Internat. J. Math. Math. Sci.* **7** (1984), 205-207.

[27] A. N. Tikhonov and V. Y. Arsenin, "Solution of Ill-Posed Problems," Winston, Wiley, New York, 1977.

[28] L. A. Zadeh and C. A. Desoer, "Linear System Theory," McGraw-Hill, New York, 1963.

Wave Equation on a Bounded Domain with Boundary Dissipation: An Operator Approach

R. Triggiani*

Department of Mathematics

University of Florida

Gainesville, Florida

1. Introduction, literature, statement of main results

Let Ω be a bounded open domain in R^n $(n \geq 2)$ subject to further conditions to be specified below. Assume that $\Gamma = \partial\Omega$, the boundary of Ω, consists of two parts Γ_0 and Γ_1 with Γ_1 non-empty ($\Gamma_1 \neq \emptyset$ sand relatively open in Γ. Consider the following second order hyperbolic problem in the solution w(t,x):

$$w_{tt} = \Delta w \qquad \text{in } (0,\infty)\times\Omega \tag{1.1}$$

$$w(0,x) = w_0(x),\ w_t(0,x) = w_1(x) \qquad \text{on } \Omega \tag{1.2}$$

$$w(t,x) \equiv 0 \qquad \text{on } (0,\infty)\times\Gamma_0 \tag{1.3}$$

$$\frac{\partial w}{\partial \nu}(t,x) = -w_t(t,x) \qquad \text{on } (0,\infty)\times\Gamma_1 \tag{1.4}$$

where Δ is the Laplacian acting on the n-dimensional space variable x, ν is the unit normal of Γ pointing toward the exterior of Ω. The assumptions on Ω are as follows [L.1]: Ω is connected and

(i) if $\Gamma_0 \neq \phi$, then Γ is of class C^2 and $\overline{\Gamma}_0 \cap \overline{\Gamma}_1 = \emptyset$;

(ii) If $\Gamma_0 = \phi$, Ω is convex and $\overline{\Gamma}_0 \cap \overline{\Gamma}_1 = \emptyset$.

Thus, Γ_0 and Γ_1 are, respectively, the reflecting and energy absorbing portions of the boundary Γ. Let

$$E(t) \equiv E(w,t) \equiv \int_\Omega [|w_t(t,x)|^2 + |\nabla w(t,x)|^2] d\Omega \tag{1.5}$$

be the 'energy' of problem (1.1) - (1.4). There has been extensive work over the past ten years or so centered on the study of energy

*Current affiliation: University of Virginia, Charlottesville, Virginia

decay as $t \to \infty$ (so called 'stabilization' problem) for problem (1.1) - (1.4) (and its natural generalization as in Remark 1.1 below); in addition, of course, to first establishing the well-posedness of (1.1) - (1.4) in an appropriate function space. In summary, the pioneering work was first performed in the mid seventies in the study aimed at achieving energy decay rates for the wave equation exterior to a bounded obstacle (the so called 'exterior' problem). See the papers of C. Morawetz, P.D. Lax; R. S. Phillips; J. Ralston and W. Strauss quoted in [C.1-2], [L-1]. These efforts brought forward several energy identities, which were then used to obtain energy decay rates, under suitable geometrical conditions on the boundary of the obstacle. In contrast, for the 'interior' problem (1.1) - (1.4) - where energy decay is sought within the bounded domain $\overline{\Omega}$ - results that become available at about the same time were simply asserting energy decay to zero for all initial data (in the space $\mathcal{H}_1$ below in (2.12)) for which $E(w,0) < +\infty$; however, no rates of decay were given. See [Q-R.1], [S.1], [Z.1]. In the modern terminology of control theory, they were strong stabilization results in the space $\mathcal{H}_1$, not uniform stabilization results in $\mathcal{H}_1$. However, they required no assumptions on the geometry of the bounded domain Ω. The 'interior' problem is more difficult than the 'exterior' problem, since the latter enjoys the advantage that the energy distributes itself over on infinite region as $t \to \infty$. Credit goes to G. Chen for having first obtained an energy decay rate (uniform stabilization) for the interior problem (1.1) - (1.4), under some natural geometrical conditions on Ω, by adapting the techniques of the literature on the exterior problem. His first paper [C.1] used a slight modification of a functional in [M.1]. His later paper [C.2] relaxed the geometrical conditions on Ω, by employing this time a new energy functional discovered by W. Strauss [S.2] in the study of the exterior problem. This same energy functional of Strauss was also employed later on by J. Lagnese [L.1], who managed to relax even further the geometrical conditions on Ω, under which an energy decay rate (in $\mathcal{H}_1$) is obtained. Lagnese's main result in [L.1] is:

Theorem 1.1 (uniform stabilization), Assume there is a vector field $h(x) = [h_1(x),..,h_n(x)] \in C^2(\overline{\Omega})$ such that

(i) $h \cdot \nu \leq 0$ on Γ_0 (1.6)

(ii) $h \cdot \nu \geq \gamma > 0$ on Γ_1 (1.7)

for some constant $\gamma > 0$;

(iii) The matrix $H(x) + H^*(x)$ is uniformly positive definite on $\overline{\Omega}$, where

$$H(x) = \begin{vmatrix} \frac{\partial h_1}{\partial x_1}, \dots, \frac{\partial h_1}{\partial x_n} \\ \vdots \qquad \vdots \\ \frac{\partial h_n}{\partial x_1}, \dots, \frac{\partial h_n}{\partial x_n} \end{vmatrix} \tag{1.8}$$

Then, there are positive constants C, δ such that

$$E(w,t) \leqslant C\, e^{-\delta t} E(w,0), \quad t \geqslant 0 \tag{1.9}$$

for every solution of (1.1) - (1.3) for which $E(w,0) < +\infty$ □

Commenting on the improvement over [C.2], paper [L.1] notes that "The key to the proof of Theorem 1.1 is the following result which may be of independent interest. This is the analog for problem (1.1) - (1.4) of a result of W. Strauss [S.2] concerning solutions of the Dirichlet problem in a region exterior to a bounded obstacle".

Theorem 1.2 For every $\varepsilon > 0$, there is a number C_ε such that for every $\beta > 0$,

$$\int_0^\infty \int_\Omega e^{-2\beta t}(w - I(w_0))^2 dQ \leqslant \varepsilon \int_0^\infty \int_\Omega e^{-2\beta t} w_t^2 dQ + C_\varepsilon E(w,0) \tag{1.10}$$

[where here and hereafter $dQ = d\Omega dt$], for every solution of (1.1) - (1.4) for which $E(w,0) < +\infty$, where

$$I(w_0) = \begin{cases} 0 & , \text{ for } \Gamma_0 \neq \emptyset \quad \text{(a)} \\ \frac{1}{\text{meas}(\Omega)} \int_\Omega w(x,0)d\Omega, & \text{for } \Gamma_0 = \emptyset \quad \square \quad \text{(b)} \end{cases} \tag{1.11}$$

"Another consequence of Theorem 1.2 - notes [L.1] - is a simple and direct proof of energy decay in the absence of restrictions on $\partial\Omega$."

Theorem 1.3 [Strong stabilization] If w is a solution of (1.1) - (1.4) with $E(w,0) < +\infty$, then

$$\lim_{t\to\infty} E(w,t) = 0 \quad \square$$

As mentioned before, this result was previously proved in [Q-R.1] using a compactness argument and the Holmgren uniqueness theorem; in [S.1], using LaSalle - Hale's invariance principle, and in [Z.1], by reducing the unbounded operator case to a bounded case. As to the well-posedness question, [L.1] quotes [Q-R.1], [C-1] to assert that problem (1.1) - (1.4) generates a strongly continuous semigroup $S(t)[w_0,w_1] = [w(t;w_0,w_1), w_t(t,w_0,w_1)]$ on the space $\mathcal{H}_1$ equipped with equivalent energy norm, if $\Gamma_0 \neq \emptyset$. In terms of this semi-group, the conclusions of Theorem 1.1 and Corollary 1.3 may be restated as $\|S(t)\| \leq Ce^{-\delta t}$, $t \geq 0$ and $\|S(t)[w_0,w_1]\| \to 0$ as $t \to \infty$ for all $[w_0,w_1] \in \mathcal{H}_1$, in the uniform, respectively, strong norm of $\mathcal{H}_1$. Thus, the plan of Lagnese's paper is as follows:

$$\text{Thm } 1.2 \begin{array}{l} \nearrow \text{ Theorem 1.1 (uniform stabilization)} \\ \searrow \text{ Theorem 1.3 (strong stabilization)} \end{array} \tag{1.12}$$

In all of the above mentioned papers [C.1-2], [L.1] on uniform stabilization for (1.1) - (1.4), the general thrust of the proof has a flavor in the Lyapunov method style, as applied to the appropriate functional taken from the exterior problem literature. In the same spirit as [L.1] in terms of the general approach - albeit with genuinely new serious technical difficulties to overcome - is a more recent paper [L-T.1]. This shows, for the first time, uniform stabilization of the wave equation with a suitable feedback in the Dirichlet B.C. (under some geometrical conditions on Ω). This result is established on a different space, $L^2(\Omega)xH^{-1}(\Omega)$ (which is the natural space for this problem). The last section of [L-T.1] presents another proof - a combination of Lyapunov method and a multiplier technique - of the conclusion of Theorem 1.1, however with assumption (iii) of Theorem 1.1 replaced by the following weaker assumption on the matrix $H(x)$ defined in (1.8):

$$\text{(iii)} \quad \int_\Omega H(x)v(x)\cdot v(x)d\Omega \geq \rho \int_\Omega |v(x)|^2_{R^n} d\Omega, \text{ for some } \rho > 0 \text{ for all } v(x) \in L_2(\Omega;R^n) \tag{1.13}$$

[J. Lagnese has kindly informed us that his proof in [L.1] uses precisely property (iii') = (1.13), even though the stronger (but more easily verifiable) property (iii) was explicitly chosen in the statement of his Theorem 1.1. Thus, henceforth, we shall consider Lagnese's Theorem, as subject to assumptions (i), (ii), (iii').]

In the present paper, we wish to examine once more well-posedness, strong stabilization and - above all - uniform stabilization for problem (1.1) - (1.4). We claim no strengthening over the statement of Lagnese's results, in particular his Theorem 1.1 (with assumption (iii') = (1.13) replacing assumption (iii)). However, we believe that our present, essentially self-contained treatment brings forth worthwile simplifications and new insight into the problems. There are two major distinctive features in our present approach.

(i) The formulation of problem (1.1) - (1.4) as an appropriate abstract evolution equation (operator model), which explicitly incorporates the boundary action of the non-homogeneous term, by means of a suitable operator ($\tilde{N}$ below). This model - in its differential forms (2.7a-b) below or in its integral form (3.36) below is a new feature over [L.1] and represents the approach to hyperbolic problems proposed in [T.1] and successfully used in many problems since. The ensuing analysis is then carried out, to the extent possible, on the abstract model, to obtain goemetry-independent results: Lemma 2.1 (generation and spectral properties), 1.3 (strong stabilization) and the key Theorem 1.2

(ii) A direct application on (1.1) - (1.4) of multiplier techniques, which reveal the role of two multipliers: $h \cdot \nabla w$ to obtain the basic initial identity (3.3), and w div h to obtain an identity for the difference "kinetic minus potential energy", see (3.13) below. This approach is entirely self-contained and shows clearly which multiplier is needed, for what end and at what stage. Moreover, it permits to keep control on the various constants. In the Lyapunov method of [L.1] , the Lyapunov functional of W. Strauss contains, among others, also these multiplier terms, but their role is less motivated. Our approach here benefits from recent understanding of the role of multipliers for the wave equation (1.1) in the context of several different problems: regularity theory of Dirichlet mixed problems [L.3] [L-L-T.1]; uniform stabilization with feedback in the Dirichlet B.C [L-T.1] as well as a direct approach to exact controllability (without passing through stabilization first) [L.4-L.6], [T.2], [L-L-T.2]

In contrast with [L.1], the plan of our paper is as follows

Lemma 2.1 (semigroup generation & spectral properties)	$\longrightarrow$	Theorem 1.3 (strong stabilization)
$\downarrow$		(1.14)
Theorem 1.2	$\longrightarrow$	Theorem 1.1 (uniform stabilization)

Thus, unlike [L.1], we obtain here a purely operator theoretic proof of (semigroup generation and) strong stabilization based on our abstract operator model, which is fully independent of the technical Theorem 1.2. As to our proof of Theorem 1.1, perhaps, the most notable and original part of it in comparison with [L.1] is a technically different - and we believe much simplified - proof of the key Theorem 1.2 which is based here on our operator model in integral form (3.36), rather than on the analysis as in [L.1], based on Fourier transforming in time problem (1.1) - (1.4): a technique, the latter, introduced by Lax-Phillips in the early seventies in the study of the <u>exterior</u> problem and used since by various authors (C. Morawetz, W. Strauss, etc.) in the exterior problem. Our operator approach to Theorem 1.2 clearly reveals a key feature behind the proof: a cancellation of the poles of $R(\lambda^2,-A)$ (on the imaginary axis), which occurs at the level of equation (3.38) below. As a result, the subsequent equation (3.39) obtained after the cancellation can be viewed <u>also on the imaginary axis</u>.

<u>Remark</u> 1.1 The techniques of this paper cover also the generalized wave equation as in [L.1] p. 167. We omit details $\square$

2. Preliminaries. Well-posedness and strong stabilization: an operator theoretic approach.

2.1 The case $\Gamma_0 \neq \emptyset$. Let $A: L^2(\Omega) \supset \mathcal{D}(A) \to L^2(\Omega)$ be the operator defined by

$$Af = -\Delta f, \quad \mathcal{D}(A) = \{f \in H^2(\Omega): f|_{\Gamma_0} = 0, \frac{\partial f}{\partial \nu}|_{\Gamma_1} = 0\} \tag{2.1}$$

Then, A is non-negative self-adjoint and has compact resolvent $R(\cdot,A)$. Moreover, the problem $\Delta h = 0$ in Ω and $h|_{\Gamma_0} = \frac{\partial h}{\partial \nu}|_{\Gamma_1} = 0$ implies $h = 0$

(by Green Theorem applied to $(\Delta h, h)_\Omega = 0$). Thus, $A^{-1} \in \mathcal{L}(L^2(\Omega))$, a distinctive feature over the case $\Gamma_0 = \emptyset$. of subsection 2.2. Next, define the operator $\tilde{N}$ by

$$h = \tilde{N}g \Longleftrightarrow \begin{cases} \Delta h = 0 & \text{in } \Omega \\ h|_{\Gamma_0} = 0 & \text{in } \Gamma_0 \\ \frac{\partial h}{\partial \nu}|_{\Gamma_1} = g & \text{in } \Gamma_1 \end{cases} \tag{2.2}$$

Then, elliptic theory gives

$$\tilde{N}: \text{ continuous } L^2(\Gamma_1) \to H^{3/2}(\Omega) \subset H^{3/2-2\rho}(\Omega) \equiv \mathcal{D}(A^{3/4-\rho}), \ \rho>0 \tag{2.3}$$

the identification on the right of (2.3) meaning equivalent norms [ ]. If now $\tilde{N}^*$ denotes the adjoint of N: $(\tilde{N}v,u)_{L^2(\Omega)} = (v,\tilde{N}^*u)_{L^2(\Gamma_1)}$, then (2.3) implies

$$A^{3/4-\rho}\, \tilde{N} \in \mathcal{L}(L^2(\Gamma_1), L^2(\Omega)) \text{ and } \tilde{N}^* A^{3/4-\rho} \in \mathcal{L}(L^2(\Omega), L^2(\Gamma_1)), \ \rho>0 \tag{2.4}$$

---

i.e. $\mathcal{D}(A)$ is the closure in $H^2(\Omega)$ of $C^2(\overline{\Omega})$ - functions f such that $f|_{\Gamma_0} = \frac{\partial f}{\partial \nu}|_{\Gamma_1} = 0$

The followig Lemma - in the style of previous work [T.1] - will be needed.

Lemma 2.0 For $y \in \mathcal{D}(A)$ we have

$$\tilde{N}^*Ay = \begin{cases} y|_{\Gamma_1} & \text{on } \Gamma_1 \\ 0 & \text{on } \Gamma_0 \end{cases} \tag{2.5}$$

Proof With $v \in L^2(\Gamma)$, we compute by Green's second Theorem where subscripts denote L^2-norms

$$-(\tilde{N}^*Ay,v)_\Gamma = -(Ay,\tilde{N}v)_\Omega = (\Delta y,\tilde{N}v)_\Omega = \tag{2.6}$$

$$= (y,\Delta(\tilde{N}v))_\Omega + (\frac{\partial y}{\partial \nu}, \tilde{N}v)_\Gamma - (y, \frac{\partial(\tilde{N}v)}{\partial \nu})_\Gamma = -(y,v)_{\Gamma_1}$$

since $y|_{\Gamma_0} = \frac{\partial y}{\partial \nu}|_{\Gamma_1} = 0$ by (2.1); $(\tilde{N}v)|_{\Gamma_0} = 0$, $\frac{\partial(\tilde{N}v)}{\partial \nu}|_{\Gamma_1} = v$ and

$\Delta(\tilde{N}v) = 0$ in Ω by (2.2). Then (2.6) yields (2.5) □

We may now introduce an abstract operator model for problem (1.1) - (1.4). Following previous work [T.1], problem (1.1) - (1.4) admits the following abstract versions: as a second order equation

$$\ddot{w} = -A[w + \tilde{N}\tilde{N}^*A\dot{w}] \tag{2.7a}$$

where $\tilde{N}^*A\dot{w}$ is the boundary term by (2.5); or else as first order system

$$\frac{d}{dt}\begin{vmatrix} w \\ \cdot \\ w \end{vmatrix} = \mathcal{A}_{\tilde{N}}\begin{vmatrix} w \\ \cdot \\ w \end{vmatrix}, \text{ on the space } W \equiv \mathcal{D}(A^{1/2})xL^2(\Omega) \tag{2.7b}$$

$$\mathcal{A}_{\tilde{N}} = \begin{vmatrix} 0 & I \\ -A & -A\tilde{N}\tilde{N}^*A \end{vmatrix} (\text{formally}); \quad \mathcal{A}_{\tilde{N}}\, z = \begin{vmatrix} 0 & I \\ -A & 0 \end{vmatrix} \begin{vmatrix} z_1 + \tilde{N}\tilde{N}^*Az_2 \\ z_2 \end{vmatrix} \tag{2.8a}$$

$$\mathcal{D}(\mathcal{A}_{\tilde{N}}) = \{[z_1,z_2]: z_2 \in \mathcal{D}(A^{1/2}) \text{ and } z_1 + \tilde{N}\tilde{N}^*Az_2 \in \mathcal{D}(A)\} \tag{2.8b}$$

Note that the fractional powers of the positive self-adjoint operator A are well defined and that $\mathcal{D}(A^{1/2})$ is topologized with the usual Hilbert norm given explicitly by the first equality below in

$$\|z\|^2_{\mathcal{D}(A^{1/2})} \equiv \|A^{1/2}z\|^2_{L^2(\Omega)} = \int_\Omega |\nabla z|^2 d\Omega, \quad z \in \mathcal{D}(A^{1/2}) \tag{2.9}$$

The second equality in (2.9) follows by first Green's theorem applied to $\|A^{1/2}z\|^2_\Omega = (Az,z)_\Omega = -(\Delta z,z)_\Omega$ with $z \in \mathcal{D}(A)$, and then arguing by density. Writing $z = A^{-1/2}A^{1/2}z$ we obtain from (2.9) the generalized Poincare inequality

$$\int_\Omega z^2 d\Omega = \|z\|^2_\Omega \leqslant \|A^{-1/2}\|^2_\Omega \|A^{1/2}z\|^2_\Omega = \|A^{-1/2}\|^2_\Omega \int_\Omega |\nabla z|^2 d\Omega, \quad z \in \mathcal{D}(A^{1/2}) \tag{2.10}$$

a distinctive feature of the present case $\Gamma_0 \neq \phi$. (Compare with section 2.2). Then, from the above

$$\mathcal{D}(A^{1/2}) \equiv H^1_{\Gamma_0}(\Omega) = \{u \in H^1(\Omega): u = 0, \text{ on } \Gamma_0\} \tag{2.11}$$

If $z \in \mathcal{D}(\mathcal{A}_{\tilde{N}})$, then by using (2.1), (2.5), (2.11) and $\overline{\Gamma}_0 \cap \overline{\Gamma}_1 = \phi$, we have that $z_1 = 0$ on Γ_0, $\partial z_1/\partial\nu + z_2 = 0$ on Γ_1 and

$$\mathcal{D}(\mathcal{A}_{\tilde{N}}) = \mathcal{H}_2 = \{[z_1,z_2] \in H^2_{\Gamma_0}(\Omega) x H^1_{\Gamma_0}(\Omega): \frac{\partial z_1}{\partial\nu} + z_2 = 0 \text{ on } \Gamma_1\} \tag{2.8c}$$

$$H^k_{\Gamma_0}(\Omega) = \{u \in H^k(\Omega): u = 0 \text{ on } \Gamma_0\}$$

and $$W \equiv \mathcal{H}_1 = H^1_{\Gamma_0}(\Omega) x L^2(\Omega) \tag{2.12}$$

<u>Lemma</u> 2.1 (i) The operator $\mathcal{A}_{\tilde{N}}$ in (2.8) is dissipative on $W \equiv \mathcal{D}(A^{1/2}) x L^2(\Omega)$ and satisfies here: range $(\lambda I - \mathcal{A}_{\tilde{N}}) \equiv W$ for $\lambda > 0$. Thus, by Lumer-Phillips theorem, $\mathcal{A}_{\tilde{N}}$ generates a s.c. (strongly continuous) <u>contraction</u> semigroup $e^{\mathcal{A}_{\tilde{N}}t}$ and the solution of (2.7) - hence of problem (1.1) - (1.4) - is given by

$$\begin{vmatrix} w(t;w_0,w_1) \\ w_t(t;w_0,w_1) \end{vmatrix} = e^{\mathcal{A}_{\tilde{N}}t} \begin{vmatrix} w_0 \\ w_1 \end{vmatrix}, \quad t \geqslant 0 \tag{2.13a}$$

and by (2.9)

$$\|e^{\mathcal{A}_{\tilde{N}}t}[w_0,w_1]\|^2_W = E(t) = \int_\Omega |\nabla w|^2 + w_t^2 d\Omega. \tag{2.13b}$$

(ii) The resolvement operator $R(\lambda,\mathcal{A}_{\tilde{N}})$ of $\mathcal{A}_{\tilde{N}}$ is given by

$$R(\lambda,\mathcal{A}_{\tilde{N}}) = \begin{vmatrix} \dfrac{I - V^{-1}(\lambda)}{\lambda} & V^{-1}(\lambda)A^{-1} \\ -V^{-1}(\lambda) & \lambda V^{-1}(\lambda)A^{-1} \end{vmatrix} \tag{2.14}$$

where we have set

$$V(\lambda) = [I + \lambda\, \tilde{N}\tilde{N}^*A + \lambda^2 A^{-1}] \tag{2.15}$$

at least for all λ with $Re\lambda > 0$; moreover $\mathcal{A}_{\tilde{N}}$ has compact resolvent on W.

(iii) The operator $R(\lambda,\mathcal{A}_{\tilde{N}})$ is well-defined (and compact) on W on the closed half-plane $Re\lambda \geqslant 0$. Thus, the spectrum (i.e. point spectrum) $\sigma(\mathcal{A}_{\tilde{N}})$ of $\mathcal{A}_{\tilde{N}}$ satisfies

$$\sigma(\mathcal{A}_{\tilde{N}}) \subset \{\lambda: \; Re\lambda < 0\} \equiv \mathbb{C}^- \tag{2.16}$$

<u>Proof</u> (i) For $y \in \mathcal{D}(\;_{\tilde{N}})$, dissipativity follows from

$$\begin{aligned} Re(\mathcal{A}_{\tilde{N}}y,y)_W &= Re\left(\begin{vmatrix} 0 & I \\ -A & 0 \end{vmatrix} y,y\right)_W - (A\tilde{N}\tilde{N}^*Ay_2,y_2)_{L^2(\Omega)} \\ &= 0 - \|\tilde{N}^*Ay_2\|^2_{L^2(\Gamma)} = -\|y_2|_{\Gamma_1}\|^2_{L^2(\Gamma_1)} \end{aligned} \tag{2.17}$$

where in the last step we have used (2.5).

(ii) With $\lambda > 0$ fixed, we solve $(\lambda I - \mathcal{A}_{\tilde{N}})y = z = [z_1,z_2] \in W$, i.e.

$$\begin{aligned} \lambda y_1 - y_2 &= z_1 \in \mathcal{D}(A^{1/2}) \\ A[y_1 + \tilde{N}\tilde{N}^*Ay_2] + \lambda y_2 &= z_2 \in L^2(\Omega) \end{aligned} \tag{2.18}$$

for $y \in \mathcal{D}(\mathcal{A}_{\tilde{N}})$. By substitution and application of A^{-1} (recall (2.15))

$$V(\lambda)y_2 = \lambda A^{-1}z_2 - z_1 \in \mathcal{D}(A^{1/2}) \tag{2.19}$$

Since the operator $V(\lambda)$ is strictly positive definite on $\mathscr{D}(A^{1/2})$ by

$$(\widetilde{N}\widetilde{N}^*Ay_2,y_2)_{\mathscr{D}(A^{1/2})} = (A\widetilde{N}\widetilde{N}^*Ay_2,y_2)_{L^2(\Omega)} = \|\widetilde{N}^*Ay_2\|^2_{L^2(\Gamma)} \tag{2.20}$$

then $V^{-1}(\lambda) \in \mathscr{L}(\mathscr{D}(A^{1/2}))$ and the unique solution of (2.18) is

$$\begin{cases} y_1 = \dfrac{z_1 + y_2}{\lambda} \\ y_2 = V^{-1}(\lambda)[\lambda A^{-1}z_2 - z_1] \in \mathscr{D}(A^{1/2}) \end{cases} \tag{2.21}$$

which can be checked to be in $\mathscr{D}(\mathscr{A}_{\widetilde{N}})$, see (2.8), since

$\lambda[y_1 + \widetilde{N}\widetilde{N}^*Ay_2] = \lambda A^{-1}z_2 - \lambda^2A^{-1}y_2 \in \mathscr{D}(A)$. Re-writing (2.21) explicitly

yields (2.14), from which compactness follows.

(iii) By contraction of the semigroup, the spectrum of the generator $\mathscr{A}_{\widetilde{N}}$ is in $\mathrm{Re}\lambda \leq 0$. We now show that it is, in fact, in $\mathrm{Re}\lambda < 0$. This will follow by showing that $V^{-1}(\lambda) \in \mathscr{L}(\mathscr{D}(A^{1/2}))$ for $\lambda = ir$, r real and, say, $\neq 0$. Indeed, letting $[I + ir\,\widetilde{N}\widetilde{N}^*A - r^2A^{-1}]x = 0$, $x \in \mathscr{D}(A^{1/2})$, we show that $x = 0$. Taking the $\mathscr{D}(A^{1/2})$ - inner product with x gives easily $\widetilde{N}^*Ax = 0$, by (2.20): hence $Ax = r^2x$. Then either $x = 0$ and we are done, or else x is an eigenvector of A, say $x = e_n$, with eigenvalue $\mu_n = r^2$. Thus,

$e_n|_{\Gamma_0} = \dfrac{\partial e_n}{\partial \nu}\Big|_{\Gamma_1} = 0$, by (2.1); moreover, $\widetilde{N}^*Ae_n = 0$

gives $e_n|_{\Gamma_0} = 0$ by (2.5). Then $e_n|_{\Gamma} = 0$, $\dfrac{\partial e_n}{\partial \nu}\Big|_{\Gamma_1} = 0$ imply $e_n = 0$,

[ ], i.e. $x = 0$ □

Corollary 2.2 With $w_t(t) = w_t(t;w_0,w_1)$, the map $\widetilde{N}^*Aw_t$: continuous $[w_0,w_1] \in W \to L_2(0,\infty; L^2(\Gamma_1))$; more precisely by (2.5) and (2.13):

$$\int_0^\infty \|\tilde{N}^*Aw_t(t)\|^2_{L^2(\Gamma)}\, dt \equiv \int_0^\infty \|w_t(t)\|^2_{L^2(\Gamma_1)}\, dt \le \|[w_0,w_1]\|^2_W = E(0) \tag{2.22}$$

Proof As usual, by (2.17)

$$\tfrac{1}{2}\frac{d}{dt}\|e^{A_{\tilde{N}}t}[w_0,w_1]\|^2_W = \mathrm{Re}\left(\mathcal{A}_{\tilde{N}}\begin{vmatrix} w(t) \\ w_t(t)\end{vmatrix}, \begin{vmatrix} w(t) \\ w_t(t)\end{vmatrix}\right)_W = -\|\tilde{N}^*Aw_t\|^2_{\Gamma_1} \tag{2.23}$$

$$= \int_{\Gamma_1} w_t^2\, d\Gamma$$

(which follows likewise by differentiating (2.13) and using Green first theorem) and (2.22) is obtained by intergrating (2.23) and using contraction of the semigroup □

Remark 2.1 As a consequence of the strong stabilization result to be proved next, we can refine (2.22) to read

$$\int_0^\infty \|w_t(t)\|^2_{L^2(\Gamma_1)}\, dt \le \tfrac{1}{2}E(0), \quad [w_0,w_1] \in W \quad \square \tag{2.24}$$

Proof of Theorem 1.3 (strong stabilization). We follow the approach carried out, say, in [L-T.1] for the case of feedback in the Dirichlet B.C., which we sketch to make this paper self-contained. Since the semigroup $e^{\mathcal{A}_{\tilde{N}}t}$ is contraction on W, Lemma 2.1(i), the Nagy-Foias-Fogel decomposition theory applies (For an excellent expository treatment of this theory, as applied to stabilization problems with distributed feedback operators, see [L.2]). Accordingly, W can be decomposed in a unique way into the orthogonal sum of three subspaces

W_{cnu}, Z_u and $Z^\perp$, all reducing for $e^{\mathcal{A}_{\tilde{N}}t}$:

$$W = W_{cnu} + Z_u + Z^\perp, \text{ such that } Z_u + Z^\perp = W_u,\ W_{cnu} + Z_u = Z, \tag{2.25}$$

where:

(i) on W_{cnu}; $e^{\mathcal{A}_{\tilde{N}}t}$ is completely nonunitary and weakly stable;

(ii) on W_u: $e^{\mathscr{A}_{\tilde{N}}t}$ is a unitary s.c. group. It follows that in our present case $W_u = \{0\}$, the trival subspace, for otherwise Stone's theorem would yield that the eigenvalues of $\mathscr{A}_{\tilde{N}}$ on W_u one on the imaginary axis, a contradiction with Lemma 2.1 (iii).

Thus, in our case, $W = W_{cnu}$ and $e^{\mathscr{A}_{\tilde{N}}t}$ is weakly stable on W, hence strongly stable on W, since $\mathscr{A}_{\tilde{N}}$ has compact resolvent.

Thus $e^{\mathscr{A}_{\tilde{N}}t} y \to 0$ as $t \to \infty$, $\forall y \in W$ □

2.2 <u>The case</u> $\Gamma_0 = \phi$. Here the usual problem arises that the operator A in (2.1) is invertible with bounded inverse not on $L^2(\Omega)$, but on $L_0^2(\Omega) = L^2(\Omega)/\mathscr{N}(A) = \{f \in L^2(\Omega): \int_\Omega f d\Omega = 0\}$, where $\mathscr{N}(A)$ is the null space of A spanned by the normalized constant function k. We then define $\tilde{w}(t,x) \equiv w(t,x) - I(w_0)$, with $I(w_0)$ as in (1.11) as usual. Thus, $\tilde{w}(0,x) \in L_0^2(\Omega)$. Moreover, if C(t) is the cosine operator generated by -A on $L^2(\Omega)$ and $S(t) = \int_0^t C(\tau)d\tau$, then $S(t)k \equiv k \sin(0t) \equiv 0$. Thus,
$S(t)\, \tilde{w}_t(0,x) = S(t)w_t(0,x) = S(t)w_{1,0}$, where $w_1 = w_{1,0} + (w_1,k)_\Omega\, k$ and $w_{1,0} \in L_0^2(\Omega)$. The problem can then be reduced to the space $L_0^2(\Omega)$, where A is boundedly invertible, in the variable $\tilde{w}$, to which the analysis as in the case $\Gamma_0 \neq \phi$ applies. We then use $E(\tilde{w},t) = E(w,t)$. Details are omitted.
An alternative and direct operator approach to handle the important special case $\Gamma_0 = \phi$ (Neumann problem) is the following. since A in (2.1) (with $\Gamma_0 = \phi$) is not boundedly invertible in $L^2(\Omega)$, we simply translate the Laplacian by, say, $\lambda = 1$; i.e. we consider the elliptic problem

$$\begin{cases} (\Delta-1)\, u=0 & \text{in } \Omega \\ \dfrac{\partial u}{\partial \nu} = g & \text{on } \Gamma \end{cases} \tag{2.26}$$

which admits a unique solution $\bar{u} \in L_0^2(\Omega)$, for each $g \in L^2(\Gamma)$. We then define the operator N_1 (depending on $\lambda = 1$) by setting

$$\bar{u} = N_1 g. \tag{2.27}$$

Second, the same argument as in Lemma 2.0, using this time problem (2.26), yields now

$$N_1^*(A+I)v = v|_\Gamma, \quad v \in \mathcal{D}(A) \tag{2.28}$$

with A the operator in (2.1) with $\Gamma_0 = \phi$. Thus, if S(t) is the corresponding 'sine' operator generated by -A (as before), then the argument in [T.1] gives that the abstract operator model for the hyperbolic problem

$$\begin{cases} y_{tt} = \Delta y & \text{in } (0,T)\times\Omega \\ y_0 = y_1 = 0 & \text{in } \Omega \\ \dfrac{\partial y}{\partial \nu} = u & \text{in } (0,T)\times\Gamma \end{cases} \tag{2.29}$$

is given in integral form by

$$y(t) = (A+I) \int_0^t S(t-\tau)N_1 u(\tau)d\tau; \tag{2.30a}$$

or in second order equation form by

$$\ddot{y} = -Ay + (A+I)N_1 g \in [\mathcal{D}(A)]' \text{ or } \ddot{y} = -A[y-N_1 g] + N_1 g \in L^2(\Omega) \tag{2.30b}$$

Thus, the corresponding abstract operation models for the feedback problem (1.1) - (1.4) are: second order equation

$$\ddot{w} = -Aw -(A+I)N_1 N_1^*(A+I)\dot{w} \in [\mathcal{D}(A)]' \tag{2.31a}$$

or as first order system

$$\begin{vmatrix} w \\ \dot{w} \end{vmatrix} = \mathcal{A}_{N_1} \begin{vmatrix} w \\ \dot{w} \end{vmatrix} \tag{2.31b}$$

$$\mathcal{A}_{N_1} = \begin{vmatrix} 0 & I \\ -A & -(A+I)N_1 N_1^*(A+I) \end{vmatrix} \tag{2.32}$$

dissipative on $\mathcal{D}((A+I)^{1/2}) \times L^2(\Omega)$; etc.

3. Proof of Theorem 1.1 (with assumption (iii') = (1.1) replacing assumption (iii))

The proof will be broken down into two major results; first Theorem 3.1 stated below and then Theorem 1.2 stated in the Introduction.

Theorem 3.1 Under the same assumptions as in Theorem 1.1 we have for any $\beta > 0$:

$$\int_0^\infty \int_\Omega e^{-2\beta t}(w_t^2 + |\nabla w|^2)dQ \leq K_1 E(w,0) + K_2 \int_0^\infty \int_\Omega e^{-2\beta t}(w - I(w_0))^2 dQ \tag{3.1}$$

$$+ K_3 \int_0^\infty \int_{\Gamma_1} e^{-2\beta t} w_t^2 d\Sigma \tag{3.2}$$

with constants K_1, K_2, K_3 (identified explicitly in (3.32) below) independent of β □

Accordingly, subsection 3.1 gives the proof of Theorem 3.1 while subsection 3.2 gives the proof of Theorem 1.2.

3.1 Proof of Theorem 3.1: a multiplier approach

We first take initial data smooth, say $[w_0, w_1] \in \mathcal{D}(\mathcal{A}_{\tilde{N}}) = \mathcal{H}_2$, which guarantees $[w(t), w_t(t)] \in C([0,\infty]; \mathcal{D}(\mathcal{A}_{\tilde{N}}))$ and $[w_t(t), w_{tt}(t)] = \mathcal{A}_{\tilde{N}} [w(t), w_t(t)] \in C([0,\infty]; W)$, and find the desired estimates with constants independent of $[w_0, w_1]$. Then, we extend by continuity, as usual. This remark will not be repeated.

Case 1: here we assume $\Gamma_0 \neq \phi$.

Step 1 With h the postulated vector field and $\beta > 0$, we multiply both sides of (1.1) by $e^{-2\beta t} h.\nabla w$ and integrate in $\int_0^\infty \int_\Omega d\Omega dt$. Letting

$$\Sigma = (0,\infty)x\Gamma; \ \Sigma_i = (0,\infty)x\Gamma_i, \ i = 0,1; \ Q = (0,\infty)x\Omega$$

and proceeding as in [L-L-T.1], [L-T.1] we obtain the following identity

$$\int_\Sigma e^{-2\beta t}(h.\nabla w)\frac{\partial w}{\partial \nu}\, d\Sigma + 1/2 \int_\Sigma e^{-2\beta t} w_t^2 h.\nu d\Sigma - 1/2 \int_\Sigma e^{-2\beta t}|\nabla w|^2 h.\nu d\Sigma \quad (3.2)$$

$$= \int_Q e^{-2\beta t} H\nabla w.\nabla w dQ + 1/2 \int_Q e^{-2\beta t}(w_t^2 - |\nabla w|^2)\text{div } h\, dQ$$

$$+ (w_1, h.\nabla w_0)_\Omega - 2\beta \int_0^\infty e^{-2\beta t}(w_t, h.\nabla w)_\Omega dt \quad (3.3)$$

where H(x) is the matrix in (1.8) (To make the present paper self-contained, we provide in Appendix A a derivation of (3.3)). We set throughout

$$2M_h \equiv \max_{\bar{\Omega}} |h| \quad \text{and} \quad 2M_{b,h} \equiv \max_{\bar{\Gamma}_1} |h|, \; (b = \text{boundary}) \quad (3.4)$$

Then, using the boundary conditions (1.3), (1.4), and hence

$$\text{on } \Sigma_0: \; w_t \equiv 0; \; \left|\frac{\partial w}{\partial \nu}\right| = |\nabla w|; \; h.\nabla w = h.\nu \frac{\partial w}{\partial \nu} \quad (3.5)$$

the left hand side (L.H.S.) of (3.3) becomes

$$\text{L.H.S. of (3.3)} = \int_{\Sigma_0} e^{-2\beta t}|\nabla w|^2 h.\nu d\Sigma + \int_{\Sigma_1} e^{-2\beta t}(h.\nabla w)(-w_t)d\Sigma$$

$$+ 1/2 \int_{\Sigma_1} e^{-2\beta t} w_t^2 h.\nu d\Sigma$$

$$- 1/2 \int_{\Sigma_0} e^{-2\beta t}|\nabla w|^2 h.\nu d\Sigma - 1/2 \int_{\Sigma_1} e^{-2\beta t}|\nabla w|^2 h.\nu d\Sigma$$

$$\text{L.H.S. of (3.3)} = 1/2 \int_{\Sigma_0} e^{-2\beta t}|\nabla w|^2 h.\nu d\Sigma + 1/2 \int_{\Sigma_1} e^{-2\beta t} w_t^2 h.\nu d\Sigma$$

$$- \int_{\Sigma_1} e^{-2\beta t}(h.\nabla w) w_t d\Sigma - 1/2 \int_{\Sigma_1} e^{-2\beta t}|\nabla w|^2 h.\nu d\Sigma \quad (3.6)$$

We now invoke asumptions (i) = (1.6) and (ii) = (1.7) on h and use

$$- \int_{\Sigma_1} e^{-2\beta t}(h.\nabla w) w_t \leq \frac{M_{b,h}^2}{\varepsilon} \int_{\Sigma_1} e^{-2\beta t} w_t^2 d\Sigma + \varepsilon \int_{\Sigma_1} e^{-2\beta t}|\nabla w|^2 d\Sigma \quad (3.7)$$

to obtain from (3.6), (3.7) for any $\varepsilon > 0$

$$\text{L.H.S. of (3.3)} \leq \left(\frac{M^2_{b,h}}{\varepsilon} + \frac{M_{b,h}}{2}\right) \int_{\Sigma_1} e^{-2\beta t} w_t^2 d\Sigma$$

$$+ (\varepsilon - \frac{\gamma}{2}) \int_{\Sigma_1} e^{-2\beta t} |\nabla w|^2 d\Sigma$$

$$\leq K_{h,\varepsilon} \int_{\Sigma_1} e^{-2\beta t} w_t^2 d\Sigma \tag{3.8}$$

choosing $\varepsilon < \gamma/2$, where.

$$K_{h,\varepsilon} = M^2_{b,h}/\varepsilon + M_{b,h}/2. \tag{3.9}$$

We have

$$(w_1, h.\nabla w_0)_\Omega \geq -M_h \int_\Omega w_1^2 + |\nabla w_0|^2 d\Omega \geq -M_h E(0) \tag{3.10}$$

$$-2\beta \int_0^\infty e^{-2\beta t}(w_t, h.\nabla w)_\Omega dt \geq -2\beta M_h \int_0^\infty e^{-2\beta t} E(t)dt \geq -M_h E(0) \tag{3.11}$$

using the contraction of $E(t)$, see Lemma 2.1 Thus, combining (3.3), (3.8), (3.10), (3.11) and the assumption (iii') = (1.13) on H, we obtain

$$K_{h,\varepsilon} \int_{\Sigma_1} e^{-2\beta t} w_t^2 d\Sigma \geq \rho \int_Q e^{-2\beta t} |\nabla w|^2 dQ - 2M_h E(0)$$

$$+ 1/2 \int_Q e^{-2\beta t}(w_t^2 - |\nabla w|^2) \text{ div } h dQ \tag{3.12}$$

<u>Step</u> 2 We examine the last integral term in (3.12).

<u>Lemma</u> 3.2 We have

$$\text{(i):} \quad \int_Q e^{-2\beta t}(w_t^2 - |\nabla w|^2) \text{div } h dQ = \int_{\Sigma_1} e^{-2\beta t} w w_t \text{div } h \, d\Sigma$$

$$+ \int_Q e^{-2\beta t} w \nabla(\text{div } h).\nabla w dQ$$

$$-(w_o,w_1 \text{ div } h)_\Omega + 2\beta \int_0^\infty e^{-2\beta t}(w,w_t \text{ div } h)_\Omega \, dt \tag{3.13}$$

$$\text{(ii)}: -(w_0,w_1 \text{ div } h)_\Omega + 2\beta \int_0^\infty e^{-2\beta t}(w,w_t \text{div } h)_\Omega dt \geq -2C_{1,h}E(0) \tag{3.14}$$

$$C_{1,h} = \max \{D_h \|A^{-1/2}\|_\Omega^2, D_h\}; \; 2 D_h \equiv \max_{\overline{\Omega}}|\text{div } h| \tag{3.15}$$

Proof of Lemma 3.2 (i). Multiply both sides of (1.1) by $e^{-2\beta t}w$ div h and integrate on Q. We obtain after integration by parts in t:

$$\int_Q e^{-2\beta t} w_{tt} w \text{ div } hdQ = -(w_0,w_1 \text{div } h)_\Omega - \int_Q e^{-2\beta t} w_t^2 \text{ div } hdQ \tag{3.16}$$

$$+2\beta \int_Q e^{-2\beta t} ww_t \text{div } hdQ$$

since $\lim e^{-2\beta T}(w(T), w_t(T)\text{div } h)_\Omega = 0$ as $T \uparrow \infty$, as it follows from

$$|(w(T),w_t(T)\text{div } h)_\Omega| \leq D_h[\|w(T)\|_\Omega^2 + \|w_t(T)\|_\Omega^2]$$

$$\leq D_h[\|A^{-1/2}\|_\Omega^2 \|A^{1/2} w(T)\|_\Omega^2 + \|w_t(T)\|_\Omega^2]$$

$$\leq C_{1,h} E(T) \leq C_{1,h} E(0) \tag{3.17}$$

where we have used $0 \in \rho(A)$, (2.9) and contraction of E( ), Lemma 2.1. On the other hand, using Green's first theorem and the identity

$$\nabla(w \text{ div } h).\nabla w = w\nabla(\text{div } h).\nabla w + |\nabla w|^2 \text{div } h \tag{3.18a}$$

we obtain

$$\int_Q e^{-2\beta t}(\Delta w)w \text{ div } hdQ = \int_\Sigma e^{-2\beta t} \frac{\partial w}{\partial \nu} w \text{ div } h \, d\Sigma$$

$$- \int_Q e^{-2\beta t} w\Delta(\text{div } h).\nabla wdQ - \int_Q e^{-2\beta t}|\nabla w|^2 \text{div } hdQ \tag{3.18b}$$

Using the B.C. (1.3) - (1.4), and equating (3.16) and (3.18) yields (3.13). Next, (3.14) follows via (3.17) used for T = 0 and T = t, and the contraction of E(.), □

Step 3 We consider the boundary term in (3.13)

Lemma 3.3 We have

$$\int_{\Sigma_1} e^{-2\beta t} w w_t \operatorname{div} h \, d\Sigma = -\tfrac{1}{2}(w_0, w_0 \operatorname{div} h)_{\Gamma_1} + \beta \int_{\Sigma_1} e^{-2\beta t} w^2 \operatorname{div} h \, d\Sigma \qquad \text{(a)}$$

$$\geq -2\, C_{2,h}\, E(0) \qquad \text{(3.19)(b)}$$

$$2\, C_{2,h} = C\, D_{b,h};\ D_{b,h} = \max_{\overline{\Gamma}_1} |\operatorname{div} h|,\ C = \text{trace constant in (3.22)} \qquad \text{(c)}$$

Proof of Lemma 3.3 Identity (3.19a) follows from

$$e^{-2\beta t} w w_t = \tfrac{1}{2} \frac{d}{dt}\left(e^{-2\beta t} w^2\right) + \beta\, e^{-2\beta t} w^2 \qquad (3.20)$$

since

$$\lim_{T \uparrow \infty} e^{-2\beta T} \int_{\Gamma_1} w^2(T) \operatorname{div} h \, d\Gamma = 0 \qquad (3.21)$$

Indeed, by trace theory, (2.9), and contraction of $E(\cdot)$

$$\left| \int_{\Gamma_1} w^2(T) \operatorname{div} h \, d\Gamma \right| \leq D_{b,h} \| w^2(T) \|^2_{\Gamma_1} \leq C\, D_{b,h} \| A^{1/2} w(T) \|^2_{\Omega}$$

$$\leq C\, D_{b,h}\, E(T) \leq C\, D_{b,h} E(0) \qquad (3.22)$$

and (3.21) follows. Moreover, (3.19a) implies (3.19b) via (3.22) □

Step 4 Combining Lemmas 3.2 and 3.3, we obtain

Corollary 3.4

$$\int_Q e^{-2\beta t}(w_t^2 - |\nabla w|^2) \operatorname{div} h \, dQ \geq \int_Q e^{-2\beta t} w \nabla(\operatorname{div} h) \cdot \nabla w \, dQ - 2C_{3,h} E(0) \qquad \text{(3.23)(a)}$$

$$C_{3,h} = C_{1,h} + C_{2,h} \quad \square \qquad \text{(b)}$$

Step 5 Set

$$4\, G_h \equiv \max_{\overline{\Omega}} |\nabla(\operatorname{div} h)| \qquad (3.24)$$

Then for any $\varepsilon_1 > 0$

$$\tfrac{1}{2} \int_Q e^{-2\beta t} w\nabla(\mathrm{div}\ h).\nabla w dQ \geq - \frac{G_h^2}{\varepsilon_1} \int_Q e^{-2\beta t} w^2 dQ$$

$$- \varepsilon_1 \int_Q e^{-2\beta t} |\nabla w|^2 dQ \tag{3.25}$$

and inserting (3.25) into (3.23a)

$$\tfrac{1}{2} \int_Q e^{-2\beta t} (w_t^2 - |\nabla w|^2)\ \mathrm{div}\ h dQ \geq - \frac{G_h^2}{\varepsilon_1} \int_Q e^{-2\beta t} w^2 dQ$$

$$-\varepsilon_1 \int_Q e^{-2\beta t} |\nabla w|^2 dQ - C_{3,h} E(0) \tag{3.26}$$

Finally, inserting (3.26) into (3.12) we obtain

<u>Lemma</u> 3.5 For any $0 < \varepsilon_1 < \rho$ and any $\beta > 0$, we have:

$$K_{h,\varepsilon} \int_{\Sigma_1} e^{-2\beta t} w_t^2 d\Sigma \geq (\rho - \varepsilon_1) \int_Q e^{-2\beta t} |\nabla w|^2 dQ$$

$$- \frac{G_h^2}{\varepsilon_1} \int_Q e^{-2\beta t} w^2 dQ - C_{4,h} E(0) \tag{3.27}$$

$$C_{4,h} \equiv 2M_h + C_{3,h} \tag{3.28}$$

<u>Step</u> 6 We return to Lemma 3.2.

<u>Corollary</u> 3.6 We have for any $\beta > 0$

$$\left| \int_Q e^{-2\beta t} (w_t^2 - |\nabla w|^2) dQ \right| \leq C_5 E(0) \tag{3.29}$$

$$C_5 = \tfrac{1}{2} \max \{ \|A^{-1/2}\|_\Omega, 1 \} = \text{independent of } \beta \ \square \tag{3.30}$$

Proof. Identity (3.13) holds for any smooth vector field, not just the postulated h. Taking there div h ≡ 1, hence $\nabla(\text{div } h) \equiv 0$ (i.e. multiplying only by $e^{-2\beta t}w$ in the proof of Lemma 3.2) and using the same argument as in (3.17) yields (3.29), see (3.15) □

Step 7 By (3.29), we obtain from (3.27)

$$K_{h,\varepsilon} \int_{\Sigma_1} e^{-2\beta t} w_t^2 d\Sigma \geq (\rho-\varepsilon_1) \int_Q e^{-2\beta t} w_t^2 dQ$$

$$-\frac{G_h^2}{\varepsilon_1} \int_Q e^{-2\beta t} w^2 dQ - [C_{4,h} + (\rho-\varepsilon_1)C_5]E(0) \tag{3.30}$$

Summing up (3.27) and (3.30) yields

$$(\rho-\varepsilon_1) \int_Q e^{-2\beta t}[w_t^2 + |\nabla w|^2]\, dQ \leq \frac{2G_h^2}{\varepsilon_1} \int_Q e^{-2\beta t} w^2 dQ$$

$$+ 2\, K_{h,\varepsilon} \int_{\Sigma_1} e^{-2\beta t} w_t^2 d\Sigma$$

$$+ [2C_{4,h} + (\rho-\varepsilon_1)C_5]E(0) \tag{3.31}$$

From (3.31), we then obtain the desired conclusion (3.1) of Theorem 3.1 with the constants K_i there given by

$$K_1 = \frac{2C_{4,h}}{\rho-\varepsilon_1} + C_5;\ K_2 = 2\,\frac{G_h^2}{\varepsilon_1(\rho-\varepsilon_1)};\ K_3 = 2\,\frac{K_{h,\varepsilon}}{\rho-\varepsilon_1} \tag{3.32}$$

at least in the case $\Gamma_0 \neq \emptyset$.

Case 2 For the case $\Gamma_0 = \emptyset$, considerations at the end of section 2 apply. We omit the details.

3.2 Proof of Theorem 1.2: an operator approach

As in [L.1], we shall employ Laplace transform techniques in the variable $\lambda = \beta + i\alpha$, α R, and obtain the needed estimates for small $|\alpha|$ and large $|\alpha|$ separately. However, unlike [L.1], we shall use our operator model for problem (1.1) - (1.4). Following [L.1], we introduce a new variable

$$u(t,x) \equiv \Phi(t)w(t,x);\ \Phi\in C^\infty(R);\ \Phi(0) = \Phi'(0) = 0;\ \Phi(t) \equiv 1,\ t\geq 1 \tag{3.33}$$

Case 1: let $\Gamma_0 \neq \emptyset$. Then, in the new variable u, problem (1.1) - (1.4) becomes

$$\begin{cases} u_{tt} = \Delta u + b & \text{in } (0,\infty)\times\Omega \equiv Q & \text{(a)} \\ u|_{t=0} = u_t|_{t=0} = 0 & \text{in } \Omega & \text{(b)} \\ u(t,x) \equiv 0 & \text{in } (0,\infty)\times\Gamma_0 \equiv \Sigma_0 & \text{(c)} \\ \frac{\partial u}{\partial \nu} = -u_t + \Phi' w & \text{in } (0,\infty)\times\Gamma_1 \equiv \Sigma_1 & \text{(d)} \end{cases} \tag{3.34}$$

where the nonhomogeneous term b is given by

$$b \equiv \Phi'' w + 2\Phi' w_t \in \begin{cases} C([0,1];L_2(\Omega)) \\ C([1,\infty];\ \mathcal{D}(A^{1/2})) \end{cases} \tag{3.35}$$

According to the operator model, for (3.34), the solution to problem (3.34) is given by the following abstract variation of parameter formula, see Lemma 2.0 and [T.1]:

$$u(t) = A\int_0^t S(t-\tau)\tilde{N}\tilde{N}^*A[-u_t(\tau) + \Phi'(\tau)w(\tau)]d\tau \tag{3.36}$$

$$+ \int_0^t S(t-\tau)b(\tau)d\tau$$

where $S(t)y = \int_0^t C(\tau)y d\tau$, $C(.)$ the cosine operator on $L^2(\Omega)$, generated

by the (negative self-adjoint) operator $-A$. Taking the Laplace transform of (3.36) with $\hat{u}_t(\lambda) = \lambda\,\hat{u}(\lambda)$ by (3.34b), and $\widehat{S(t)} = R(\lambda^2,-A)$, we obtain:

$$\hat{u}(\lambda) = AR(\lambda^2,-A)\tilde{N}\tilde{N}^*A[-\lambda\hat{u}(\lambda) + \widehat{[\Phi' w]}(\lambda)] + R(\lambda^2,-A)\hat{b}(\lambda)$$

i.e.

$$[I + \lambda AR(\lambda^2,-A)\tilde{N}\tilde{N}^*A]\hat{u}(\lambda) = AR(\lambda^2,-A)\tilde{N}\tilde{N}^*A\widehat{\Phi' w} \tag{3.37}$$

$$+R(\lambda^2,-A)\hat{b}(\lambda)$$

But

$$I + \lambda AR(\lambda^2,-A)\tilde{N}\tilde{N}^*A = R(\lambda^2,-A)A[I + \lambda\tilde{N}\tilde{N}^*A + \lambda^2A^{-1}]$$

$$\text{(by (2.15))} \qquad = AR(\lambda^2,-A)V(\lambda)$$

inserted in (3.37) yields

$$AR(\lambda^2,-A)V(\lambda)\hat{u}(\lambda) = AR(\lambda^2,-A)\tilde{N}\tilde{N}^*A\widehat{[\Phi'w]}(\lambda) + AR(\lambda^2,-A)A^{-1}\hat{b}(\lambda) \tag{3.38}$$

valid at least for all $\lambda = \beta + i\alpha$, $\beta \geq 0$, with $\lambda^2 = \beta^2 - \alpha^2 + 2i\alpha\beta \neq \{-\mu_n, n = 1,2...\}$, $\mu_n > 0$ eigenvalues of the (positive self-adjoint operator A (as in section 2); i.e. except $\beta = 0$ and $\alpha^2 = \mu_n$, i.e. $\lambda = \lambda_n = \pm i\sqrt{\mu_n}$, where $R(\lambda^2,-A)$ is <u>not</u> defined. Then (3.38) yields

$$\hat{u}(\lambda) = V^{-1}(\lambda)\ \widehat{[\tilde{N}\tilde{N}^*A\Phi'w + A^{-1}b]}(\lambda) \tag{3.39}$$

But, because of Lemma 2.1(iii) we know that $V^{-1}(\lambda) \in V^{-1} \in \mathscr{L}(L^2(\Omega))$ in the <u>closed</u> right half-plane Re $\lambda \geq 0$, including the imaginary axis $\beta = 0$ and is holomorphic in Re$\lambda > 0$ [K1, p. 365 bottom]. Moreover, for any λ in the <u>closed</u> rectangle $\mathscr{R}_{\alpha_0}$: $0 \leq \text{Re}\lambda \leq 1$, $|\text{Im}\lambda| \leq \alpha_0$, with $\alpha_0 > 0$ arbitrary, we have

$$\|V^{-1}(\lambda)\|_{\mathscr{L}(L^2(\Omega))} \leq C_{\alpha_0}, \quad \lambda \in \mathscr{R}_{\alpha_0} \tag{3.40}$$

Thus, with $\lambda = \beta + i\alpha$, β fixed, $0 < \beta \leq 1$, recalling the definition of b in (3.35) and using Parseval equality, we obtain from (3.39), (3.40)

$$\int_{|\alpha|\leq\alpha_0} \|\hat{u}(\beta+i\alpha)\|_\Omega^2\, d\alpha \leq C_{\alpha_0}\{\int_{|\alpha|\leq\alpha_0} \|\widehat{[\tilde{N}\tilde{N}^*A\Phi'w]}(\beta+i\alpha)\|_\Omega^2\, d\alpha$$

$$+ \int_{|\alpha|\leq\alpha_0} \|\widehat{[A^{-1}b]}(\beta+i\alpha)\|_\Omega^2\, d\alpha\}$$

$$\leq 2\pi C_{\alpha_0} \{\int_0^\infty e^{-2\beta t} \|\Phi'(t)\widetilde{N}\widetilde{N}^*Aw(t)\|_\Omega^2 \, dt$$

$$+ \int_0^\infty e^{-2\beta t} \|\Phi''(t)A^{-1}w(t) + 2\Phi'(t)A^{-1}w_t(t)\|_\Omega^2 dt\} =$$

(using (3.33) and (2.3)

$$\leq C_\Phi 2\pi C_{\alpha_0} \{\|\widetilde{N}\widetilde{N}^*A^{1/2}\|_\Omega^2 \int_0^1 \|A^{1/2} w(t)\|_\Omega^2 dt$$

$$+ \|A^{-3/2}\|_\Omega^2 \int_0^1 \|A^{1/2} w(t)\|_\Omega^2 dt$$

$$+ 2\|A^{-1}\|_\Omega^2 \int_0^1 \|w_t(t)\|_\Omega^2 dt\}$$

$$\leq C_{6,\alpha_0} \int_0^1 E(t)dt \leq C_6 E(0) \tag{3.41}$$

where in the last two steps we have used (2.9), the contraction of $E(t)$, and

$$C_{6\alpha_0} \equiv C_\phi 2\pi C_{\alpha_0} \max \{\|\widetilde{N}\widetilde{N}^*A^{1/2}\|_\Omega^2, \ \|A^{-3/2}\|_\Omega^2, \ 2\|A^{-1}\|_\Omega^2\} \tag{3.42}$$

The case $\lambda = \beta + i\alpha$, $0 < \beta \leq 1$, $|\alpha| > \alpha_0 > 0$ proceeds as in [L1]: $\hat{u}_t(\lambda) = \lambda\hat{u}(\lambda)$ and $\frac{1}{|\lambda|^2} \leq \frac{\alpha^2}{\alpha_0^2}\frac{1}{|\lambda|^2} \leq \frac{1}{\alpha_0^2}$ so that Perseval equality

$$\int_{|\alpha|>\alpha_0} \|\hat{u}(\lambda)\|_\Omega^2 d\alpha = \int_{|\alpha|>\alpha_0} \frac{1}{|\lambda|^2} \|\lambda\hat{u}(\lambda)\|_\Omega^2 d\alpha \leq \frac{1}{\alpha_0^2} \int_{|\alpha|>\alpha_0} \|\lambda\hat{u}(\lambda)\|_\Omega^2 d\alpha$$

$$\leq \frac{2\pi}{\alpha_0^2} \int_0^\infty e^{-2\beta t} \|u_t(t)\|_\Omega^2 dt \tag{3.43}$$

Choosing $1/\alpha_0^2 = \varepsilon$, we obtain from (3.41), (3.43) by Perseval equality

$$\int_0^\infty e^{-2\beta t}\|u(t)\|_\Omega^2 dt = \int_{-\infty}^\infty \|\hat{u}(\beta+i\alpha)\|_\Omega^2 d\alpha \leqslant K_\varepsilon E(0)+\varepsilon \int_0^\infty e^{-2\beta t}\|u_t(t)\|_\Omega^2 dt \tag{3.44}$$

where $K_\varepsilon = C_{6,\alpha_0}|2\pi$, which is an inequality of the desired type, but for u, not w. We return now from u to w: since $u \equiv w$ for $t \geqslant 1$, see (3.33)

$$\int_0^\infty e^{-2\beta t}\|w(t)\|_\Omega^2 dt \leqslant \int_0^1 e^{-2\beta t}\|w(t)\|_\Omega^2\, dt + \int_1^\infty e^{-2\beta t}\|u(t)\|_\Omega^2 dt$$

(by (3.44))

$$\leqslant \|A^{-1/2}\|_\Omega^2 \int_0^1 \|A^{1/2} w(t)\|_\Omega^2\, dt + K_\varepsilon E(0)+\varepsilon \int_0^\infty e^{-2\beta t}\|w_t(t)\|_\Omega^2 dt \tag{3.45}$$

and (1.10) of Theorem 1.2 follows by (2.9), with

$$C_\varepsilon = \|A^{-1/2}\|_\Omega^2 + C_{6,\alpha_0}/2\pi, \quad \alpha_0^2\,\varepsilon = 1 \;\square \tag{3.46}$$

Appendix A: sketch of proof of (3.3)

The identity: $\mathrm{div}\,(\phi h) = h.\nabla\phi + \phi\,\mathrm{div}\,h$ and the divergence theorem give

$$\int_\Omega h.\nabla\phi d\Omega = \int_\Gamma \phi h.\nu d\Gamma - \int_\Omega \phi\ \mathrm{div}\ h d\Omega \tag{A.1}$$

Multiply both sides of (1.1) by $e^{-2\beta t}h.\nabla w$ and integrate $\int_0^\infty \int_\Omega d\Omega dt$. As to the left hand side, we integrate by parts in t, use $w_t h.\nabla w_t = 1/2\, h.\nabla(w_t^2)$, and identity (A.1) with $\phi = w_t^2$. We obtain

$$\int_0^\infty e^{-2\beta t}\int_\Omega w_{tt} h.\nabla w d\Omega dt = -\int_\Omega w_1 h.\nabla w_0 d\Omega + 2\beta \int_0^\infty e^{-2\beta t}\int_\Omega w_t h.\nabla w d\Omega dt$$
$$- 1/2 \int_0^\infty e^{-2\beta t}\int_\Gamma w_t^2\ h.\nu d\Gamma dt$$
$$+ 1/2 \int_0^\infty e^{-2\beta t}\int_\Omega w_t^2\ \mathrm{div}\ h d\Omega dt \tag{A.2}$$

As to the right hand side, we use Green's first theorem, the identity

$$\nabla w.\nabla(h.\nabla w) = H\nabla w.\nabla w + \tfrac{1}{2}\, h.\nabla(|\nabla w|^2)$$

and identity (A.1) with $\psi = |\nabla w|^2$. We obtain

$$\int_0^\infty e^{-2\beta t} \int_\Omega \Delta w h.\nabla w d\Omega dt = \int_0^\infty e^{-2\beta t} \int_\Gamma \frac{\partial w}{\partial \nu} h.\nabla w d\Gamma dt$$
$$- \tfrac{1}{2} \int_0^\infty e^{-2\beta t} \int_\Gamma |\nabla w|^2 h.\nu d\Gamma dt$$
$$- \int_0^\infty e^{-2\beta t} \int_\Omega H\nabla w.\nabla w d\Omega dt$$
$$+ \tfrac{1}{2} \int_0^\infty e^{-2\beta t} \int_\Omega |\nabla w|^2 \text{ div } h d\Omega dt \qquad (A.3)$$

Equating the left hand side (A.2) with the right hand side (A.3) results in (3.3) □

REFERENCES

[C.1] G. Chen, Energy decay estimates and exact boundary value controllability for the wave equation in a bounded domain, J. Math. Pures et Appliques (9) 58 (1979), 249-274.

[C.2] G. Chen, A note on the boundary stabilization of the wave equation SIAM J. Control & Optim. 19 (1981), 106-113.

[L.1] J. Lagnese, Decay of solutions of wave equations in a bounded region with boundary dissipation, J. Diff. Eqts. 50(1983), 163-182.

[L.2] N. Levan, The stabilization problem: A Hilbert space operator decomposition approach, IEEE Trans. Circuits & Systems (AS-2519) (1978), 721-727.

[L.3] J. L. Lions, Contrôle des systemes distribués singuliers, Guthier-Villars, 1983.

[L.4] J. L. Lions, Controlabilite exacte de systemes distribues CRAS 1986.

[L.5] J. L. Lions, Controlabilité exacte de systemes distribues: remarques sur la theorie generale et les applications, Proceedings 7th International Conference on Analysis & Optimization of Systems, Antibes, France, June 25-27, 1986. Lecture Notes CIS, 1-13.

[L.6] J. L. Lions, Von Neumann Lecture July 1986, SIAM Meeting, Boston.

[L-T.1] I. Lasiecka and R. Triggiani, Uniform exponential energy decay of the wave equation in a bounded region with $L_2(0,\infty;L_2(\Gamma))$- feedback control in the Dirichlet B.C., J. Diff. Eqts, to appear.

[L-L-T.1] I. Lasiecka, J. L. Lions, and R. Triggiani, Nonhomogeneous boundary value problems for second order hyperbolic operators, Journal de Mathematiques Pure et Appliquees, 1986.

[L-L-T.2] I. Lasiecka, J. L. Lions, and R. Triggiani, Work in progress on exact controllability for the wave equation with control in the Neumann B.C.

[M.1] C. S. Morawetz, Energy identities for the wave equation, NYU Courant Institute Math. Sci. Res. Rep. NoIMM 346, 1976.

[Q-R.1] J. Quinn and D. L. Russell, Asymptotic stability and energy decay rates for solutions of hyperbolic equations with boundary damping, Proc. Royal Soc. Edinburgh 77A (1977) 97-127.

[S.1] M. Slemrod, Stabilization of boundary control systems, J. Diff. Eqts, 22(1976), 402-415.

[S.2] W. Strauss, Dispersal of waves vanishing on the boundary of an exterior domain, Comm. Pure Appl. Math. 28(1976), 265-278.

[T.1] R. Triggiani, A cosine operator approach to modeling $L_2(0,T;L_2(\Gamma))$- boundary input problems for hyperbolic systems, Proceedings 8th IFIP Conference on Optimization Techniques, University of Würzburg West Germany 1977, Lectures Notes CIS M6 Springer-Verlag (1978), 380-390.

[T.2] R. Triggiani, Exact boundary controllability on $L_2(\Omega)xH^{-1}(\Omega)$ of the wave equation with Dirichlet boundary control acting on a portion of the boundary $\partial\Omega$, and related problem.

[Z.1] J. Zabczyk, Stabilization of boundary control systems, International Symposium on Systems Optimization and Analysis, December 1978, Lecture Notes in Control and Information Sciences, #14, edited by A. Bensoussan and J. L. Lions, Springer-Verlag.

Index